TETRAHEDRON ORGANIC CHEMISTRY SERIES

Series Editors: J E Baldwin, FRS & P D Magnus, FRS

Volume 6

Modern NMR Techniques
for
Chemistry Research

D1225959

Related Pergamon Titles of Interest

BOOKS

Tetrahedron Organic Chemistry Series:

DAVIES: Organotransition Metal Chemistry: Applications to Organic Synthesis
DESLONGCHAMPS: Stereoelectronic Effects in Organic Chemistry
GAWLEY: Asymmetric Synthesis*
GIESE: Radicals in Organic Synthesis: Formation of Carbon-Carbon Bonds
GIESE: Radicals in Organic Synthesis 2: Formation of Carbon-Hydrogen and Carbon-
    Heteroatom Bonds*
HANESSIAN: Total Synthesis of Natural Products
HASSNER: Name Reactions and Unnamed Reactions*
PAULMIER: Selenium Reagents and Intermediates in Organic Synthesis
PERLMUTTER: Conjugate Addition Reactions in Organic Synthesis*
SIMPKINS: Sulfones in Organic Chemistry*
WILLIAMS: Synthesis of Optically Active Alpha-Amino Acids
WONG: Enzymes in Synthetic Organic Chemistry*

JOURNALS

EUROPEAN POLYMER JOURNAL
JOURNAL OF PHARMACEUTICAL AND BIOMEDICAL ANALYSIS
TETRAHEDRON
TETRAHEDRON: ASYMMETRY
TETRAHEDRON COMPUTER METHODOLOGY
TETRAHEDRON LETTERS

Full details of all Pergamon publications/free specimen copy of any Pergamon journal available on request from your nearest Pergamon office

*In preparation

Modern NMR Techniques
for
Chemistry Research

ANDREW E. DEROME✝

The Dyson Perrins Laboratory,
University of Oxford, UK

PERGAMON PRESS

OXFORD • NEW YORK • SEOUL • TOKYO

UK	Pergamon Press plc, Headington Hill Hall, Oxford OX3 0BW. England
USA	Pergamon Press, Inc, 395 Saw Mill River Road, Elmsford, New York 10523, USA
KOREA	Pergamon Press Korea, KPO Box 315, Seoul 110-603, Korea
JAPAN	Pergamon Press Japan, Tsunashima Building Annex, 3-2-12 Yushima, Dunkyo-ku, Tokyo 113, Japan

First edition 1987
Reprinted 1987, 1988 (twice), 1990, 1991

Library of Congress Cataloging in Publication Data

Derome, Andrew E.
Modern NMR techniques for chemistry research.
(Organic chemistry series, v. 6)
Includes index.
1. Nuclear magnetic resonance spectroscopy.   I. Title
II. Series: Organic chemistry series (Pergamon Press); v. 6.
QD96.N8D47   1987   543′.0877   86–20510

British Library Cataloguing in Publication Data

Derome, Andrew E.
Modern NMR techniques for chemistry research.—(Organic chemistry series, v.6)
1. Nuclear magnetic resonance spectroscopy
I. Title   II. Series
541.2′8   QD96.N8

ISBN 0–08–032514–9 (Hardcover)
ISBN 0–08–032513–0 (Flexicover)

Printed in Great Britain by BPCC Wheatons Ltd, Exeter

Foreword

The development of commercially available high resolution NMR spectrometers in the 1950's provided the chemist with a new tool of enormous power. The direct relationship between spectral symmetry and the molecular symmetry of the sample enabled the routine solution of structural problems which previously required a level of intellectual power and chemical insight given to few. All this was achieved in the 100 milligram range, with instruments based on powerful electro- or permanent magnets.

A second revolution has now occurred in the NMR field. This has come about through the development of reliable superconducting magnets coupled to the application of the pulse technique and its associated Fourier transform. Now dispersion and sensitivity have increased in leaps and bounds, so that sample size can be in the microgram range. Perhaps more important have been developments consequent on the pulse technique which permit enormously greater control and manipulation of the sample's magnetisation. As a result, the structural information now available to the chemist, through pulse NMR, is probably greater and more readily obtained than by any other single technique.

In this book Andrew Derome takes the practising chemist through the practical aspects of this new method. Through this work a chemist, not trained as a physical NMR spectroscopist, can understand and use the enormously powerful tools for structural investigation available through this new generation of superconducting FT machines. It is strongly recommended to chemists and biologists both in academe and industry, who wish to realise the full possibilities of the new wave NMR.

Jack E. Baldwin, FRS

The Dyson Perrins Laboratory
Oxford University

Andrew E. Derome, 1957–1991

The tragic death of Andy Derome earlier this year has robbed the field of chemical research of an outstanding teacher in the application of NMR spectroscopy to chemical problems. Throughout his brief but highly productive career he was able to bring together his knowledge of chemistry and his expertise in NMR to create the most powerful analytical techniques used by organic chemists.

This book has served as a catalyst for the conversion of a multitude of chemists throughout the world into capable practitioners of modern NMR methods. As such it has significantly advanced the field of chemical research. Although he is no longer with us his *magnum opus* will, through its use by chemists, be a living memorial to his outstanding teaching ability.

Jack E. Baldwin, FRS

Dyson Perrins Laboratory
University of Oxford
August, 1991

Preface

If you have just picked up this book with the thought of buying it, you may be asking yourself 'why go to all this trouble'. Here is a text which, although set in a qualitative framework, does spend a good deal of time discussing the physical principles behind NMR experiments; if you are a chemist or biologist you may shy away from such material. There is also a lot of experimental detail, which may seem superfluous if you are used to placing your spectroscopic problems in the hands of specialists. Presumably you are curious about the new developments in NMR, but maybe you would feel more comfortable with a survey which concentrated on spectral interpretation, leaving the technical details for somebody else. I believe there are several reasons why the application of NMR to chemical and biological problems cannot be made properly using such a 'black box' approach, and I hope the following paragraphs will convince you of this. You can find a detailed description of what this book contains in the first chapter.

Why is NMR such a useful technique? There are many reasons, but a central theme is that it can identify *connections* between entities. The gross features of chemical shift and signal intensity have, essentially, the same character as other spectroscopic methods such as infra-red absorption, and if that was all there was to NMR it would not outpace other methods to such a degree. But there is more: the fine structure in spectra, which arises from the coupling between nuclei, and various other interactions such as the nuclear Overhauser effect, depend on the relationships *between* nuclei; this is what gives NMR its special usefulness. Whether the aim is to probe the structure of an isolated, pure compound, or to measure proton-proton distances in a protein, or to extract the signal of a labelled metabolite from some biological soup, it is to those properties relating one nucleus with another that we turn.

Over the last ten years or so the peculiar advantages of pulse NMR have come to be fully appreciated, leading to the development of many ingenious new ways to exploit these connections. Simply glancing at a list of experiments can be daunting; there are so many that it seems impossible to even begin to understand them. However, *it is not so difficult*. A few basic principles have been applied in various permutations; once these principles have been grasped any new experiment can be fitted easily into the overall picture. This is an important reason for making the effort to understand what is happening during pulse experiments; you will be better able to judge whether some new method is applicable to your particular problem. A nice feature of pulse NMR is that it is, on the whole, easier to understand than continuous wave NMR, even though the latter may be more familiar. In a pulse experiment the system spends most of its time evolving independently of outside stimuli, so there are less things to think of at once.

Modern pulse NMR is performed exclusively in the Fourier transform mode. The reasons for this are set out in detail later in the text, but the use of the FT method is another important motivation for thinking about the mechanics of NMR experiments. Of course, it is useful to appreciate the advantages of the transform, and particularly the spectacular results which can be achieved by applying it in more than one dimension, but it is also essential to understand the limitations imposed by digital signal analysis. The sampling of signals, and their manipulation by computer, often limit the accuracy of various measurements of frequency and amplitude, and may even prevent the detection of signals altogether in certain cases. These are not difficult matters to understand, but they often seem rather abstract to newcomers to FT NMR. Even if you do not intend to operate a spectrometer, it is irresponsible not to acquire some familiarity with the interaction between parameters such as acquisition time and resolution, or repetition rate, relaxation times and signal intensity. Many errors in the use of modern NMR arise because of a lack of understanding of its limitations.

Another reason for studying NMR, and especially its practical side, is that it is *fun*. Many chemical and, nowadays, biological projects culminate in an NMR experiment. After a year-long slog to synthesise some labelled substrate, it may only take a few days to perform the *real* experiment by NMR, whether it be following the mechanism of an organic reaction or watching a metabolic pathway inside whole cells. It is not fair that the excitement associated with such experiments should be reserved for specialist NMR operators: *learn to do it yourself*. The investment is the initial effort required to become familiar with a spectrometer (which may require a few late nights, unless you are very lucky), and the struggle to understand the experiments. The return is a broad understanding of the most important spectroscopic technique, and a refreshing and stimulating change from everyday lab. routine.

Many people have helped during the preparation of this book, and I will now try to thank them; to anyone who has been inadvertently omitted I extend my apologies. First, I am indebted to Jane MacIntyre, for originally persuading me that the project was possible. Several of my colleagues in the Dyson Perrins and Inorganic Chemistry Laboratories have read the text, from the diverse perspectives of graduate student, NMR operator, electronics engineer, university lecturer and postdoctoral researcher, and provided many helpful suggestions. My sincere thanks to: Barbara Domayne-Hayman, Elizabeth McGuinness, Dermot O'Hare, Mike Robertson, Mike Robinson, Chris Schofield and Nick Turner. I also especially have to thank Elizabeth McGuinness and Tina Jackson for bearing the brunt of the NMR requirements of hundreds of research workers in the Oxford chemistry departments, thus leaving me sufficiently undisturbed to work on the text. Mike Robertson carried out extensive modifications to a commercial instrument to permit some of the more unusual experiments to be performed, while Tina Jackson provided technical assistance.

A number of people have been kind enough to contribute figures, or to permit me to perform devilish experiments on their compounds. I thank, from the Dyson Perrins Laboratory, Oxford: John Brown (for Figures 9.6, 9.7), George Fleet (for compound **1**, Chapters 8-10), Nick Turner (for compound **6**, Chapter 5) and Rob Young (for compound **6**, Chapter 8); from the Inorganic Chemistry Laboratory, Oxford: Dermot O'Hare (for Figures 8.45, 9.10, 10.11); from the Physical Chemistry Laboratory, Oxford: Chris Bauer (for Figure 7.19), Ray Freeman (for Figure 8.21) and Peter Hore (for Figure 2.23); from the Biochemistry Department, Oxford: Jonathan Boyd (for Figure 8.36); from the University Chemical Laboratory, Cambridge: James Keeler (for Figures 8.24, 10.13); from the Ruđer Bošković Institute, Zagreb: Dina Keglević (for compound **2**, Chapter 8). Oxford Instruments Ltd. and Bruker Spectrospin also provided a number of figures, which are acknowledged in the relevant captions. Figure 5.10 was reproduced with permission from: J. H. Noggle and R. E. Schirmer, *The Nuclear Overhauser Effect - Chemical Applications*, Academic Press. 1972.

In common with the other books in the Organic Chemistry series, this text was reproduced photographically from camera-ready copy. However, in order to achieve an acceptable appearance, I typeset the material myself on a Monotype Lasercomp phototypesetter at Oxford University Computing Service. I thank the Computing Service for access to this device, and Catherine Griffin for advice on its use. Christine Palmer, of the Inorganic Chemistry Laboratory, Oxford, miraculously produced the line drawings from my semi-legible sketches, often at high speed; I thank her for this vital contribution. Final paste-up of the text was performed at Pergamon, under the expert supervision of Colin Drayton, who also managed to avoid making me feel bad even when the whole project was months behind schedule.

Andrew Derome

Contents

3 Basic Experimental Methods   31

4 Describing Pulse NMR   63

6 Polarisation Transfer and Spectrum Editing   129

7 Further Experimental Methods   153

Introduction

Nuclear resonances are affected by a variety of weak interactions between the nuclei and the electrons of molecules, between nuclei within the molecules and between nuclei in neighbouring molecules. If these interactions can be disentangled and interpreted, they are found to contain an extraordinarily rich mine of information about the structures and conformations of the molecules of the sample, about interactions between molecules and about molecular motion.

All this is made possible by the very long relaxation times that characterise NMR spectra of spin ½ nuclei in mobile liquids; line widths of 0.1 Hz or less at 500 MHz are by no means uncommon. This means that even resonances which lie very close together can be resolved.

High resolution spectra, even of molecules of modest molecular weight, are therefore often quite complicated, but because the interactions are weak, cross terms are not very important if the measurements are made in strong magnetic fields, and the spectra are relatively easy to interpret. It is not surprising, therefore, that high resolution NMR has become such an indispensable tool for the organic chemist.

As the size of the sample molecules becomes greater, the complexity of the NMR spectra increases rapidly, lines overlap, and their interpretation becomes more difficult. At this stage much more sophisticated sequences of experiments must be made, the analysis of which can be technically demanding. Here the organic chemist is faced with a dilemma. How much effort should he or she invest in grappling with the not inconsiderable technicalities of the very powerful NMR techniques now available? It is certain that without some understanding of these methods he may not even realise how NMR can help him solve his particular problems, and even if he thinks help is available, he must beware of using NMR as an imperfectly understood black box.

Most organic chemists to-day need to understand what NMR can do for them, and in this book Dr Derome sets out the principles and provides a practical guide to the use of NMR spectroscopy in terms accessible to organic chemists; it is a long journey but one which will be well worth while. The reader who perseveres with the book will be considerably enlightened. I hope the reader will also be sufficiently fired by the enthusiasm running through the text to read at some later stage the further references provided and achieve still greater depths of understanding of these elegant experiments.

Sir Rex Richards, FRS
University of Oxford

1

What This Book Is About

1.1 INTRODUCTION

Figure 1.1 is a proton NMR spectrum of cholesteryl acetate, run at 60 MHz on a continuous wave spectrometer. During the sixties and early seventies, a period when much research into compounds of this type was carried out, such a spectrometer would have been regarded as a top rank research instrument. It is a remarkable reflection on the power of NMR to illuminate chemical problems (and on the prowess of those who pioneered its use) that the availability of spectra even as crude as this revolutionised organic chemistry.

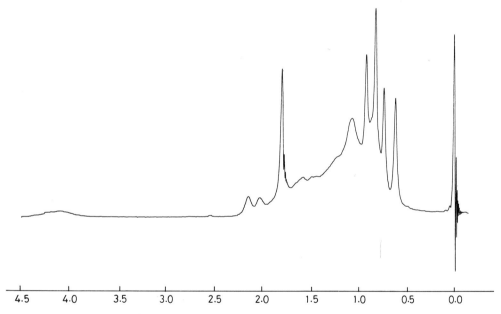

Figure 1.1 A low-field, continuous wave NMR spectrum.

Nowadays we can expect much more from our spectrometers (Figure 1.2), with instruments operating at almost ten times the frequency (and costing at least thirty times as much). The improvements in this spectrum are obvious, but the point I wish to make by showing it to you is perhaps less so. *It is not good enough.* The intrinsic limitations of NMR are such that, however strenuous our efforts to improve spectrometer technology, we will not be able to interpret the proton spectra of relatively simple compounds at sight. This will remain so into the foreseeable future, barring some entirely unexpected development in magnet construction. It is worth bearing in mind that, while the switch from electromagnets to superconducting magnets allowed an immediate jump in field strength of a factor of 3-4, further efforts in superconducting magnet design over almost twenty years have resulted in only another doubling of routinely available field.

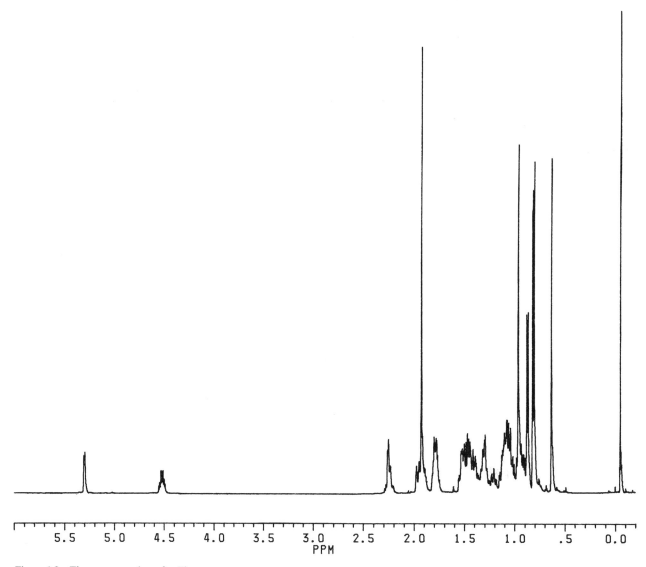

Figure 1.2 The same sample as for Figure 1.1, but run at 500 MHz in Fourier transform mode.

We cannot rely on progress in instrument design to solve our spectroscopic problems for us, but fortunately we do not need to do so. In parallel with technological development of NMR spectrometers, there has been an astonishing growth in our understanding of the properties of nuclear spin systems. Since about 1980, this has reached some kind of critical point, with the sudden development of a wide range of new experimental techniques. Everyone involved in chemistry research will certainly be aware by now that something unusual is happening to NMR, but still the impact of these developments has been fairly restricted. The technical nature of most NMR papers, and their spread through an unusual range of chemical, physical and biological journals, means that the non-specialist has little hope of keeping abreast of developments. In the teaching of NMR to students of chemistry, it sometimes seems that pulsed FT NMR has yet to supplant CW NMR in many courses pitched at organic chemists; modern NMR methods are almost never treated. This cannot remain so for much longer, because understanding how to use NMR to solve problems is fundamental to success in chemical research.

In this book I hope to provide an introduction to modern NMR accessible to the non-specialist. It is intended to be self-contained, aside from some basic background knowledge requirements described in the next section. The reader I have in mind is the advanced student or beginning research worker, or those with more experience who wish to learn about new techniques. The bare skeleton of the material (without experimental details or such a wide range of methods) derives from a short course for graduate students at Oxford, but despite this I think that much of the contents should be accessible to undergraduates. The application of modern NMR methods is rarely taught at undergraduate level at present, but this reflects the speed with which we have been overtaken by events, rather than any extreme conceptual difficulties in the subject.

1.2 WHAT YOU NEED TO KNOW

I take it you have already encountered 'traditional' NMR. Thus, I assume a good deal of familiarity with the use of proton NMR for structure elucidation. You will find no discussion of the variation of chemical shifts or coupling constants with structure here. This is partly because the information is readily available elsewhere, but more importantly because modern NMR experiments can reduce our reliance on such empirical correlations. We can move away from 'the spectrum is consistent with structure X' and much closer to 'the proof of structure X follows from $\cdots$'. The only proton NMR 'method' I expect you to have met is homonuclear decoupling, which is mentioned several times but not explained in detail. You might also have come across the nuclear Overhauser effect, but if not never mind, as it merits a whole chapter of its own.

As to other nuclei, you should be aware of the possibility of observing ^{13}C, and of the consequences of its low natural abundance. If you have used carbon spectra, then you must have seen the 'off-resonance' decoupling experiment, which helps to determine the number of protons borne by each carbon. If you are an inorganic chemist, or a biologist, you might make more use of ^{31}P NMR. It is not really important, though, as the treatment is entirely multi-nuclear. While examples will often be drawn from proton and carbon spectroscopy, as a natural reflection of the fact that I work in an Organic Chemistry department, the only real distinction is between nuclei with spin $\frac{1}{2}$ and those with greater spin (quadrupolar nuclei). Most of the discussions in this book centre on spin-$\frac{1}{2}$ nuclei, with extensions to quadrupolar nuclei mentioned when appropriate.

Some rudimentary mathematics crops up from time to time, but fear not, because it is never essential to understanding the material. This is not a book for physical chemists, and there are no derivations or detailed mathematical analyses, but the occasional formula does help sometimes. All you need to understand the physical model introduced in Chapter 4 is a knowledge of basic trigonometry and Cartesian coordinates. Some first-order rate equations also occur in a few places, but no calculus more difficult than that (apart from a small burst of vector calculus in Chapter 4, which can safely be ignored). In one or two places I assume slight familiarity with computer terms such as 'memory' or 'byte'.

1.3 WHAT IS IN THE BOOK

This book is *not* a comprehensive review of recent NMR methods. Any attempt to include a review of this kind in a reasonably short text would leave little room for explanation or examples. Instead, I have chosen to survey several broad classes of experiment, to try to demonstrate the principles involved in - and the practical aspects of - modern NMR. In discussing new experiments, I have tried very hard to convey the physical essence of how they work, taking the explanations as far as I could subject to the restriction of no mathematics. Naturally, the point at which such an approach should be abandoned as vague and unenlightening is a matter of personal taste, and if I sometimes seem to oversimplify, or to push physical analogy too far, I hope you will bear with me.

The starting point for our discussions has to be an introduction to Fourier methods (Chapter 2) because, although these have been in widespread use for many years, chemists still often have little appreciation of their merits or limitations. This situation has been tolerable (if less than ideal) until recently, because in running simple one-dimensional spectra we are seldom limited by the requirements of the Fourier technique. This ceases to be so for the *two-dimensional* experiments which are the subjects of Chapters 8, 9 and 10. I know that the temptation not to worry about the more technical aspects of spectroscopy is strong, particularly when they seem vaguely mathematical, but it has to be resisted in this case. NMR is so important to chemistry that we cannot treat it as a 'black-box' technique.

We also need some model of the events during a pulse experiment, so that we can talk about them; this is developed in Chapter 4 in a pictorial way. Much of what is said in the remainder of the book depends on the material in Chapters 2 and 4, so these need to be read first. The remaining chapters, however, are more loosely interdependent. Chapter 5, on the nuclear Overhauser effect, stands on its own. Chapter 6 introduces the idea of *polarisation transfer* in a specific context, before we encounter it in more generalised form in Chapters 8 and 9. The concept of two-dimensional NMR data is introduced in the first part of Chapter 8; this forms an essential introduction to the last three chapters. It seemed overly abstract to talk about two-dimensional Fourier techniques separately from their application, so Chapter 8 also discusses a particular class of two-dimensional experiment (homonuclear shift correlation). This is extended to the heteronuclear case in Chapter 9, and to various experiments to do with multiplet structure (*J*-spectroscopy) in Chapter 10.

There remain two chapters (3 and 7) on experimental techniques. I think it is most important to gain experience of experimental NMR whenever possible, because then apparently dry and abstract discussions take on an entirely different light. Nobody who has made serious use of an FT spectrometer has much difficulty comprehending the importance of digital resolution, acquisition time, window functions and the like. Since it is impossible to condense a comprehensive manual of experimental NMR[1] into two chapters, I have had to be selective, and I am conscious that the selection offered may seem rather obscure. My choices were determined by observations of many researchers trying to use high-field instruments for the first time, and the difficulties they encountered.

Thus, the inclusion of very basic material in the first part of Chapter 3 reflects the peculiar observation that, faced with an extremely expensive and rarely available spectrometer, many people nevertheless are content to stick a sample in it scarcely fit for a 60 MHz instrument. High-field NMR spectrometers are precision instruments, and samples for them should be handled with care. Most of the remainder of Chapter 3 is concerned with more understandable operational difficulties relating to resolution and sensitivity. There is also a critical discussion of the tests used to measure the performance of spectrometers, which I hope might be useful to anyone involved in instrument purchasing.

In Chapter 7 I tackle some more advanced experimental problems. Beginners quickly discover that it is no good at all understanding, say, the theory of heteronuclear shift correlation backwards and inside out if you can't measure the pulse width on the decoupling channel. This type of knowledge tends to be passed down by word of mouth at present; in gathering together a few simple procedures perhaps I can make these new experiments seem a little more accessible. Chapter 7 also includes a discussion of how to select experimental parameters so as to obtain optimum sensitivity or quantitative results, and various procedures (such as peak suppression, composite pulses and tailored excitation) which do not fit in to other chapters, but seem important enough to include.

1.4 WHAT CAN BE DONE WITH NMR

Many spectroscopy books include a chapter on 'tackling the problem', or sets of case studies or problems. While there are plenty of examples throughout this text, I have quite deliberately avoided such an explicit attempt at telling you what to do with the new experiments. There are two reasons for this. The first is an entirely personal bias; I find I never read those parts of other books anyway. The trouble is, spectroscopic problems are either trivial, in which case they are unenlightening, or it requires so much effort to get in to them that most people will not bother. You need the stimulus of the compound being your own creation before you can rack your brains properly. This is a fairly frivolous reason; the second is much more important.

At this time, many experiments have not been around long enough for their usefulness to be established. I have my own views, but these are naturally conditioned by the type of compounds I am asked to investigate, and by what the instrumentation at my disposal is suited to perform. This latter point is particularly significant; if you asked me "what do you think of triple-quantum filtration", my reply last year would have been "I don't know, we haven't got a good enough phase shifter", this year it would be "seems to work, but we haven't solved a real problem yet", as for next year, who can say. The point is, I do not want to impose my current ideas about a fast-moving subject on you, but rather to develop sufficient understanding for you to be able to assess the position yourself.

That said, it does seem appropriate to try to give some kind of perspective on the recent developments in NMR. The applications of NMR cover a very wide range, from simple structure elucidation, through studies of enzyme conformations in solution to *in vivo* monitoring of metabolism and medical diagnostics. Despite this diversity, there are in fact rather few basic themes behind the whole panoply of experiments. The following paragraphs may make more sense much later, when you have read the rest of the book, but seem to belong at the beginning nevertheless.

The exploitation of *coupling* is the basis for all the experiments described in this book. Scalar or *J*-coupling is familiar because it causes line splittings in ordinary spectra; all experiments other than those based on the nuclear Overhauser effect (nOe) involve *J*-coupling in some way. The nOe itself also depends on a coupling, but it is the *dipolar* coupling between nuclei, of which more in Chapter 5.

The usefulness of *J*-coupling in structure determination is clear enough from one-dimensional proton spectra. By examining multiplet patterns we can often work out how many neighbours a proton has, and we may even be able to trace chains of protons by matching up splittings. If we can do homonuclear decoupling experiments the potential for identifying neighbouring protons is still greater. The reason that detection of coupling is so informative is that it occurs in very predictable circumstances; for protons, 2 and 3 bond couplings are nearly always in the range 2-20 Hertz, while longer range couplings are usually very small. The predictable nature of couplings, and the fact that they indicate *pairwise relationships* between nuclei, make them a sensitive probe of molecular structure. Chemical shifts, in contrast, give only a crude indication of the environment of *individual* nuclei.

Increased understanding of how we can manipulate coupled systems is a major theme of 'modern' NMR. This leads to better ways to make familiar measurements, in the form of experiments which map out spin systems in a fashion similar to decoupling (Chapters 8, 9). It also leads to a range of methods which aid in the observation of low-frequency nuclei, by using coupling with protons as a 'lever' to boost their signal strength (Chapter 6). Both these applications do things which have been possible for a long time, but do them better (i.e. faster, with better powers of discrimination, or in a more general way). There are also experiments which are entirely novel, such as those that select spin systems according to the number of nuclei they involve (*multiple quantum filtration*, Chapters 6 and 8). The various experiments are approached in later chapters from the point of view of their applications, which can make them seem quite distinct, but it is as well to keep in mind that the manipulation of coupling, and of the energy levels it creates, underlies them all.

The second great theme of modern NMR is the application of Fourier techniques to the data analysis. We will see in the next chapter how this accelerates the information retrieval process, an essential requirement in view of the low intensity of NMR signals. Extension of the concept of data measurement as a function of time to experiments involving two (or more) time variables can give extraordinary dividends, both in terms of speed and resolving power (Chapters 8, 9, 10). The two-dimensional Fourier transform is essentially an adjunct to the ideas about coupling which form the basis for new experiments; it helps us get the answers we want as efficiently as possible. It also enables the measurement in an indirect way of phenomena which otherwise have no physical manifestation, such as transitions between energy levels 'forbidden' by quantum mechanical selection rules.

1.5 A SHORT TOUR ROUND AN NMR SPECTROMETER

So that we understand what we are talking about in later chapters, let's take a look at a high-field spectrometer and its important components (Figure 1.3). To achieve proton frequencies greater than 100 MHz, super-conducting magnets are required. The solenoid (Figure 1.4), wound from alloys based on niobium, is immersed in a bath of liquid helium contained within an efficient cryostat (large cylindrical structure at the right of Figure 1.3). Careful cryostat design, involving use of an outer jacket of liquid

Figure 1.3 A 500 MHz FT spectrometer.

nitrogen to cool a 'radiation shield', means that fresh helium need be added only rarely (every 2 to 9 months, depending on model). The spectrometer in the figure observes protons at 500 MHz, so the field at the centre of its magnet is 11·7 Tesla.

A set of field gradient coils is mounted inside the bore of the magnet (the 'shims' - see Chapter 3), and inside them sits the NMR probe (Figure 1.5). This is the most crucial part of the whole system, and contains the apparatus for transmitting pulses to the sample and receiving the NMR signals they elicit. In the spectrometer we are looking at, the probes (which are exchangeable to suit the experiment at hand) are inserted up from the base of the cryostat into the active region of the magnet. Samples, held in familiar cylindrical NMR tubes, descend from the top of the cryostat to rest in the upper part of the probe (Figure 1.6). The sample is spun about its vertical axis by an air turbine. Note that the gradient coils, probe and sample are all at room temperature, despite their close proximity to the 4K liquid helium.

Figure 1.4 The solenoid from a 470 MHz NMR magnet (courtesy of Oxford Instruments Ltd.).

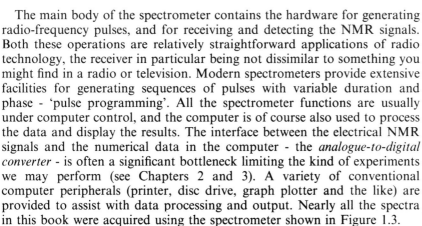

Figure 1.5 A collection of NMR probes of various shapes and sizes, for varying diameter samples and varying magnet bores.

The main body of the spectrometer contains the hardware for generating radio-frequency pulses, and for receiving and detecting the NMR signals. Both these operations are relatively straightforward applications of radio technology, the receiver in particular being not dissimilar to something you might find in a radio or television. Modern spectrometers provide extensive facilities for generating sequences of pulses with variable duration and phase - 'pulse programming'. All the spectrometer functions are usually under computer control, and the computer is of course also used to process the data and display the results. The interface between the electrical NMR signals and the numerical data in the computer - the *analogue-to-digital converter* - is often a significant bottleneck limiting the kind of experiments we may perform (see Chapters 2 and 3). A variety of conventional computer peripherals (printer, disc drive, graph plotter and the like) are provided to assist with data processing and output. Nearly all the spectra in this book were acquired using the spectrometer shown in Figure 1.3.

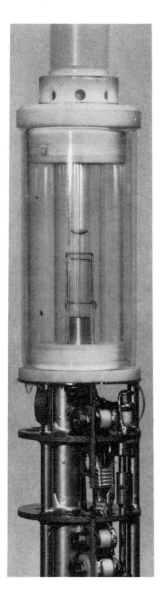

Figure 1.6 Inside the probe, the sample hangs inside the transmit/receive coil. Below can be seen some of the components of the resonant circuit (see Chapter 3).

REFERENCE

1.   For a detailed survey of traditional aspects of experimental NMR, see: M. L. Martin, J-J. Delpuech and G. J. Martin, *Practical NMR Spectroscopy*, Heyden, 1980.

2

Why Bother With Pulse NMR

2.1 INTRODUCTION

It somehow seems natural to record spectra in the *continuous wave* mode. The idea of applying monochromatic radiation to a sample and varying its frequency to locate absorption maxima is simple and straight-forward. In optical spectroscopy, such as infra-red (IR) or ultra-violet (UV) absorption measurements, this is still the commonest approach. Why then do we need to consider the apparently obscure alternative of pulsed excitation for NMR? To find out, we need to examine some NMR data closely (Figure 2.1). Superficially this spectrum looks nice enough, but we only have to amplify it by a factor of 4 (inset) to discover the bane of all NMR spectroscopists: noise.

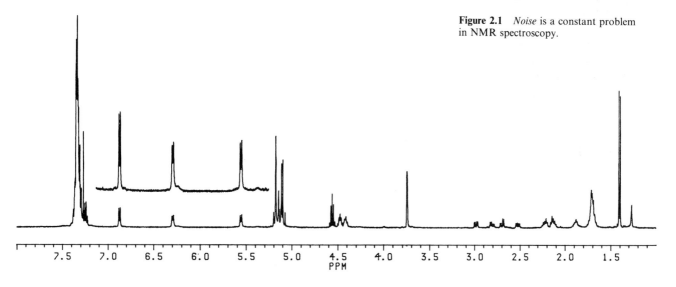

Figure 2.1 *Noise* is a constant problem in NMR spectroscopy.

The trouble is, with physically practical magnets an NMR transition has very low energy. So low that it is small by comparison with kT (k = Boltzmann's constant) at room temperature, which leaves very little popul-ation excess in the lower energy state. The signals we obtain are corre-spondingly weak, and in many cases are not substantially larger than the inevitable electrical noise generated by the spectrometer. Careful design minimises sources of electrical noise, but compared with NMR signals even the thermal motions of electrons in a piece of wire are significant. However much trouble is taken over spectrometer construction there is an unavoid-able base level of noise, which is inconveniently high from an NMR point of view. We need therefore to find some other way to improve the comparative sizes of NMR signal and noise - the *signal-to-noise ratio*.

One way to improve the signal-to-noise ratio beyond the natural limit-ations of the spectrometer is by *signal averaging*. We take advantage of the fact that the noise contribution is random by recording the spectrum more than once. The NMR signals occur in exactly the same place each time, and therefore build up over a number of repetitions. The fate of the noise is slightly more complicated; it does not 'average out', as is often erroneously believed, but builds up *more slowly* than the signal. In fact, over n repetitions of the experiment, during which the signal clearly increases n times, the noise amplitude only increases by $\sqrt{n}$. Thus the signal-to-noise *ratio* improves by a factor of $\sqrt{n}$. Proving that the noise increases as the square root of the number of experiments is not trivial; if you are interested, look up the 'central limit theorem' in a statistics textbook.

In NMR spectroscopy we almost always want to carry out signal averaging to improve the signal-to-noise ratio. This is where NMR diverges from other common spectroscopic techniques; the sensitivity of UV and IR is such that single experiments are usually sufficient. This requirement for multiple experiments is not in itself a justification for the use of pulsed excitation however; signal averaging could in principle be applied to CW NMR (and was, in historical times - the Computer Averaged Transients, or CAT, technique). To discover why this is not a very viable approach, we need to examine the problems of acquiring multiple spectra in CW mode. The stumbling block proves to be the *length of time* it takes to acquire each spectrum.

The speed at which we can extract NMR data in CW mode is limited by very fundamental considerations. NMR lines (of spin-$\frac{1}{2}$ nuclei) are usually 'sharp', by which I mean that the corresponding transition energies vary little from one contributing nucleus to the next. Often we want to take advantage of this by making measurements capable of discriminating between closely spaced lines. Suppose the smallest line spacing we would like to resolve is 1 Hz. This is equivalent to measuring an energy difference ΔE of h Joules (h = Planck's constant; $E = hv$). The uncertainty principle tells us:

$$\Delta E \Delta t \sim h \qquad (2.1)$$

So if $\Delta E = h$ we need to perform the measurement during a time interval of the order of 1 s. What does this mean in terms of a CW NMR experiment? The measurements are made by varying the irradiation frequency across some band where we expect the resonances to fall. The requirement that we spend around 1 s measuring each 1 Hz interval restricts the *sweep rate* to 1 Hz/s. Now, the typical width of a proton spectrum is 10 p.p.m., which would be 1000 Hz on a 100 MHz spectrometer, so a complete scan across the spectrum would require *one thousand seconds* (about 15 minutes). Since we require 4 repetitions to double the signal-to-noise ratio and 16 to quadruple it (because it depends on the square root of the number of experiments), the first doubling takes an hour and the second can't be achieved before lunch. Thus, CAT applied to CW NMR is not a very helpful technique.

We should note at once that the trouble arises because we require 'high resolution' measurements. If we relax the requirements on frequency discrimination, we can acquire data more quickly. We will encounter this trade off between speed and resolution constantly when thinking about NMR (and indeed, all spectroscopy; it is simply a more pressing problem in NMR). It happens that, for solution NMR of spin-$\frac{1}{2}$ nuclei, the CW mode is at a considerable disadvantage compared with the method discussed later. This might not be so were the lines of interest much broader, as we would find in solid state NMR for instance, but that is outside the scope of this book. For the kind of spectra of most consequence to everyday chemistry, we need to find a quicker way of extracting the data so we can signal average to better effect. First, though, let us turn to the apparently irrelevant question of bells $\cdots$

2.2 THE TUNING OF BELLS

I am given to understand that a bell (I am thinking of a big one, like a church bell), when it first emerges from the foundry, is not much use. The casting process is not accurate enough to ensure a harmonious product; it has many discordant resonances. Careful tuning is required, which is brought about by shaving slivers of metal from the bell to change the relative contributions of the various harmonics which may be present. Suppose that we, as physical scientists, have been called in to examine the possibility of automating the measurement of the bell's harmonic content.

We might propose a scheme something like this. Mount a sound source on the bell, such as a loudspeaker, and also some kind of receiving transducer (a microphone). With a signal generator, stimulate the bell with audio frequencies, varying them from zero up to the limit of human hearing. The rate at which we vary the frequency will be limited by the required accuracy of the measurement, and by physical properties of the bell. The signal from our transducer will then vary as we pass through the characteristic resonances, and we can connect up the output to a graph plotter to yield the spectrum of responses as a function of frequency. Once we have the spectrum, we can shave off a bit of metal and repeat the whole procedure until the desired response is obtained. This might work, but it would be inconveniently slow, *because it is a continuous wave measurement.*

The bell tuner with years of experience, meanwhile, would be expressing quiet amusement in the corner of the room, because he knows of a much faster way to do the experiment. He has his trusty hammer at the ready, and when we have finished messing around he gives the bell a good clout with it. It is immediately clear to anyone who is not tone-deaf that the required information is now directly available. The response of the bell *to an impulse* (the hammer-blow) contains all the characteristic frequencies together, and we can analyse them directly with our ears. We have switched from a continuous wave to a pulse method, and the advantage is obvious.

2.3 PULSE NMR

2.3.1 Introduction

The possibility of extracting the complete frequency response of a system in one go is just what we need to accelerate NMR data collection. How to implement the 'hammer-blow' in NMR terms is not self-evident, and neither is the method for unscrambling the frequency information from the resulting response, but the potential advantage is readily estimated. In the CW experiment, requiring 1 Hz resolution over a 1000 Hz spectral width led to a 1000 s experiment time. Supposing it proves possible to analyse the response of the sample to an impulse instead, we can evidently complete this alternative experiment in just 1 s, because while the requirement for spending 1 s on the measurement of *each* frequency remains, we are measuring *all* frequencies simultaneously instead of one after another. I will say that again, because it is so important. We can change from sequential measurement (the CW experiment) to simultaneous measurement (the impulse experiment), and thereby make the total experiment time smaller.

It seems at this point as though the advantage we have gained (known as the *Felgett advantage*) is 1000-fold. In fact, although this is true for a single experiment as described here, it is a gross overestimate of what we can achieve in multi-experiment signal averaging. So far we have only examined fundamental considerations of accuracy; we will find when we look into the pulse experiment more closely that other properties of the nuclear system regulate how often we may repeat the measurement. Nevertheless, it proves that very significant time savings can be achieved using pulse methods to excite NMR signals. This was the original stimulus that lead to the widespread adoption of pulse NMR. In the rest of this chapter we are going to look at how to implement this technique, and in particular how to analyse the data it generates. Pulse methods have numerous other interesting properties aside from speed, and these form the subject of the remainder of the book.

2.3.2   Stimulating the Sample

In the 'continuous wave' measurement of a bell, we apply a weak mechanical stimulus, while to perform an experiment in 'pulse' mode we need a strong one. Since CW NMR uses irradiation with a 'weak' radio-frequency (*rf*) source, it is easy to imagine that the required stimulus in a pulse experiment is 'strong' *rf*. This is true, but it is not completely obvious what 'weak' and 'strong' mean in this context, or how we decide what constitutes a short enough 'impulse' in an NMR experiment. For a proper understanding of these questions we need to wait until Chapter 4, when we will know enough to analyse exactly what happens inside a sample when it is subject to a pulse. However, we can argue on fundamental grounds that the *rf* 'impulse' must have certain characteristics.

What we require is an excitation of the NMR signals that is effective over a range of frequencies. For practical reasons the *rf* source at our disposal is invariably monochromatic, so we have to work out a way of using a single frequency to excite multiple frequencies. To see how this can be done, we take our cue from the uncertainty principle again. If the irradiation is applied for a time Δt, then arguing exactly as before we have:

$$h\Delta v \Delta t \sim h \qquad\qquad (2.2)$$

So the nominally monochromatic irradiation is uncertain in frequency by about $1/\Delta t$ Hz if we only apply it for Δt s. In pulse NMR experiments Δt is the *pulse width*, so for our 1000 Hz wide spectrum we estimate that we need a pulse no longer than 1 ms to excite all the signals. Once again, this is only an order-of-magnitude estimate from first principles, and later we will discover reasons why the pulse should be much shorter, μs rather than ms. Still, we now have at least an idea of what kind of 'impulse' is required in NMR.

This is the basic recipe for doing a pulse NMR experiment. The sample is subject to a short burst of *rf*, certainly lasting less than 1 ms. This should stimulate signals, which we measure for a time depending on the accuracy required, 1 s in the example we have been discussing. Just as the bell will not sound forever, due to physical damping of its vibration, the NMR response fades over a period following the pulse, so it is often referred to as a *Free Induction Decay* (FID). Subsequently we can repeat the experiment, so as to improve the signal-to-noise ratio. When sufficient repetitions have been made, we will be left with data containing information about all the frequencies in the NMR spectrum, but in an unfamiliar form. In the case of the ringing bell, it is intuitively obvious that the 'unfamiliar' type of data contains the same information as the CW spectrum, and we can pick out the details we want using our senses. We must now turn to the question of how to extract the required frequency spectrum from NMR data obtained in the pulse mode.

2.3.3   Time and Frequency

What is the difference between the signals we obtain in CW and pulse experiments? When we sweep the irradiation across the spectrum, the measurement we make is signal amplitude as a function of frequency (the measurement is in the *frequency domain*). In recording data after a pulse, however, we measure amplitude as it evolves with *time* (i.e. in the *time domain*) (Figure 2.2). The reciprocal nature of time and frequency suggests that there may be a direct relationship between the two forms of data, and this is found to be the case. The *Fourier transform* allows us to interconvert them, and is the usual means for analysing the results of pulse experiments. Fourier analysis is a complete branch of mathematics in itself, and we hardly have time to investigate it, but we can at least persuade ourselves that interconverting the time and frequency representations of the data is a viable operation.

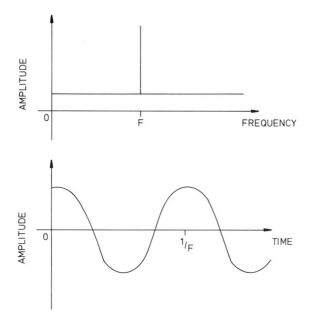

Figure 2.2 Two different ways of viewing the same thing: a wave (amplitude as a function of *time*) and a frequency spectrum (amplitude as a function of *frequency*).

This requires careful consideration of the nature of each form of data. The frequency spectrum consists of a set of 'peaks', whose intensities represent the proportions of each frequency component present. With perfect resolution, each peak would be perfectly sharp, but in practice they contain a spread of frequencies. In time domain form, all the resonances act together, and the oscillation we observe is the sum of the contributing frequencies, in the proportions of their amplitudes in the frequency domain. To ask whether we can interconvert the two forms is equivalent to asking whether it is reasonable to search for the combination of frequencies and amplitudes which would add up to the observed time domain oscillation. It is not difficult to imagine doing this in a direct fashion, simply by simulating various combinations of frequencies numerically and comparing the result with the experimental data. While it is not clear that such an approach would yield results in a practical length of time, it does sound like it should work in principle (in fact, it *can* be done like this, but the computational requirements are much greater than for the Fourier transform approach - see section 2.6).

A more detailed analysis of the problem leads to this famous expression relating the two types of data[1]:

$$f(\omega) \;=\; \int_{-\infty}^{\infty} f(t)e^{i\omega t}\mathrm{d}t \qquad (2.3)$$

Here $f(t)$ represents the time domain data, and $f(\omega)$ the required frequency spectrum. This formula has a number of alarming features, being an integral and also involving complex numbers, but fortunately we do not need to worry about it too much. Provided the NMR signals can be converted to numerical form, the integral can be approximated as a sum, and an efficient algorithm is available for its evaluation (the Cooley-Tukey fast Fourier transform). If you accept that this relationship works, we can move on to examine the more immediate question of how it can be applied in practice.

2.4 PRACTICAL IMPLEMENTATION OF PULSE NMR

2.4.1 Introduction

In this section I want to examine how we can put the rather abstract ideas of section 2.3 into practice. So far we have concluded that it is advantageous to measure the NMR response following a pulse (the FID), because the experiment can be completed more quickly. I have argued that it is reasonable to expect to be able to extract the familiar frequency spectrum from this data, and told you that the Fourier transform is the most common way of doing it. For newcomers to FT NMR this idea of interconverting two kinds of data tends to be the biggest conceptual difficulty. Disquiet about this process can be dispelled in no better way than sitting at a spectrometer and watching it do the calculation; if you get a chance to do this, please take it. Here is the kind of thing you might see.

Figure 2.3 shows, at the top, the FID resulting from a sample with only one line (an H_2O/D_2O mixture). It has the features we might expect; an oscillation, corresponding with the chemical shift of the line, which fades out over a period of a few seconds (we will see *why* this happens in Chapter 4, but of course it would be physically unreasonable for the signal not to decay). Below we see the result of numerical transformation into the frequency domain; because the FID lasted a finite time there is some uncertainty in the calculated frequency, so the line has a characteristic shape. The shape we see here, known as a *Lorentzian,* is the result of transforming an FID which decayed *exponentially,* and is typical of NMR signals obtained in the liquid state.

Clearly there are a number of problems involved in implementing this scheme for pulse NMR. NMR signals are electrical, being high-frequency oscillations, but we need to analyse them numerically. The process of getting the NMR data into a form suitable for numerical analysis imposes restrictions on the experiment which it is most important to understand. We have to take an electrical oscillation, convert it into a stream of numbers, and store these somewhere so that the process can be repeated for signal averaging. Each stage introduces its own problems and limitations. Until quite recently, the computations required to execute the Fourier transform would have taken a considerable time on available computers, but this is now an insignificant problem for one-dimensional experiments. However, it may still cause trouble in the two-dimensional methods described later, where the data sets are generally much larger.

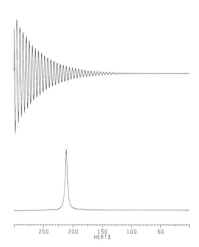

Figure 2.3 The Fourer transform converts a time domain signal (above) into the corresponding frequency spectrum (below).

2.4.2 Detection and Acquisition

Introduction

Converting electrical signals into numerical form is a common problem, and devices to do it are readily available (*analogue-to-digital converters,* or ADC's). The property of an ADC is that, presented with a voltage at its input, it outputs a binary number representing the size of the voltage. An ADC is characterised by the *time* this process takes, and by the number of bits used in the binary representation (the *ADC resolution*). Both these properties affect the kind of NMR experiment we can do; the effect of the speed of conversion is discussed below, and that of the ADC resolution in Chapter 3 (section 3.4.3 - *Dynamic range and ADC resolution*).

NMR signals emerge from the receiver of our spectrometer, and our first idea might be to sample them directly with an ADC (Figure 2.4); however,

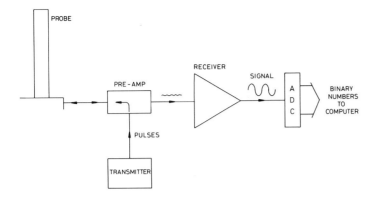

Figure 2.4 Our first idea for carrying out FT NMR experiments (but *not* the best way!).

a moment's thought reveals that this would be unnecessarily difficult. Consider a proton spectrum recorded at 500 MHz. The various *rf* signals arising from the sample are all in the region of 500 MHz, differing only by the chemical shift range present. For protons this would typically be 10 p.p.m. or 5000 Hz, so the frequencies encountered might run from 500,000,000 Hz to 500,005,000 Hz. There is no objection in principle to attempting direct digitisation of these signals, but life can be made much easier by noting that it is only the chemical shift differences that need to be measured. If we *subtract* some reference frequency from the NMR signal, say 500 MHz, then we are left only with the chemical shifts to digitise, in this case frequencies from 0 to 5000 Hz. In essence, all the MHz are uninteresting because they are just the carrier on which the chemical shifts ride. This is exactly the same as the manner in which radio waves are used to carry audio signals from one location to another in broadcasting: radio programmes would seem exceedingly peculiar if we did not subtract the carrier frequency first!

So a practical scheme for a pulse spectrometer involves a *detection* process (Figure 2.5). In Chapter 4 (section 4.3.5) we will find that there are various approaches to this step, but for the time being we can take it as amounting to subtraction of a frequency less than the smallest frequency we expect to encounter. The detector output, consisting of frequencies from 0 to 5000 Hz in our example of a 500 MHz proton spectrum, is passed to the ADC; it is necessary to decide how frequently and for how long to sample this signal.

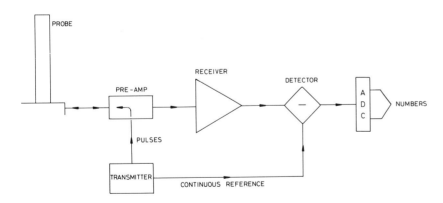

Figure 2.5 A more sensible scheme for FT NMR involves detection against a reference frequency.

Sampling the signals

The question of how long to spend sampling an FID has already been considered; it was the whole basis for preferring a pulse method. If we want to discriminate between features separated by Δv Hz, then sampling should continue for at least $1/\Delta v$ s. This presupposes that the NMR signals persist this long; if they do not then we *can't* resolve such close features because the intrinsic linewidth due to the sample is too great. The length of time spent sampling is referred to as the *acquisition time* A_t (the subscript t, which may seem redundant at the moment, will come into play later when we see experiments involving more than one acquisition time).

In practice, proton signals are often sampled long enough for almost the whole FID to be recorded (typically 2-3 s), because we require the resolution to be limited by the sample and not by our measuring technique. For other nuclei we often sample only the first part of the FID, because resolution is less important than speedy data acquisition. *Truncating* the data in this way can cause some interesting effects, discussed later.

While the duration of sampling depends on the *minimum* frequency difference of interest, its *rate* depends on the total spectral range. Clearly, some entirely arbitrary waveform might not be represented accurately by samples taken at finite intervals; anything might happen between each point. NMR signals, however, are periodic oscillations, and for a given experiment we know the highest frequency likely to be present. This means that it is possible to calculate a sampling rate sufficient to characterise the data; if frequencies up to $\mathcal{N}$ Hz are present, the signal must be sampled every $1/2\mathcal{N}$ s.

The highest frequency which can be characterised by sampling at a particular rate (i.e. $\mathcal{N}$) is known as the *Nyquist frequency*, but in the context of NMR it is often just called the spectral width. By analogy with CW NMR, it is also common to hear the spectral width, as defined by the sampling rate, called the 'sweep width', but this usage is very misleading, since there is no sweep in a pulse experiment. I will always say 'spectral width', 'spectral range' or 'Nyquist frequency' when referring to this quantity. Notice that the maximum possible spectral width a particular spectrometer can handle is determined by its ADC, because there will be an upper limit to the rate at which it can sample the signal. For spectrometers designed for high-resolution work the shortest ADC conversion time is typically in the range 10 μs to 3 μs, giving maximum spectral widths from 50 kHz to 150 kHz.

It would be too much of a digression to try to derive the relationship between sampling rate and spectral width, but it is easy to convince yourself that it has the right order of magnitude by imagining the sampling process (Figure 2.6). If we think of trying to characterise a sine wave by sampling it at discrete intervals, it is evidently inappropriate to sample much below its frequency, because each cycle is then characterised by less than one point. Likewise, it is redundant to sample very many times per cycle, because the periodic nature of the oscillation allows us to interpolate the complete waveform from only a few (in fact, 2) points.

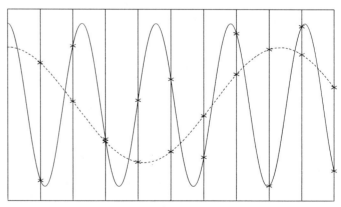

Figure 2.6 Digital sampling of two waveforms at regular intervals (represented by the vertical lines). The interval between sample points might be 100 μs, in which case this picture represents the first ms of sampling with a Nyquist frequency of 5 kHz, and the two waves represented here have frequencies of 4·5 and 1·2 kHz.

It is also interesting to consider what happens to a waveform with frequency higher than the Nyquist frequency (Figure 2.7). Here we see two signals, one with frequency F ($F < \mathcal{N}$), and another with frequency $2\mathcal{N} - F$ (clearly $> \mathcal{N}$). Note that they *both* pass through the same sample points; the higher frequency goes through an extra part cycle between

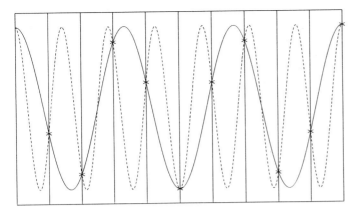

Figure 2.7 Attempting to digitise a waveform with frequency greater than the Nyquist frequency; see text for details.

points, but after digitisation we know nothing of this. This is a most important property of spectroscopy based on digital sampling - frequencies outside the band defined by the sampling rate are still detected, but at false positions. Signals occurring in the wrong place in this way are said to be *aliased* or, more commonly when speaking of NMR, *folded*. For an example of how they appear in a real spectrum, see section 2.5.5.

These considerations of sampling have some important practical consequences for experimental FT NMR. Suppose we desire 0·2 Hz resolution from an experiment, which implies $A_t = 5$ s. If we are observing protons at 500 MHz the spectral width is likely to be 5000 Hz, so according to the Nyquist criterion the signal must be sampled every $1/10,000$ s ($= 0·1$ ms). Sampling every 0·1 ms for 5 s generates *fifty thousand* data points, which have to be stored and subject to Fourier transformation. On most modern spectrometers such a quantity of data could be handled with little difficulty, but when we come to two-dimensional experiments (in which the number of data points may be *squared*) digitisation on this scale becomes unthinkable.

Controlling the bandwidth

The fact that digital sampling of a signal admits *all* frequencies, by representing them in incorrect places, is a major disadvantage of pulse NMR. In fact, the more you think about it, the worse it seems. The most obvious problem is that uninteresting peaks, such as intense solvent resonances, cannot be excluded from the spectrum. In CW NMR, if a region of the spectrum contains an inconveniently large peak, we simply choose not to look at it by selecting the sweep range accordingly. In pulse NMR, however, the pulse brings about non-selective excitation; together with the properties of digital sampling this means that we get all peaks whether we want them or not. The problems that this can cause are discussed in Chapter 3, section 3.4.3, and various ways round them in Chapter 7, section 7.7.2.

Even if we arrange that all *peaks* are within the spectral range characterised by our choice of sampling rate, something will still lie outside it. This is the electrical noise we have been striving to eliminate, which contains an essentially infinite range of frequency components (*white* noise). This seems at first like a fatal flaw in the scheme for pulse NMR: a virtually unlimited amount of noise could be folded into the spectrum, completely negating any sensitivity advantage gained through signal averaging. To avoid this catastrophe, it is necessary to limit the *electrical* bandwidth of the spectrometer, by placing a bandpass filter before the ADC (so we have a third experimental scheme - Figure 2.8).

Since we presumably want to be able to select different spectral ranges for different experiments, this filter needs to be adjustable, with bandwidths from a few Hz upto the maximum spectral range determined by the ADC. In operating a spectrometer the setting of the filter is often invisible, being

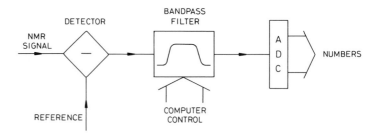

Figure 2.8 A yet more refined scheme for FT NMR adds a bandpass filter before the ADC.

carried out by the control software once the sampling rate has been chosen, but it is important to be aware of its existence. Figure 2.9 illustrates some permutations on settings of spectral width and filter width. Making a filter suitable for this application is non-trivial, and various spectral distortions resulting from it are mentioned in other chapters.

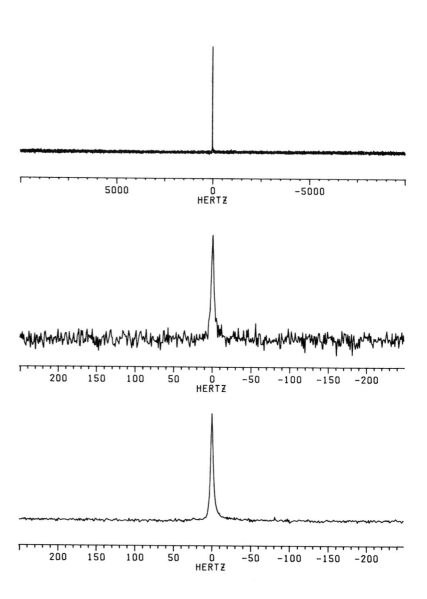

Figure 2.9 Showing the importance of the bandpass filter. In the uppermost trace, both the spectral width and the filter width are large. In the centre, the spectral width has been reduced, but the filter remains unchanged; all the noise visible in the top spectrum is folded into the new observed region. By correcting the setting of the filter to suit the new spectral width (bottom) the signal-to-noise ratio is greatly enhanced.

2.4.3 Transformation

Once sufficient time averaging has been performed to give an adequate signal-to-noise ratio, the digital data must be transformed into the frequency domain. Practical aspects of this process, and some very interesting manipulations that can be performed on the time domain data before it is transformed, are discussed in section 2.5. Here I want to investigate briefly the meaning of the terms in the Fourier transform, as a prelude to a more extensive discussion of detection methods in Chapter 4. I would recommend omitting this section if it seems to confuse the issue, aside from looking at the two forms of the Lorentzian line.

Recall the formula for the Fourier transform:

$$f(\omega) = \int_{-\infty}^{\infty} f(t)e^{i\omega t}\mathrm{d}t \tag{2.3}$$

$f(t)$ could be a complex function, but in the experimental set up we have been discussing it is real: the time dependence of the amplitude of the NMR signal. Despite this, it appears that $f(\omega)$ is going to be complex, because of the complex exponential in the integral. This seems a little obscure, but it turns out to have a simple interpretation. It helps to represent the exponential in its alternative form as a combination of trig functions:

$$e^{i\omega t} = \cos\omega t + i\sin\omega t \tag{2.4}$$

The transform is then seen to have real and imaginary parts:

$$\mathrm{Re}[f(\omega)] = \int_{-\infty}^{\infty} f(t)\cos\omega t\,\mathrm{d}t$$

$$\mathrm{Im}[f(\omega)] = \int_{-\infty}^{\infty} f(t)\sin\omega t\,\mathrm{d}t \tag{2.5}$$

Each of these parts contains a representation of the spectrum, but with different forms of the Lorentzian line. Provided certain experimental conditions are satisfied (Chapter 4), the real part of the transform contains *absorption mode* lines and the imaginary part *dispersion mode* lines (Figure 2.10). It is conventional to display NMR spectra in absorption mode, so the real part of the spectrum is used for plotting.

Figure 2.10 Absorption mode (left) and dispersion mode (right) forms of the Lorentzian line; note the wide wings of the dispersive line.

This generation of *two* forms of spectrum by the transform is a manifestation of the fact that there is another variable in the time domain we have not yet considered. Each NMR signal has its characteristic amplitude and frequency, but a wave also has *phase* (i.e. where in the waveform it starts - Figure 2.11). The signals may all have constant phase, different from zero, or they may have different phases according to their frequencies, and this

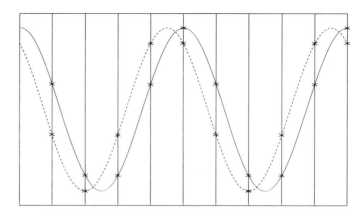

Figure 2.11 Two waves with equal frequency and amplitude may still differ in phase (in this case by $\pi/5$ radians).

fact is reflected in the relationship between the real and imaginary parts of the transform. We will look at this more thoroughly in Chapter 4, where we will also discover an experimental scheme which requires that the function $f(t)$ is complex (i.e. the time domain signal has two components). For the time being, note that the effect of phase changes in the time domain is to mix the real and imaginary parts of the frequency domain spectrum, which gives rise to lineshapes containing part absorptive and part dispersive character (Figure 2.12).

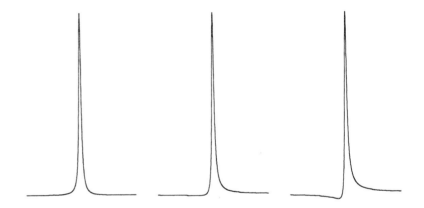

Figure 2.12 As the phase of the time domain signal changes (here in 10° steps), the absorption and dispersion parts of the frequency spectrum intermingle, leading to lineshape changes of this kind.

2.5 WORKING WITH FT NMR

2.5.1 Introduction

The complete machinery of detection, digitisation, data storage and transformation discussed in Section 2.4 is often ignored by casual users of FT spectroscopy. For many simple applications it is possible to get away with this, because the limitations imposed by the technique are well below the level of interpretation of the results. For example, proton spectra may be run over 10 p.p.m. and then plotted on a piece of paper 50 cm long. On a 500 MHz spectrometer this would mean the plot scale was 100 Hz/cm. The data points representing the spectrum might be 0·4 Hz apart, so every cm of plot is characterised by 500 points. At this level the influence of the digitisation is quite invisible, and you need never know that the spectrum was anything other than a continuous line. This approach is fine for routine assays or purity checks, but as soon as we tackle a real structural problem it becomes inadequate.

In dealing with real problems we want to push the spectrometer to its limits. There is no sense having a powerful and expensive FT spectrometer and treating it like a CW machine with a bit of extra dispersion and no annoying wiggles after the lines. The following sections illustrate various features of FT spectra; there are both advantages and limitations.

2.5.2 Digital Resolution and Acquisition Time

Introduction

So far we have discussed the sampling parameters mainly in terms of *time*. Thus, the spectral range determines the interval between measurements of the signal, and the required resolution the total duration of the measurement. This is a convenient viewpoint to take when setting up the experiment, because measurements are made in the time domain. However, once the data has been transformed it becomes more natural to think in terms of frequencies. If we call the interval between data points in the frequency spectrum $\mathscr{R}_d$ Hz (the *digital resolution*), then:

$$\mathscr{R}_d = 1/A_t \qquad (2.6)$$

(because to characterise a spectral width F we sample every $1/2F$ s, so the total number of sample points N is $2F.A_t$; since half of these points represent the real part of the spectrum the digital resolution is $2F/N$).

This abstract consideration has a very concrete impact on spectra, as we can see by examining one closely (Figure 2.13). For proton spectra such as this $\mathscr{R}_d$ is typically 0·3–0·4 Hz/point, but proton linewidths in small molecules can be 0·1 Hz or less. Thus as soon as you wish to interpret fine structure in a proton spectrum it becomes necessary to improve the digital resolution (note that $\mathscr{R}_d$ should be significantly less than the linewidth to characterise a line properly). This is achieved by increasing A_t, either by reducing the spectral width or increasing the number of data points used to acquire the spectrum. The inevitable cost of this is reduced sensitivity, because more time is spent acquiring each spectrum so less signal averaging is possible.

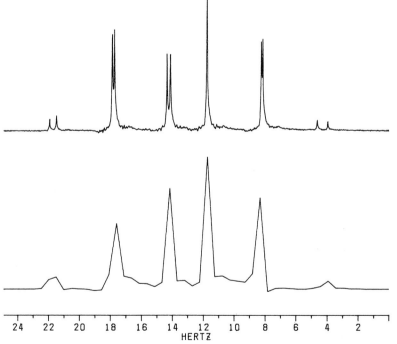

Figure 2.13 Inadequate digital resolution can completely mask spectral features. In the lower spectrum, acquired under fairly 'normal' conditions for proton NMR ($A_t = 2$ s, $\mathscr{R}_d = 0·5$ Hz/point), the individual data points can clearly be seen (the spectrum is composed of straight line segments). Improving the digitisation (upper spectrum, $A_t = 65$ s, $\mathscr{R}_d = 0·015$ Hz/point) reveals the true spectrum, limited only by the natural linewidths and the resolution of the spectrometer (this is a test sample with extremely narrow lines).

In heteronuclear spectroscopy, even of nuclei like ^{13}C that have very narrow lines, it is common to work with quite low digital resolution (2-3 Hz/point or worse). This is natural, because the concern is usually with sensitivity; many heteronuclei are of low abundance and show no fine structure in their spectra. Amongst the common nuclei, a notable exception to this statement is ^{31}P. In phosphorus NMR we have the undesirable combination of high frequency, wide chemical shift range and homonuclear coupling to contend with, so achieving adequate digital resolution can be tricky.

Zero-filling

If all we are interested in is *distinguishing* lines, then setting the acquisition time according to their separation will automatically ensure the digital resolution is suitable. However, there is a big difference between just being able to tell two lines apart and completely characterising their shapes by covering them with data points. To achieve the latter requires far higher digital resolution, more than we may be inclined to obtain by increasing A_t. The process of *zero-filling* can then be used to improve $\mathscr{R}_d$, provided the FID has decayed close to zero by the end of A_t. The idea is that, if this condition is met, increasing the acquisition time further would only amount to recording no signal anyway. Thus we might as well not bother with the measurement, but simply use the computer to append zeros on to the data before the transform (Figure 2.14). This takes advantage of the fact that computer memory is readily available, while time for acquiring spectra may not be.

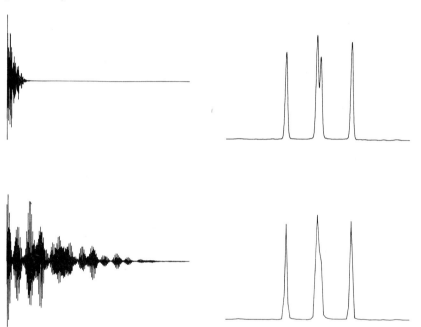

Figure 2.14   Zero-filling the time domain data interpolates extra points into the frequency spectrum, improving its appearance.

The improvement resulting from zero-filling is cosmetic (no extra information is added to the spectrum), but often worthwhile nonetheless. A typical 1D application would be to zero-fill in conjunction with the resolution enhancement discussed in section 2.5.4, to improve the appearance of a complex multiplet. In 2D experiments, where acquisition times are sometimes disconcertingly short, zero-filling can often be used to good effect to improve the appearance of at least one of the dimensions (this is discussed at some length in Chapter 8). In the 2D case, or anytime that the assumption that the FID has decayed nearly to zero during A_t is not true, great care must be taken to ensure correct *apodisation* - see below.

2.5.3   Truncation and Apodisation

Some quite strange things happen to digitised NMR lines if A_t is short relative to the natural decay time of the FID. We could regard data acquired in this fashion as being the product of a 'complete' FID and a step function (i.e. a function with value 1 between $t = 0$ and A_t and 0 thereafter), which chops off the signal at A_t, particularly if we imagine the data being zero-filled to some extent (Figure 2.15). The Fourier transform of the product of two functions turns out to be a kind of mixture (called a *convolution)* of their individual Fourier transforms; it is a sort of running

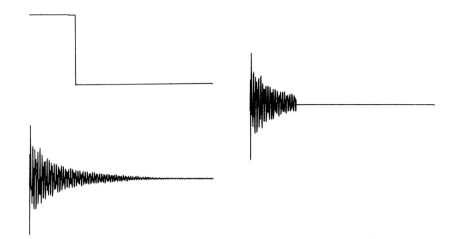

Figure 2.15 One way to view data acquired with A_t too short: the product of a complete FID and a step function.

average of the two curves. The transform of a step is the function $(\sin x)/x$ (often called sinc x - Figure 2.16); when this is convoluted with a Lorentzian line the result has 'wiggles' at its base (Figure 2.17).

Since this kind of lineshape distortion only occurs when A_t is relatively short, it is not particularly common in routine 1D proton spectroscopy, but it may sometimes be seen at the base of solvent resonances. In heteronuclear and 2D NMR, however, eliminating truncation 'wiggles' becomes a matter of some importance. The key to doing this is to note that it is the 'sharp edge' at the end of the FID that is responsible for the problem; if this is smoothed out the wiggles disappear. The edge can be smoothed by multiplying the FID by a function which begins at 1 but decays smoothly to zero by the end of A_t; this is an example of the use of a *window function* prior to Fourier transformation (Figure 2.17). When the object of this manipulation is reduction of truncation wiggles it is referred to as *apodisation* ('cutting the feet off'); for examples of the kind of function we might use to achieve this, see the following section.

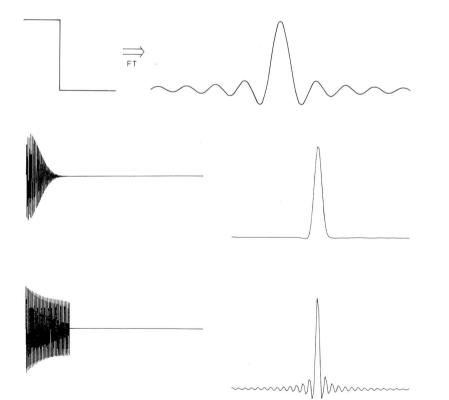

Figure 2.16 The Fourier transform of a step crops up quite often in FT NMR; it is the function sinc x.

Figure 2.17 The transform of truncated data (bottom) is a 'convolution' of a Lorentzian line and a sinc function; smoothing out the step in the FID by apodisation cures the wiggles, but broadens the line (top).

2.5.4 Window Functions

Introduction

Apodisation, as described in the previous section, is just one example of a range of effects that can be brought about by manipulating the FID prior to transformation. In essence, by adjusting the decay envelope of the FID, we can play with the balance of signal-to-noise ratio and resolution in the transformed spectrum. This possibility is by no means specific to FT NMR, but the availability of the data in time domain form makes the required computations rather straightforward (and of course CW spectrometers do not usually have computers attached). Use of window functions is an essential part of the process of analysing a spectrum; whether the aim is to optimise sensitivity or resolution or simply to apodise the data, the limits of the spectrometer have not been reached until the best window function has been found and applied. Out of a wide range of functions which have been suggested for this application I will illustrate two; one designed to improve sensitivity and the other designed to improve resolution.

Sensitivity - the matched filter

Since the NMR signal decays during each scan, while the noise amplitude remains constant, decreasing the relative contribution of the tail end of the FID might improve the signal-to-noise ratio. This can be achieved by multiplying the data by a decreasing exponential function. If the natural decay of the signal is described by:

$$\mathscr{S} = Ae^{-t/T_2} \tag{2.7}$$

(it will become clear in Chapter 4 why the symbol T_2 is used for the time constant of this decay), and we multiply the FID by a window function $\mathscr{E}$:

$$\mathscr{E} = e^{-t/a} \tag{2.8}$$

(where a is positive) the desired reduction in the size of the tail occurs (Figure 2.18). However, we have to be rather careful. Multiplication in this fashion also speeds the apparent decay of the signal; in the frequency domain this corresponds with broadening the line (because we have apparently been able to measure the signal for less time). The half-height width δv of a Lorentzian line is related to the time constant T_2 by:

$$\delta v = \frac{1}{\pi T_2} \tag{2.9}$$

Application of the window function $\mathscr{E}$ decreases the apparent T_2 to a value T_2' given by:

$$\frac{1}{T_2'} = \frac{1}{T_2} + \frac{1}{a} \tag{2.10}$$

Now, broadening the line in this way reduces its height, so the ratio of peak height to noise amplitude is not necessarily improved by the manipulation. Careful analysis of the problem indeed shows that excessive line broadening (i.e. choice of too small a value for a) degrades the sensivity, while of course large values of a have little effect. There is an optimum balance between the noise reduction and the line broadening effects, reached when $a = T_2$, or in other words when the window function *doubles the linewidth* in the frequency domain. This window function is known as the *matched filter*, and is the appropriate choice to use when the best sensitivity is required (Figure 2.18). Note that 'matched' means 'matched to the decay envelope of the FID', so the ideal matched filter is only an exponential if the FID envelope is as well.

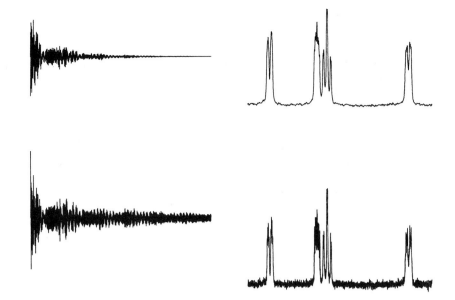

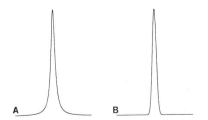

Figure 2.18 Application of the matched filter improves sensitivity.

In practice it is easy to determine the required value for a; simply transform the spectrum without use of the window function, examine the linewidth of the interesting signals (making sure $\mathcal{R}_d$ is adequate first by zero-filling if necessary), and compute T_2 (and hence a) according to the above equation (2.9). The transform can then be repeated after application of the window function to the original FID. Actually, spectrometer software systems often specify the parameter of this window function as a line-broadening factor anyway, and if this is the case there is no need for mental arithmetic. Application of a window fuction of this kind (often called *exponential multiplication*, for obvious reasons) may help with apodisation, but usually the amount of line broadening required is such as to be far from the matched filtration condition. If this is the case, careful use of the function described next is likely to give better results.

Resolution - the Lorentz - Gauss transformation

Since forcing the FID to decay more quickly by exponential multiplication broadens the lines in the frequency domain, we might expect the opposite effect to be achieved by suppressing its decay. In other words, reversing the sign of a in equation 2.8 should give us a *resolution enhancement* function. This is true, but there are problems; the resulting amplification of the tail of the FID increases the noise level, and large wiggles due to truncation are likely to arise. A better result is obtained by using a function which cancels the decay of the early part of the FID, but then falls smoothly towards zero by the end. There are many possible candidates for functions with this property; one of the most popular is the Lorentz-Gauss transformation:

$$\mathcal{G} = e^{-t/a}e^{-t^2/b} \qquad (2.11)$$

This time a is chosen equal to $-T_2$, and b is positive. Adjustment of the parameter b can then be made according to the result required (again, some spectrometers may not provide direct input of b, but there will be a parameter related to it - you need to discover what the relationship is from the software manual). This function improves the resolution in two ways. First of all, some values of b will actually reduce the half-height linewidth; the price for this will be considerably reduced signal-to-noise ratio. Perhaps more important than this, though, is the change in the *shape* of the line, which is converted to a Gaussian (Figure 2.19). Gaussian lines are much narrower at the base than equivalent Lorentzians (for example, at 1% of

Figure 2.19 Lorentzian (A) and Gaussian (B) lines of equal half-height width; the Gaussian is much narrower at the base.

the peak amplitude a Gaussian is nearly 5 times narrower than a Lorentzian with equal half-height width), and this is the property of the Lorentz-Gauss transformation which makes it so useful.

The best way to apply this function in practice is by trial and error; a can be determined as before, and then b (or its equivalent) varied. Each time the parameters are changed, the data processing is repeated and the frequency domain spectrum examined to see whether the desired improvement has occurred. Since a depends on the original width of the lines, it may not be possible to choose an optimum value for every peak simultaneously; processing with several different versions of the window function is often necessary in a complex spectrum. This is a case where persistence is definitely worthwhile, since quite startling extraction of fine structure from apparently hopeless lumps and bumps may be possible (Figure 2.20). This kind of result can only be achieved when the 'lumps' in fact consist of many overlapping sharp lines; attempts to narrow signals which are genuinely broad (for instance due to exchange) will only result in loss of sensitivity. With large values of a, and b chosen so as to reduce the FID smoothly to zero by the end of A_t, the Lorentz-Gauss transformation is also well suited to reduction of truncation wiggles; the example in Figure 2.17 was processed in this way.

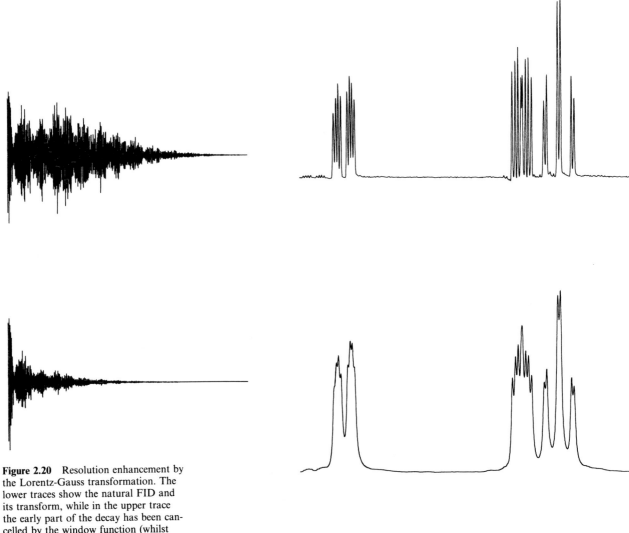

Figure 2.20 Resolution enhancement by the Lorentz-Gauss transformation. The lower traces show the natural FID and its transform, while in the upper trace the early part of the decay has been cancelled by the window function (whilst still ensuring apodisation). Such strong manipulation as this can only be applied to data with a very high signal-to-noise ratio.

2.5.5 Spotting Folded Peaks

The phenomenon of folding, described in section 2.4.2, can be a considerable nuisance. Even when working with 'familiar' nuclei like ¹H it is not always possible to ensure that the spectral width is adequate to encompass every peak; surprises such as protons involved in hydrogen bonding or attached to metals crop up from time to time. Therefore it is important to be able to determine whether a peak is folded. There are several clues to look out for which may indicate this.

The typical circumstance that arises in proton NMR is that most peaks are within the spectral window, while one or two unusual ones fall outside it. In this case it is easy to identify the folded peaks, because they show different *phase* to the others (Figure 2.21). This method falls down when the spectrum only contains one peak, as often happens with heteronuclei, because then there is no phase comparison to be made. In such cases, if there is any reason to think that the spectral window may not encompass the signals (e.g. when examining unusual nuclei), a test must be made to check for folding. This is done by moving the window slightly, say by 100 Hz. If peaks are not folded, then naturally they appear to move in the opposite direction to the window, but by the same amount (Figure 2.22). If they are folded, however, they will show some combination of moving in the wrong direction and/or by the wrong amount (depending on by how much their true frequencies exceed the Nyquist frequency).

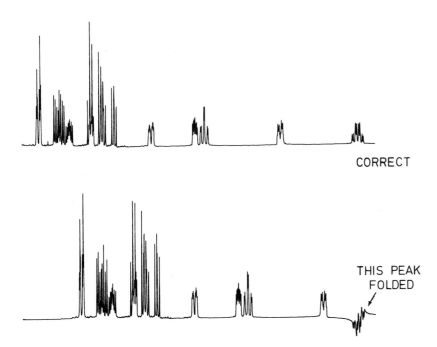

CORRECT

THIS PEAK
FOLDED

Figure 2.21 Identifying folded peaks by their phase properties.

2.6 FT IS NOT THE ONLY WAY

All the spectra we look at in this book will be processed using the Fourier transform. However, as I indicated in section 2.3.3, this is not necessarily the only way to extract the frequency spectrum from time domain data. An alternative technique, known as the *maximum entropy method* (MEM), has attracted so much attention and comment recently that it seems appropriate to mention it, even though we will not use it again. The idea behind this is the apparently rather crude notion of reconstructing the FID by simulating possible combinations of oscillators (i.e. NMR lines) and synthesising the corresponding time domain signal numerically. Obviously such a simulated signal can be compared against

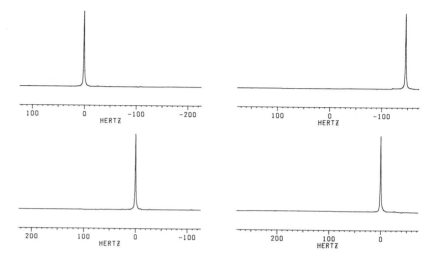

Figure 2.22 In the absence of other peaks against which to make the phase comparison, try moving the spectral window. At the left is the result for a correctly characterised peak, while at the right that for one folded from just outside the window (the result here will depend on precisely how the spectrometer detects signals, but will differ from the correct case).

the experimental data, and some measure of closeness of fit evaluated, such as a least squares calculation. Variation of the simulated spectrum would then allow a search to be made for that which best models the experimental data. The tricky thing (aside from any practical considerations of how long the search might take) is that no unique solution will exist, because the experimental data is not complete - it contains noise, and it was only measured for a finite time.

How should a solution be selected from a range of possibilities which all seem equally 'close' to the experimental data? The maximum entropy criterion specifies that the best spectrum to pick is that which contains *minimum* information (i.e. *maximum* entropy), because then there is no risk that the experimental data is being overinterpreted. This is an approach used widely in other areas such as radio astronomy and optical image enhancement, where noisy data must be analysed. The initial excitement generated by its application to NMR[2] was the hope that, since MEM provides a criterion for identifying peaks free from human bias, much greater effective sensitivity would be obtained in spectra processed this way. This point is still being argued, but it seems that any advantage here is marginal. The question is whether processing using MEM is a better way of selecting peaks from noisy data than simply examining an FT spectrum processed with a matched filter, and choosing a threshold (e.g. some multiple of the noise amplitude) below which peaks should be rejected. This has not been answered quantitatively yet, but there is a widespread feeling that there is little difference between the two.

Where MEM does come into its own is in the processing of very incomplete data, such as spectra accumulated with short acquisition time. Using the Fourier transform in this case we have to ensure apodisation, and whatever window function we choose this will inevitably broaden the lines. With MEM, however, the the simulated time domain signal is fitted to however much real signal was recorded, and the problem of truncation simply does not arise in the same way (Figure 2.23). This seems likely to be particularly useful in the processing of 2D experiments, in which acquisition time are often rather short[3]. To date most spectrometer software packages do not provide MEM processing (and the computational requirements are much greater than for FT), but no doubt in due course this will become available.

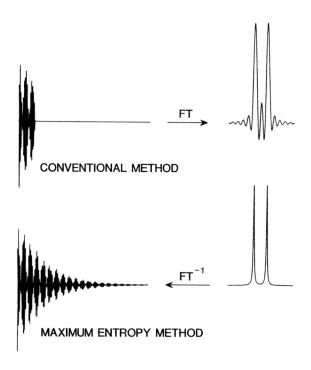

CONVENTIONAL METHOD

MAXIMUM ENTROPY METHOD

Figure 2.23 Processing a highly truncated FID using MEM (compare this with Figure 2.17).

REFERENCES

1.   In accord with the non-mathematical approach of this text, no attempt will be made to explain the origin of the Fourier transform or to discuss its mathematical properties. Many other NMR or physics texts include such discussions; see, for example (in increasing order of difficulty): K. Müllen and P. S. Pregosin, *Fourier Transform NMR Techniques: A Practical Approach*, (Chapter 1), Academic Press, 1976; D. Shaw, *Fourier Transform NMR Spectroscopy*, 2nd. edition, (Chapter 3), Elsevier, 1984; G. Arfken, *Mathematical Methods for Physicists*, 3rd. edition, (Chapters 14, 15), Academic Press, 1985.

2.   S. Sibisi, *Nature (London)*, **301**, 134, (1983); S. Sibisi, J. Skilling, R. G. Brereton, E. D. Laue and J. Staunton, *Nature (London)*, **311**, 446, (1984).

3.   P. J. Hore, *J. Mag. Res.*, **62**, 561-567, (1985).

3

Basic Experimental Methods

3.1 INTRODUCTION

This is the first of two chapters about experimental methods (the other is Chapter 7). Obviously they will be of most interest if you want to operate a spectrometer yourself, but some of the matters discussed here are relevant to bench chemists (sample preparation) or those involved in instrument purchasing (performance tests). On the subject of testing a spectrometer, manufacturers have a natural tendency to slant the results in their favour, so it is worth being familiar with the loopholes available in common test procedures. Some less common tests that can be informative are discussed in Chapter 7. As it is clearly impossible to provide a complete survey of experimental techniques in two short chapters, I have selected for inclusion topics which seem to confuse newcomers to FT instruments, and general procedures required when carrying out the experiments described in the rest of the book.

3.2 SAMPLE PREPARATION

3.2.1 Introduction

In this section I want to draw your attention to some of the commonest problems relating to sample preparation. Many of the factors mentioned are most critical when observing proton spectra, but still it is better to get into the habit of thinking about these questions whatever the experiment to be performed. If the early paragraphs seem obvious or trivial, I apologise; however it is my experience that neglect of the simple precautions described below is as common a source of failure in the practice of NMR as misunderstanding of the more complex matters discussed in the rest of the book. There is a natural excitement associated with the approach of the NMR experiment, perhaps after months of labour to prepare materials, that can lead to excessive haste in getting the sample into the tube and the tube into the spectrometer. A few minutes spent planning these operations can save hours of wasted instrument time.

3.2.2 Choice of Solvent

Obviously the sample must be soluble in the chosen solvent! However, bear in mind that it need not be *very* soluble, at least for proton work, where 1 mg in 0·4 ml would be a strong solution for a medium to high field instrument. Beyond this obvious requirement, the solvent used may influence the results obtained in a number of ways. In proton and carbon work, the solvent signals will obscure regions of the spectrum. The solvent viscosity will affect the resolution obtainable, particularly in proton spectra. Some solvents, like water and methanol, have exchangeable protons which will prevent the observation of other exchangeable protons in the solute. If experiments at other than room temperature are contemplated,

Table 3.1 Properties of some common deuterated NMR solvents. The symbol ~ before a value indicates it is for the corresponding protiated material. The 'Cost' column shows approximate prices (£/10g) for small quantity purchase in the U.K. in 1985.

Solvent	M.P. °C	B.P. °C	$\delta\ ^1H$	$\delta\ ^{13}C$	Cost
acetic acid	15·8	115·5	11·53, 2·03	178·4, 20·0	15
acetone	−93·8	55·5	2·05	206·0, 29·8	10
acetonitrile	~ −48·0	80·7	1·95	118·2, 1·3	15
benzene	6·8	79·1	7·16	128·0	10
chloroform	−64·0	60·9	7·27	77·0	1
cyclohexane	~6·5	78·0	1·38	26·4	75
dichloromethane	~ −97·0	40·0	5·32	53·8	20
dimethylformamide	~ −61·0	~153·0	8·01, 2·91, 2·74	167·7, 35·2, 30·1	130
dimethylsulphoxide	~18·0	~190·0	2·50	39·5	10
dioxan	~12·0	~100·0	3·53	66·5	110
methanol	~ −98·0	65·4	3·31	49·0	20
nitrobenzene	~5·0	~210·0	8·11, 7·67, 7·50	148·6, 134·8, 129·5, 123·5	35
nitromethane	~ −29·0	100·0	4·33	62·8	25
pyridine	~ −42·0	114·4	8·71, 7·55, 7·19	149·9, 135·5, 123·5	20
tetrahydrofuran	−106·0	65·0	3·58, 1·73	67·4, 25·2	150
toluene	~ −93·0	110·0	7·09, 7·00, 6·98, 2·09	137·5, 128·9, 128·0, 125·2, 20·4	20
trifluoroacetic acid	~ −15·0	75·0	11·30	163·8, 115·7	5
water	3·8	101·4	4·63	-	3

Sources: Aldrich Catalogue Handbook of Fine Chemicals 1985; Merck, Sharp and Dohme NMR reference data; Aldrich Library of NMR Spectra, 1983; E. Pretsch, J. Seibl, W. Simon and T. Clerc, *Tabellen zur Strukturaufklärung organischer Verbindungen*, Springer-Verlag, 1981.

then the melting and boiling point of the solvent must be taken into account, together with any temperature dependence of the solubility of the solute. Solvents containing aromatic groups such as benzene and pyridine may cause large changes in the observed chemical shifts of the solute when compared with spectra run in non-aromatic solvents. The strength and sharpness of the deuterium signal available from a solvent may affect the outcome of certain experiments such as difference spectroscopy. Finally, the cost of deuterated compounds varies enormously, which might be an important consideration when selecting a solvent for some routine assay to be performed many times daily. Careful attention to these factors can play a large part in determining the success or failure of an experiment.

A large number of fully or partially deuterated compounds are available commercially. Table 3.1 lists some of the physical properties of those most commonly used in NMR spectroscopy. Generally speaking the cheapest solvents are chloroform and water, of which chloroform has by far the most favourable properties from an NMR point of view as we shall see shortly. Other solvents increase in price according to difficulty of preparation in deuterated form; thus most of the common organic solvents such as benzene, toluene, dimethylsulphoxide (DMSO), acetone, acetonitrile, methanol, dichloromethane, dimethylformamide (DMF) and pyridine are in a similar price range, while tetrahydrofuran (THF) and cyclohexane are among the most expensive.

When proton spectra are required on samples of less than a few mg, the position of the residual peaks in the solvent must be carefully considered. There are three sources of interfering signals: residual protons in the deuterated solvent, dissolved water and other dissolved impurities. Normally only the level of the first of these is specified by the solvent manufacturer, but the other two, and particularly the water, may be very troublesome.

Typically solvents are available with levels of deuteration ranging from 99·5% to 99·995%, with the purer forms often being supplied in small ampoules containing 0·5 ml. Naturally the high purity grades are considerably more expensive. It is important to remember that the specification is referring to deuteration of the solvent, and may or may not imply greater all-round purity; water or other substances may sometimes be found in equal or higher concentration than the residual solvent in the more highly deuterated grades. The only way to find out whether it is worth investing in a highly deuterated solvent is to try it in the application of interest.

The ideal solvent would have residual signals that did not clash at all with those of the solute. Of course, this cannot be achieved and a compromise choice must be made. The chemical shift of the solvent resonance will be known, and it may be expected that a region of the spectrum around this value will be obscured. The obscured region can be quite wide, as generally the residual peak will be a multiplet due to deuterium coupling; the only exceptions to this being chloroform and water. For solvents like acetone and dimethylsulphoxide the residual peak is a quintuplet due to coupling to deuterium, and can easily obscure a region several tenths of a p.p.m. wide. The residual solvent peak in 'normal' grade (i.e. 99·5-99·9% deuterated) solvents starts to be larger than the solute peaks around the level of 2 mg in 0·4 ml for medium molecular weight compounds.

Often even more annoying than the residual solvent peak is that due to water. Almost all NMR solvents contain water when supplied, and also most are quite hygroscopic. The water peak in chloroform, for instance, is usually bigger than the residual $CHCl_3$ peak for conventionally handled samples, and in addition is broad and comes in an inconvenient place (around 1·6 p.p.m.). Some solvents, for instance DMSO, need to be handled under an inert atmosphere by syringe techniques if they are to remain dry enough for very dilute samples. Even water (deuterated) is hygroscopic, and should be kept in a desiccator! If the sample can stand it, the water level can be greatly reduced by filtration through a drying agent. This process can be combined with the necessary filtration direct into the NMR tube to remove particles, and so need not be unduly troublesome. Most normal agents can be tried, activated alumina often being quite suitable. Molecular sieves are not so good as they contain very fine dust which can be difficult to remove afterwards and is detrimental to resolution.

The viscosity of the chosen solvent can have a considerable effect on the observed linewidth, and also on relaxation parameters of the sample, a point discussed further elsewhere. For the time being, the common NMR solvents can be loosely classified as viscous (benzene, DMF, DMSO, pyridine, toluene and water) and non-viscous (acetone, acetonitrile, chloroform, dichloromethane and methanol). The highest resolution can only be obtained in a non-viscous solvent, of which acetone has generally the most favourable properties and is usually chosen for resolution test samples. In routine NMR applications this variation due to viscosity is unlikely to cause many problems, however for precise work, particularly on degassed samples, it becomes very important.

As an aid to solvent selection, Figure 3.1 shows high field NMR spectra (500 MHz) of a range of solvents. These spectra were obtained on fresh samples, usually of solvents supplied in 5g or 10g ampoules and at the lowest specification for deuterium content offered by Aldrich. Solvents which have been stored for any length of time are likely to contain a good deal more water. Low level impurities detected in certain solvents are plotted on a ×128 vertical scale; there is no intention to imply that these are always present, or that this was a typical batch of solvent, but simply that it is necessary to check for contamination before running spectra of very dilute samples.

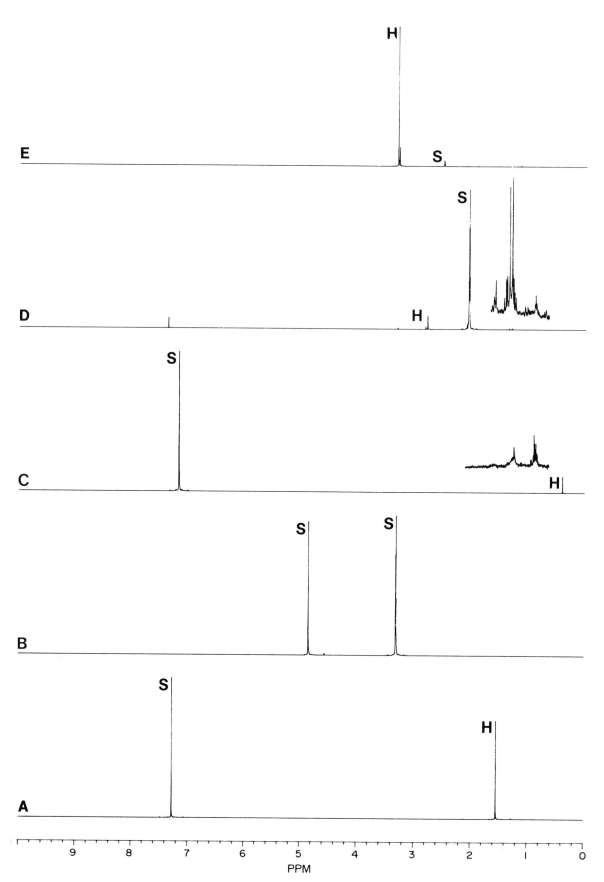

Figure 3.1 500 MHz proton spectra of some common NMR solvents: A, chloroform; B, methanol; C, benzene; D, acetone; E, dimethylsulphoxide; F, dichloromethane; G, water; H, toluene; I, acetonitrile; J, pyridine. Solvent resonances are marked 'S' and dissolved water 'H' in each case; other small impurities are also sometimes visible (see text).

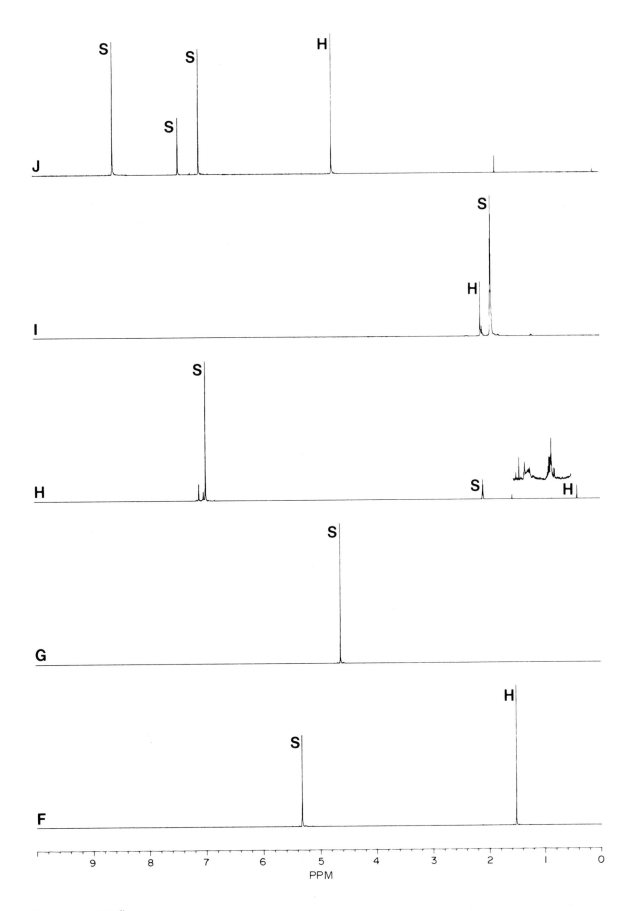

Figure 3.1 (Contd)

3.2.3   The NMR Tube

Wilmad, the principal manufacturer of NMR glassware, offer a very
wide range of different sample tubes with varying specifications. Which
tube is necessary must be the subject of experiment, as it is not possible to
predict the effects of, say, changing specified concentricity of a tube with
any confidence. A general guideline is that the higher the field the more
worthwile it becomes to buy more tightly specified tubes. The symptom of
inadequate tube selection is often unusually bad spinning sidebands, as
most of the potential defects in the tube relate to its cylindrical symmetry.

However good the chosen tube, high quality results will not be obtained
unless it is kept clean and free from dust and scratches. A good way to
clean tubes without scratching them is soaking in a proprietary degreasing
solution (after first rinsing out the contents); scrubbing with small test-tube
brushes is best avoided. Subsequently they can be taken out of the solution
prior to use and rinsed with distilled water and acetone; drying can then be
achieved by blowing filtered nitrogen into the tube through a pipette (the
filtration being conveniently achieved by a small plug of cotton wool). The
more common approach of rinsing followed by oven drying gives plenty of
opportunity for dust to enter, and degrades the specifications of the tube
dimensions.

A point worth being aware of is that brand new NMR tubes straight
from the packet are not usually very clean. A circumstance I have observed
too frequently is that of the owner of an unusually precious sample who
decides to obtain a new tube just for the occasion, fails to clean it and then
gets upset by the huge peaks due to grease or di-octyl phthalate or
whatever that obscure the spectrum. Such traumatic experiences are easily
avoided by good housekeeping.

3.2.4   Sample Volume

When an essentially unlimited amount of material is available, the
volume of solution used can be made fairly large and the precise quantity
is then non-critical. However when the limits of instrumental sensitivity are
approached, as they often are even with the highest field machines, optimising the sample volume becomes important. The reason is simple: the
detection coil in the NMR probe only receives signals from a finite volume,
and any sample outside this volume may as well not be there. Unfortunately, it is not usually possible simply to concentrate all the solution
inside the sensitive volume, because then the field distortions caused by the
solution/air interface degrade lineshape and resolution and consequently
sensitivity. The effect of sample depth on resolution is discussed further
below.

From a practical point of view, it is advisable to experiment to find the
minimum depth of sample which gives adequate resolution on a particular
probe, and also the optimum setting of the tube position in the spinner
turbine to use with a sample of such depth. These values can then be used
when sensitivity is really critical, and say 50% greater sample volume used
normally to make resolution adjustments easier.

Various gadgets are available to help when working with minimal
volume samples. The most important of these is the so-called vortex
suppressor, which is a poly-tetrafluoroethylene (PTFE) or glass plug designed to fit tightly inside the NMR tube. This is pushed down until it
contacts the solution surface, the original intention being to prevent
'whirlpools' forming in large diameter tubes. It also provides a less drastic
field distortion than passing direct from solution to air, and so is worth
experimenting with for short samples even in 5 mm tubes. An important
point to remember when using these devices is that PTFE has a very high
coefficient of thermal expansion, and so it is not suitable for variable
temperature work (at low temperatures the plug drops into the solution,
while at high ones the tube cracks).

PTFE parts are also available to fill out the lower curved end of the tube, in order to concentrate the sample into the sensitive region of the probe. This process is carried to an extreme in the form of various schemes for micro sample cells, which are usually small glass spheres or cylinders designed to fit inside a normal NMR tube, the idea being to put the sample inside the cell so it is all in the sensitive volume, but surround it with solvent so as to avoid field distortions. With probes of 10 mm diameter and greater this works well, but not so well as obtaining a smaller diameter probe, a point discussed later (section 3.4.3). In 5 mm probes, at least with proton observation at high field, the price paid in terms of lost resolution is usually too high in my experience. However, this may not be true of certain probe and magnet combinations so again micro-cells are worth experimenting with.

3.2.5 Sample Handling

High resolution spectra can only be obtained on solutions which are completely free from suspended dust or fibres. To achieve this it is best to get into the habit of always filtering solutions for NMR direct into the sample tube. A small plug of cotton wool at the neck of a Pasteur pipette is a surprisingly effective filter; however it is necessary to pre-rinse it with a little of the solvent to be used to flush out any loose fibres. Fresh medical cotton wool is free from organic soluble contaminants; however the same cannot be said of odd scraps of the material which have stood around the lab. for months. Keep it in an airtight box to avoid this problem. Some solutions may react with cotton wool; in this case glass fibre is an alternative but it does not seem to make such a good filter. Really fine suspended matter can be removed by filtering through a pad of celite. The whole operation can advantageously be combined with drying of the solution as mentioned above.

A further contamination problem occurs in handling dilute aqueous solutions, when sweat from the fingertips can easily find its way into the sample. When working with less than a few hundred μg of material it is necessary to wear gloves continuously to avoid this. The symptom of 'fingertip contamination' in proton NMR is the presence of a 7 Hz doublet around 1·4 p.p.m. and a corresponding quartet around 4 p.p.m. (these are probably due to alanine or lactic acid).

Anticipating the discussion of shimming in section 3.3.4, it is very helpful always to prepare samples with a consistent depth of solution. If this is done by topping up a more concentrated solution to the required depth in the tube, it is important to ensure that the sample is then mixed thoroughly. A surprising cause of apparently inexplicable loss of resolution is the difficulty of mixing layers of solution in 5 mm NMR tubes. The resulting swirls of variable concentration are very detrimental, especially at high field.

3.2.6 Nuclei Other Than Protons

Residual solvent signals cease to be a problem for heteronuclei other than carbon. In this case it is reasonable to make solutions containing only a sufficient proportion of deuterated solvent to provide enough signal for the deuterium lock system (see section 3.3.4). For carbon observation the use of completely deuterated solvent is usually desirable, as the intensity of the solvent signal is greatly reduced owing to the loss of the nuclear Overhauser effect from protons (see Chapter 5) and the division into several lines due to deuterium coupling. For really critical cases of solvent interference, some ^{13}C depleted solvents are available.

3.3 ACHIEVING RESOLUTION

3.3.1 Introduction

Adjusting a spectrometer to give the best performance possible can be a very satisfying process. It can also be a source of extreme frustration sometimes, particularly when the machine seems to choose to be difficult only when containing the most precious of samples or when the head of department is looking over your shoulder. The latter part of this section (3.3.4) attempts to show you how to approach shimming a high-field instrument logically, so as to give consistent and quick results. It is a mistake to suppose that it is only the linewidth that depends on the quality of the shimming. Sensitivity also is affected, and experiments such as difference or two-dimensional spectroscopy that depend on a highly stable field/frequency ratio. Anyone making serious use of a high-field NMR spectrometer needs to be confident of making the adjustments described here accurately on every sample. The first part discusses how resolution is measured, which will only really be important to you if you are involved in buying or maintaining a spectrometer. Nevertheless, if you are learning to do NMR and have the luxury of substantial access to an instrument it can be fun to see if you can equal or exceed the performance specifications for your machine; once you can do this you can be confident that you are getting the best out of real samples.

3.3.2 Criteria of Resolution

One of the fundamental aspects of spectrometer performance is the 'resolution' obtainable with a given combination of probe and magnet. It is important to understand what is meant by this term, and how a large number of factors interact to control the results obtained. The commonest parameter used to specify resolution is the width at half-height of a particular line measured in Hertz. This is a simple enough concept, but in fact is neither the most difficult parameter to optimise nor the most crucial to overall system performance. Instead, it is the extent to which the lineshape deviates from an ideal Lorentzian (or alternatively, the extent to which the time domain decay deviates from exponential) which is critical.

Unfortunately, spectrometer specifications do not quote this directly, but rather measure the half-height linewidth on one sample, and the linewidth at two specific points near the base of the line (0·55% and 0·11% of the peak amplitude) on another. It is necessary to interpret such a measurement carefully in order to infer the true spectrometer performance. The tests described below relate to proton observation; test samples are available for other nuclei (particularly carbon), but resolution of heteronuclear probes can often be adjusted and measured by observing proton lines through the decoupling coil.

The standard test sample for proton linewidth measurement is a 10-15% solution of *o*-dichlorobenzene (ODCB) in acetone. Such samples are available degassed and sealed in tubes of various diameters. In common with many NMR test samples, the choice of ODCB was made a long time ago when low-field CW machines were all that were available. There are several reasons why this is a non-ideal sample on modern high-field instruments, but nonetheless its use seems to be ingrained.

The aromatic protons of ODCB form an AA'BB' system which gives rise to up to 24 lines, any of which can in principle be taken as the test line for half-height linewidth measurement. However, at higher fields with such a strong solution, the phenomenon of radiation damping causes appreciable broadening of the signals, and in this case only the weak outer lines of the group can be used to make the measurement reliably. (Radiation damping is acceleration of the decay of the NMR signal by its coupling to the resonant circuit of the probe; it is less significant for weaker signals, and can be alleviated by deliberate de-tuning of the receiver coil). Alternatively,

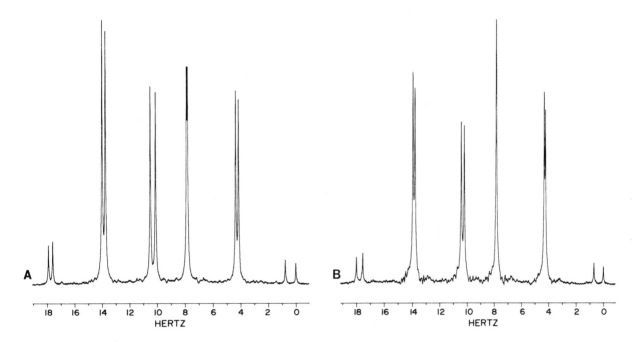

1% solutions are available and should be preferred at 400 MHz and above. Figure 3.2 shows ODCB resolution tests at 250 MHz and 500 MHz; the evident variation in appearance of the second-order spectrum with field is another inconvenience. In 5 mm probes it is usually easy to obtain linewidths of 0·1 Hz or slightly less; spectrometer manufacturers normally specify about double this.

Figure 3.2 ODCB proton resolution tests at (A) 250 MHz and (B) 500 MHz; the linewidth in the latter spectrum is 0·06 Hz.

The lineshape measurement (i.e. the linewidth at 0·55% and 0·11% heights) is made on a solution of chloroform in acetone. Again the commonly available sealed samples are too concentrated to give very narrow half-height linewidths on high-field instruments. Unfortunately it is necessary in this case to use a fairly concentrated sample to give adequate signal to noise ratio for the measurement in a few scans.

The manufacturers specify some maximum value for the width at the two heights (the percentages are chosen because they are the height of the ^{13}C satellites of the $CHCl_3$ line and one-fifth this height); however this is only really meaningful when compared with the expected value for a Lorentzian line of the same half-height width. This is easily calculated from the equation for the Lorentzian (Figure 3.3) as being 13·5 times the linewidth at 0·55% and 30 times at 0·11%. This would correspond to 3·4 Hz at 0·55% and 7·5 Hz at 0·11% for a half-height linewidth of 0·25 Hz. Common specifications are 10-15 Hz and 20-30 Hz at the same points; considerably greater than the theoretical value.

A subtle point to watch if the measurement is made on a spectrum acquired with more than one scan is the difference in T_1 between the main chloroform line and its ^{13}C satellites (see Chapter 4 for an explanation of T_1). The satellites have a T_1 about half that of the central line (15 s and 30 s are typical values), so it is easy accidentally to saturate the main line so that the satellites appear taller than they should. This in turn leads to a flattering lineshape measurement; to avoid this the amplitude of the satellites should be checked to be 0·55% of the ^{12}C line.

In practice this specification is often exceeded, and in fact it is not unusual to see lineshapes which are narrower than a Lorentzian should be at 0·55%. This is just as much an incorrect lineshape as one which is too broad, and should be avoided; therefore it is necessary to measure the half-height linewidth of the lineshape test sample to determine the desired width at the base. The lineshape specification would be better expressed as the maximum degree of deviation from Lorentzian shape achievable for a given linewidth, but there seems no prospect of this happening.

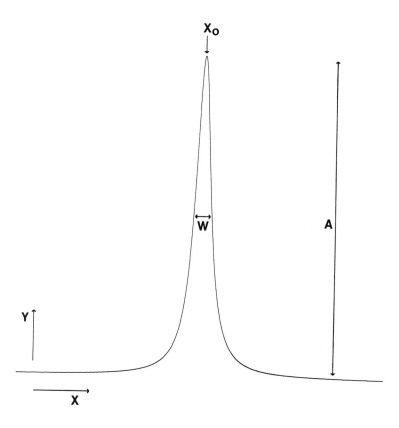

Figure 3.3 A Lorentzian line can be expressed in terms of its amplitude A and half-height linewidth W: $y = \dfrac{AW^2}{W^2 + 4(x_0 - x)^2}$

A second parameter measured with the lineshape test sample is the height of spinning sidebands. These are satellite lines occurring at multiples of the spinning speed (expressed in Hertz) on either side of the central line. Unless there is something drastically wrong with the adjustment of the magnetic field, no more than two pairs of sidebands (separated from the central line by the spinning speed and double the spinning speed) will be visible. Sidebands are usually specified as having an amplitude less than 1% of the main peak; the achieved value is highly variable from magnet to magnet, but at medium to low field sidebands can often be almost undetectable. The origin of spinning sidebands is discussed in more detail below. Figure 3.4 shows a line shape and sideband test on a 5 mm proton probe at 500 MHz.

3.3.3 Factors Affecting Resolution

Rather a large number of factors conspire to degrade the shape of NMR lines from the ideal Lorentzian with small linewidth. Some of these can be controlled by the operator of the spectrometer, and knowing how to do this efficiently constitutes an important skill. Others are in the hands of the instrument maker, and this is one of the several competitive areas between different manufacturers. As many other aspects of instrument performance depend on the quality of the lineshape, this should be carefully tested when considering the purchase of a new spectrometer.

NMR lines have a natural width, determined by the relaxation processes to which the nuclei are subject. However, for observation of spin one-half nuclei in low molecular weight molecules in non-viscous solution the contribution to the observed linewidth by the natural linewidth is almost negligible (for instance the natural width of a ^{13}C line can be 0·02 Hz or less, while the observed linewidth is usually at least ten times this). Instead, a variety of instrumental defects cause deviations from the ideal. The three most important elements involved here are the static magnetic field, the NMR probe and the sample itself.

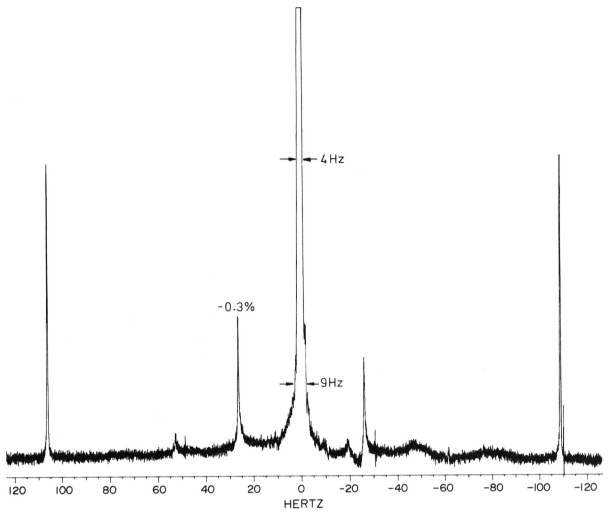

Figure 3.4 An example of a lineshape test (proton observation at 500 MHz). The width of this line at half-height was 0·3 Hz, so the expected widths at the test points for a Lorentzian are 4 and 9 Hz, which have been achieved. 1st order sidebands are visible, at an acceptable level for a magnet of this field, and also some low humps to high field of the line which might be improved by adjustment of the high order non-spinning gradients.

An NMR line 0·1 Hz wide may be observed on a 500MHz spectrometer. This implies that the static field is homogeneous through the volume of the sample to an accuracy of better than 1 part in 10⁹. As the data for any spectrum is acquired over a finite time interval, the field must also be constant in time as well as through space. Superconducting solenoids as used in high-field instruments have favourable intrinsic properties from both points of view; however corrections to the homogeneity and stabilisation through time are both still required. These are achieved through the use of gradient coils (so-called shims) and the lock system respectively. A good understanding of both these parts of the spectrometer is an essential prerequisite to carrying out high quality experiments; both are discussed below and the lock system in more detail in Chapter 5.

The NMR probe necessarily is made out of something, and parts of it, notably the coil, must approach the sample quite closely. If the magnetic susceptibility of the materials used differs from zero, distortions of the static field will occur as a result. Very often it is this distortion which proves to be the limiting factor for resolution and lineshape, rather than intrinsic inhomogeneity of the static field. Such distortions are usually more difficult to correct with the gradient coils than those arising from the magnet itself, as they vary more rapidly over a short distance. In recent years the development of materials with low magnetic susceptibility which also have suitable electrical properties has allowed a great reduction in effects of this type, with a concomitant improvement in lineshape and, therefore, sensitivity. For this reason a spectrometer built some time ago would not be expected to show equal performance to one of the same field constructed recently; however upgrading is usually straightforward simply by buying newer probes. Naturally, other aspects of the total system design may also have changed, but it is usual that the most improvement originates from changes in the probe.

A second aspect of probe construction which influences lineshape, along with several other elements of probe performance, is the degree to which the coil transmits and receives equal radiofrequency (*rf*) signals to or from different parts of the sample. This is referred to as the *rf* or B_1 field homogeneity (in contrast with the static or B_0 field). This is a parameter with great significance for the success of the more complex multi-pulse experiments described later, and is mentioned in several other contexts in this book. Techniques for determining the B_1 homogeneity of a probe are beyond the scope of a book for the non-specialist, but we will encounter some examples of experiments which are quite sensitive to an inhomogeneous B_1 field, and these might be used as a qualitative test. It is not normal for quantitative limits for B_1 inhomogeneity to be given as part of a probe specification, unfortunately.

The sample itself will also inevitably induce distortions of the static field. These will originate from any region where the susceptibility of the sample changes; this must of course happen at the top and bottom of the liquid column where there is a transition from solution to air or glass (or the PTFE of a vortex suppressor). For this reason, the highest resolution is only attainable when the sample length is substantially greater than the length of the sensitive region of the probe; examine standard test samples and you will see that the liquid column is many centimetres long. As discussed above, when sensitivity is critical it is not acceptable to dilute the sample in this way, and then only careful experimentation will determine the minimum usable depth. The field may also be distorted by changes in the solution, for instance concentration or temperature gradients or floating particles; again these factors, which can readily be eliminated, have been discussed in the previous section. Temperature gradients are particularly objectionable as they induce convection currents; in variable temperature work it is often necessary to wait a considerable time after changing the sample temperature for these to die down.

3.3.4 Shimming

Introduction

This curious term is derived from the engineering practice of using thin bits of metal (called shims) as spacers to make mechanical parts fit together exactly. In the context of NMR the shims are small magnetic fields used to cancel out errors in the static field. Every time a new sample is introduced into the magnet it is necessary to re-adjust the shims, and this process requires a degree of skill if it is to be carried out effectively. The difficulty of the adjustment increases (not necessarily linearly) with increasing field, increasing frequency of the observed nucleus and increasing sensitive volume; thus the worst case would be proton observation at 600 MHz (or even higher by now?) in a large diameter probe. Magnet and probe combinations vary enormously in the effort required and the results obtainable. The suggestions I make here are most relevant to difficult cases such as 500 MHz spectrometers or very large probes. Only experience will tell you how many gradients you need to adjust and to what extent to achieve a given result on your spectrometer.

The shims

In a superconducting magnet the homogeneity of the field is adjusted at several levels. The most basic is the actual design of the solenoid, which should be such as to leave only simple gradients which can easily be corrected. There are then two sets of gradient coils, a superconducting set

Figure 3.5 Various room temperature shim assemblies and their components; on the two at the right of the picture the outermost gradient coils are visible (courtesy of Oxford Instruments Ltd.).

adjusted during magnet installation and then left alone, and a room temperature set adjustable for each sample. The normal user of a spectrometer would only be concerned with the third level, the room temperature shims, and then most often only with a small number of the available gradients.

For a conventional vertical bore magnet the shims are made out of printed coils wrapped round a cylindrical former which is inserted into the magnet (see Figure 3.5). The NMR probe is then mounted inside the shim assembly so that the sensitive volume is at the centre of the Z gradient coil; the shim assembly itself must also align with the superconducting gradients inside the cryostat. A coil whose field is aligned along the vertical axis of the magnet is called a Z gradient (e.g. Z, Z^2, Z^3 etc.) while those aligned along the other two orthogonal axes are various orders of X and Y gradients (also often called non-spinning or horizontal gradients, see below). The shim assembly will also contain a coil for altering the total magnetic field to some extent (sometimes called the sweep or Z^0 coil); the range of adjustment usually being around 100 - 200 p.p.m.

If you are already familiar with older spectrometers using permanent or electromagnets you will notice that the gradient names used in a superconducting magnet are different; this is because the static field direction is along the vertical axis of the sample rather than across it. The two most important gradients Z and Z^2 on a superconducting magnet are similar in application to those called Y and R^2 (or sometimes 'curvature') on an iron magnet. The gradient names used in this section correspond with Oxford Instruments conventions.

There is an important difference between gradients aligned along the axis of the sample and those aligned across it. The sample will be spun at a speed sufficient to average many defects of the magnetic field in the x-y plane; however the spinning does not affect the vertical gradients. This means that the vertical homogeneity of the field is more critical, and more care is therefore needed to correct it. Superconducting magnets are invariably fitted with room-temperature Z gradients up to the fourth order (i.e. Z, Z^2, Z^3 and Z^4) and sometimes also Z^5 (usually in 500 MHz or widebore magnets only). The adjustment of these gradients strongly affects the lineshape and linewidth obtained. In contrast, the horizontal field is usually only corrected to third order, with X and Y (first order), XZ, YZ, XY and X^2-Y^2 (second order) and XZ^2 and YZ^2 (third order) gradients being common. High field magnets may also have X^3, Y^3, ZXY, $Z(X^2$-$Y^2)$ or others in addition. Adjustment of these gradients affects most strongly the spinning sidebands mentioned previously, but the higher order horizontal gradients also affect the lineshape.

Being faced with such a plethora of adjustments, which often take the form of an intimidating bank of knobs on older instruments, may cause feelings of desperation. Indeed, adjusting a magnet from scratch is an arduous task well avoided if possible. Fortunately once the initial adjustment for a probe/magnet combination has been made the differences observed from one sample to another are relatively minor. In the following description of an approach to shimming I am going to assume that, while you do not have the need for a procedure that takes you from scratch to a shimmed probe, you may be tackling 'difficult' problems such as very high field instruments or observing spectra over a large sample volume. Thus I hope to avoid being tedious while at the same time not just suggesting that you twiddle Z and Z^2 and leave it at that. An excellent description of a more detailed shim procedure first appeared in Nicolet Instruments (now GE NMR Instruments) house magazine some years ago and circulated in pirated form in the NMR community; happily this has now been published and I refer you to it for further information[1].

Ideally, shimming would be completely straightforward. The gradients are designed so that each will have some effect independent of the rest, so it should be possible to adjust each in turn to maximise resolution and then that would be that. Unfortunately it is not possible to realise fully the ideal shapes for the magnetic fields with physically practical coils (and also the sample is not spherical, which ideally it should be), and inevitably each gradient contains some of the others as 'impurities'. This is the fundamental obstacle to shimming: adjusting one gradient renders other previously optimised gradients incorrect again. If every gradient contained every other gradient as impurities it would be wellnigh impossible to make any progress at all with the adjustment, but fortunately things are not quite that bad. In fact it is possible to identify pairs or small groups of gradients that influence each other strongly. The procedure for dealing with an interacting pair of gradients can be illustrated with the Z/Z^2 pair which needs very frequent adjustment, but first we need to consider how to assess whether altering a gradient is improving the homogeneity.

The deuterium lock

The most fundamental criterion of the homogeneity is the observed lineshape, of course. However it is hardly practical to obtain a new spectrum and make a lineshape measurement every time a gradient is changed slightly. There are usually two more immediate measures of homogeneity available on a modern spectrometer in the form of the deuterium lock signal and the free induction decay. It is common to rely most on the first of these in routine spectrometer operation; certain advantages which arise from using the free induction decay as well are discussed below.

The deuterium lock is the means by which long term stability of the magnetic field is achieved. A more detailed discussion of how the lock system operates is reserved until Chapter 5, by which time some further necessary terminology will have been introduced. The basic idea is quite simple though: observe an NMR line and compare its frequency with a constant reference, then make adjustments to the magnetic field to maintain that frequency. This converts the problem of producing a very stable magnet field into one of producing a very stable frequency, which you may realise (particularly if you own a quartz watch) is fairly straightforward. It is convenient to observe the deuterium line of the solvent for this purpose. Modern spectrometers invariably do this, and will provide some indicator of the strength of the deuterium resonance, which may take the form of a lock level meter or some kind of graphical display. The intensity of the lock signal depends, amongst other things, on the field homogeneity, and so can be used as an indication of progress when shimming. In order for this measurement to be meaningful, and indeed in order for the spectrometer to operate correctly overall, it is important that the lock conditions are correctly set up.

The lock channel is in essence an extra deuterium spectrometer operating in parallel with whatever other nucleus you happen to be observing. As such, there are likely to be at least three parameters that need to be set for optimum operation. These are the amount of *rf* power used to stimulate the deuterium signal (the lock transmitter power), the amount of gain applied to the received signal (the lock amplitude or gain), and the receiver reference phase (see Chapter 5 for an explanation of this term). Terminology used for these items varies from manufacturer to manufacturer, the terms in brackets being intended only as a guide; find out what they are called on your spectrometer. Of these three, it is particularly important that the transmitter power and the reference phase are correctly set. The adjustments described below should be made after the lock condition has been established; the procedure for achieving this varies a good deal from one instrument to another, so consult your instrument's operating manuals for details.

The lock transmitter power and the lock gain controls may appear superficially similar, in that altering them will make the amount of lock signal go up and down. However, there is a vital distinction. The lock gain is altering only the amplification of the received signal. Thus it does not affect what is happening inside the sample, and may be adjusted quite freely. The penalty for using high lock gain will be excessive noise, which for many applications will not be a bad problem (but see Chapter 5). The transmitter power, on the other hand, can only be increased up to a certain point before *saturation* sets in. This is the situation in which more energy is being put into the sample by the applied *rf* field than it can dissipate through relaxation processes (Chapter 4). One manifestation of this is a broadening of the observed line, another is erratic variation in its amplitude. Both are highly detrimental to the operation of the lock.

It is essential not to operate the lock in saturation; on the other hand it is desirable to use reasonably high transmitter power so as to obtain the best lock signal to noise ratio. Gross saturation can be detected as large random variation of the lock signal, and is therefore obvious. The more subtle initial onset of saturation can be detected by the following experiment. Increase the lock transmitter power by a small increment, and observe the lock signal. It will initially increase, but if saturation is occurring it will then fall back to a lower level over a short period. Only if it increases to some point and stays there is no saturation occurring. You can do the same experiment backwards, by *reducing* the power and watching to see if the signal stays reduced or drifts back up. In this way it is easy to stop just clear of saturation. The lock gain control can then be used to set the level of the signal appropriately.

The significance of the lock reference phase is discussed in Chapter 5. It can usually be adjusted simply by altering it for maximum lock signal. However this will not work correctly unless the shape of the deuterium line is reasonably Lorentzian. A good procedure is to make an initial adjustment after obtaining lock, then check again after making first adjustments to the Z and Z^2 gradients. Symptoms of incorrectly set lock phase range from erratic variations in the lock signal level (which may easily be confused with saturation) to 'unresponsive' shim adjustment (but clearly you need to be quite familiar with a spectrometer to detect this). Completely incorrect phase will make it impossible to lock at all. Adjusting the phase without first obtaining lock is an instrument specific procedure, so again consult operating manuals.

Adjusting the shims

The lock system, once correctly optimised, provides some feedback about the state of the magnetic field which allows the shims to be adjusted. The amplitude of the signal depends, approximately at least, on the homogeneity, and so the object is to maximise it. The obstacle to making systematic adjustment of the gradients is the existence of interactions between certain combinations. Part of the skill of shimming is knowing how to cope with such interactions, and also knowing which gradients need to be treated as interacting. Another important skill is knowing how much it is necessary to alter a gradient in order to have a significant effect and, for certain gradients, how *fast* they can be altered without disrupting the lock. All these things vary a good deal from one instrument to another, so there is a definite sense in which it is necessary to get the feel of a new magnet before you can shim it reliably. Practise with the Z/Z^2 combination, which needs to be adjusted all the time anyway. Very often these will be the only gradients you will need to change, but still it is not always trivial even to adjust only this pair.

The essential assumption which must be made when adjusting a pair of gradients is that the outcome of adjusting the higher order component (e.g. Z^2) is not meaningful alone. Rather, it is essential to re-optimise the lower order component (Z) and then judge whether the combined effect is an improvement or not. Thus the following procedure can be used for Z and Z^2.

Vary the Z gradient; the lock level will go up and down. Choose the Z value which gives the highest lock level, and note that level. It is convenient if you can adjust the lock gain to set the reading against some reference point such as a division on a level meter or a grid line on a display. Now alter Z^2 until something happens to the lock level *without regard for what it is*. In other words, do not worry if the level falls or rises at this stage, as long as *something* is happening. Once the lock level has definitely changed one way or the other, readjust Z for maximum signal. Having noted the previous best reading, you will be able to determine whether or not the new combination is better. If it is, then continue adjusting Z^2 *in the same direction*, each time reoptimising Z, until no further improvement is found. If the combined result was worse than before, then *reverse* the direction of Z^2. Beginners find it hard to remember which way Z^2 was turned, but this rapidly becomes instinctive with practice. (This procedure differs somewhat from that described in reference 1, where a distinction is made between two different kinds of interaction, but in my experience it conveys the essence of what needs to be done without adding undue complications).

In many cases this sequence of operations will seem over-elaborate, as the straightforward approach of trimming each gradient for improved lock level will give satisfactory results. The reason for this is that when the shims are close to correct adjustment already, the lowest order gradients tend to home in on the right settings directly. If your spectrometer is correctly maintained, and you use the recommended sample volume, then this circumstance may often apply. Do not be misled by this into thinking that it is unnecessary to learn how to shim properly, as inevitably there will come a time when the correct procedure is needed. For instance, you may need to run an unusually short sample for sensitivity reasons, and it is the sample depth which most strongly affects the required shim values.

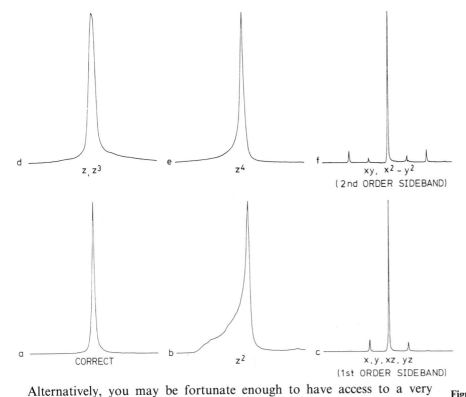

Figure 3.6 Some lineshape defects.

Alternatively, you may be fortunate enough to have access to a very high field instrument. For reasons related to the change from single to double solenoid magnet construction, there is a disproportionately large change in difficulty of shimming in the step from 400 MHz to 500 MHz spectrometers at present. At 500 MHz and higher it is generally necessary to adjust the first four Z gradients on each sample to obtain reasonable results. At the same time, one's perception of what is 'reasonable' tends to vary with the price of the spectrometer, so at high field you may feel more inclined to put greater effort into getting the best out of the instrument anyway. Third order and higher gradients almost never have a directly improving effect on the lock level, so using the procedure is essential.

The other gradients which it may sometimes be beneficial to adjust in routine spectrometer operation are Z^3, Z^4 and the low order horizontal gradients X, Y, XZ, YZ, XY and X^2-Y^2. It must be said, however, that it is quite rare for this to be necessary, the most common cause being use of a very high field spectrometer as just mentioned. Another common situation where the need for more stringent adjustment arises is when you try something unorthodox, such as observing proton spectra on the decoupling coil of a heteronuclear probe. To judge whether or not such adjustments are needed, it is helpful to have an idea of what symptoms are produced by errors in each gradient. Figure 3.6 illustrates lineshapes obtained on a 500 MHz 5 mm proton probe with systematic misadjustment of shims. Bear in mind that in real life all the errors are likely to be present simultaneously in different proportions, and the resulting lineshape may have a correspondingly obscure appearance.

All the Z gradients affect the width of the line, the higher order gradients causing effects progressively further down the line. Odd order gradients (Z, Z^3, Z^5) cause *symmetrical* broadening of the line, while even order gradients (Z^2, Z^4) cause *unsymmetrical* broadening (Figure 3.6 B, D and E). Normally the higher the order of the gradient the larger the adjustment needed to produce any effect (but obviously this depends on how your spectrometer is built). In the examples of Figure 3.6 rather large misadjustments have been used to produce an obvious effect; closer to the correct values or at lower field such obvious lumps will not be produced, just a thickening of the line.

When adjusting the Z gradients, in theory even order gradients contain largely even order impurities, and likewise odd order gradients odd order impurities. Thus it is expected that the strongest interactions will be between Z and Z^3 and Z^2 and Z^4 respectively. In practice it is rash to rely on this, and a better assumption is that whenever a gradient is altered all lower order gradients will need to be readjusted. A consequence of the presence of strong even order impurities in even order gradients is that Z^2 and Z^4 contain Z^0 (the *field*), so adjusting these too rapidly is liable to make the lock drop out. These are the gradients mentioned previously for which it is necessary to determine by experiment how fast they can be altered without causing problems.

The lower order horizontal shims affect the height and shape of the spinning sidebands, errors in gradients odd in X and Y (i.e. X, Y, XZ and YZ) leading mainly to 1st order sidebands and errors in gradients even in X and Y (i.e. XY, X^2-Y^2) mainly to 2nd order sidebands (Figure 3.6 C and F). As the field defects which are corrected by these gradients are averaged by the sample spinning *they must be adjusted without spinning*; hence the term non-spinning shims.

Amongst the gradients mentioned above, each one of second order should be considered as interacting with X and Y together, so groups of three need to be adjusted. In addition, those containing a Z component (XZ, YZ) require that Z be reoptimised if they are changed. It is also a good idea to retrim Z immediately after turning off the spinning, as this often alters the height of the sample slightly relative to the probe.

In order to reduce the adjustment to a practical level of complexity, certain interactions should be considered dominant (Table 3.2) and these groups adjusted together; other lower order gradients are changed only when a large movement has been made in the higher order gradients, or during final trimming. Spinning sidebands may also originate from other causes, for instance misshapen sample tubes, vortexing in a large diameter tube, floating particles in the sample or a dirty or scratched spinner turbine, so check for these things before wasting a lot of time trying to shim all the non-spinning gradients.

The highest order gradients (Z^5 and the third order non-spinning gradients mentioned previously) tend to have less obvious effects than the

Table 3.2 Principal shim interactions. When each gradient in the first column is changed, those in the second column will be most strongly affected. If large changes are made then gradients in the third column are also likely to need readjustment.

Adjusted gradient	Principal interactions	Other interactions
Z	-	-
Z^2	Z	-
Z^3	Z	Z^2
Z^4	Z^2	Z, Z^3
Z^5	Z, Z^3	Z^2, Z^4
X	Y	Z
Y	X	Z
XZ	X	Z
YZ	Y	Z
XY	X, Y	-
X^2-Y^2	XY	X, Y
XZ^2	XZ	X, Z
YZ^2	YZ	Y, Z
ZXY	XY	X, Y, Z
$Z(X^2$-$Y^2)$	X^2-Y^2	X, Y, Z
X^3	X	-
Y^3	Y	-

others. XZ^2 and YZ^2 often cause a symmetrical broadening reminiscent of Z^3, although further down the line. Also, in combination with Z^3 and Z^4, they may cause a change between a narrow based line with relatively large sidebands and a broader line without sidebands. Finding a compromise between these generally requires a lot of trial and error during initial shimming, and can hardly be attempted on everyday samples. The remaining gradients tend to cause very low (below 0·1%) humps if misadjusted, which may be very far from the main line (e.g. *hundreds* of Hertz). Such defects can be highly objectionable and are not detected by the normal lineshape test. I mention this only so you can complain if you encounter such problems; eliminating them should be left to specialists.

Shimming using the FID

Although the lock level is a convenient indicator of the homogeneity, and is perfectly satisfactory for adjusting routine samples, it can sometimes be misleading. This becomes clear if you consider *why* the height of the lock signal reflects what is happening to the magnetic field. The lock is derived by observing a single line, normally that of the solvent, in the deuterium spectrum. The *area* under this line is independent of the homogeneity, but of course the *width* of it varies. Thus, as the line becomes narrower it must also become taller so that it encloses a constant area, and it is the height of the line which is displayed as the lock level. Now, it is not difficult to imagine lineshape variations which cause the height of a line to increase without bringing its shape closer to Lorentzian; for instance narrowing the upper part of the line while leaving a constant broad hump at the base would have this effect. In fact it would probably be true to say that the lock level largely reflects line*width* changes of this type. For more rigorous adjustments of the lineshape another criterion is needed.

Many spectrometers are capable of displaying the free induction decay (FID) resulting from a single pulse in real time. If this facility is available it can be used to good effect when shimming. Two features of the appearance of the FID are informative: its duration and its shape. The duration of the decay gives information about the eventual linewidth akin to the lock level; however information about the lineshape is also present in the appearance (exponential or otherwise) of the decay envelope. Naturally this is a fairly subjective judgement, but with practice it becomes a more reliable means of adjusting the shims than observing the lock level alone. Figures 3.7 shows 'before' and 'after' FID's obtained during shimming.

This kind of simple exponential is only observed with samples whose signal is dominated by a strong single line (in this case, the chloroform line of a lineshape test sample). The more complex modulation patterns observed with real samples have an overall decay envelope easily related to this. Use of the FID for shimming in this way requires a slightly different

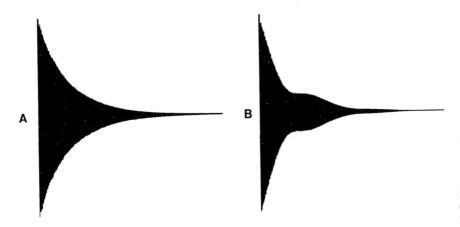

A   B

Figure 3.7 Correct (A) and incorrect (B) FID shapes, such as may be observed during shimming.

style of adjustment of the gradients, as you need to wait until the end of a scan before complete information about the effect of changing a shim becomes available. The ideal circumstance is to be able to observe the lock level and the FID together, so that you have immediate information together with slower but more accurate information. Rather remarkably this apparently simple facility is not always available even on recent instruments from some manufacturers.

Shimming unlocked

It sometimes happens that it is necessary to run samples unlocked, because the usual lock nucleus is to be observed or irradiated, or circumstances prevent the use of a solvent containing it. Alternatives to deuterium as the lock nucleus (most commonly fluorine) are available, but in the absence of such a facility unlocked operation over medium time spans (e.g. up to a few hours) is usually quite feasible with superconducting magnets. In this case the lock level is obviously unavailable for shimming, and the FID shape must be used.

An added element of confusion enters here in the form of the field impurities present in the Z^2 and Z^4 gradients. In the absence of lock, changes in the field brought about by changing these gradients are not corrected. The short term effect of this is bizarre changes in the appearance of the FID, particularly if the gradients are altered during a scan. These can simply be ignored. More importantly, the field may easily be changed enough so that the spectral window selected is no longer in the right region, so before starting a long accumulation and going off to have lunch you need to check the observe offset.

3.4 ACHIEVING SENSITIVITY

3.4.1 Introduction

In fact a rather large proportion of the book could be placed under the heading 'Achieving Sensitivity', as it is a problem that besets NMR spectroscopists all the time. What I want to look at in this section is the more instrumental side of optimising sensitivity, that is to say how to choose the right combination of spectrometer components and how to ensure they are working together properly. Choice of experimental parameters such as pulse widths and delays, which also is of great importance of course, is covered in Chapter 7. Another very significant contribution to overall performance comes from the *lineshape*, since NMR lines have constant *area* for a given amount of signal and dissipating this into broad lumps reduces the height of peaks; the whole of the previous section is therefore also important here.

3.4.2 Criteria Of Sensitivity

Sensitivity specifications for NMR spectrometers are a prime area of competition between the instrument manufacturers, which one might suppose would lead to some effort to standardise the means used to measure them. Unfortunately this has not taken place, and instead we are faced with confusing lists of test data combined with aggressive claims for high performance, which have become increasingly vigorous in recent years as all the companies have made genuine progress towards higher sensitivity. In reality it is fairly straightforward to get meaningful comparisons between instruments, provided you are clear about exactly what you are trying to measure.

I want to survey this problem with the specific perspective of the end-user of an instrument, who is most concerned with total system performance. Thus the tests I will be concerned with are those that are most influenced by all aspects of an instrument acting in combination. There has been a tendency lately to try to invent test samples (such as the ASTM test for carbon sensitivity) which eliminate various influential factors (in this case proton decoupling efficiency and to some extent quality of shimming) from the measurement. This is perfectly admirable if the object is to try to find the origins of variable sensitivity, as we might want to do if we were instrument designers, but it is not very relevant to the chemist who wants to know whether he can get a carbon spectrum out of his 2 mg sample.

The first very important point to be clear about is that the sensitivity or *signal-to-noise ratio* measurement is a statistical one. We measure the height of a peak under a standard set of conditions, and compare it with the noise level in some way, but the latter is obviously a random quantity. Making a set of measurements and picking the best one, a not uncommon practice, is an unconventional way to interpret statistical data; we should rather take the average value. Misuse of the data in this way is one possible origin of dubious claims regarding sensitivity. Much more subtle possibilities exist however, as we shall see. A principal source of confusion is the difference between sensitivity at constant concentration, which is the conventionally measured quantitity, and sensitivity for constant amount of sample, which would be a much more interesting measurement from the chemist's point of view. Before investigating the difference between these, we need to discuss the practical aspects of the signal-to-noise ratio measurement.

As I have just said, what one does is to measure a spectrum of a standard sample under defined conditions, and take the ratio of the height of a specified peak to the size of the noise. The first problem here is what is meant by the 'size of the noise'. The *mean* value of the noise amplitude over a large enough number of samples is obviously zero, as it contains positive and negative components equally (we assume it to have a Gaussian distribution centred on zero). The statistical value of interest is the *root-mean-square* noise amplitude (which is, if you prefer, the standard deviation of the noise amplitude distribution), obtained by averaging the squares of the noise values over a set of samples (i.e. a set of points in the digitised spectrum) and taking the square root of the result. This is then sometimes multiplied by 2 in recognition of the fact that the noise lies both sides of zero; always check whether or not this has been done if you are supplied with a calculated *rms* noise value.

Unfortunately, as evaluation of this quantity requires a degree of computational complexity it is rarely done (once again we are in the grip of historical precedent here, as modern instruments have computers that could easily calculate *rms* noise; the procedures for noise measurement date from the days of CW instrumentation). Instead it is usual to measure the *peak-to-peak noise* and relate this to the *rms* noise (which must be smaller) by use of a conversion factor. This immediately introduces an extra element of uncertainty, as if the noise is truly random there cannot be any real meaning to the idea of 'peak' noise, since we should always be able to find a larger piece simply by looking some more. In practice the most important thing is to adopt a consistent approach to the measurement; reasonably accurate comparative values can then be obtained.

It is essential to pick a large enough sample of noise (i.e. a wide enough signal-free region of the spectrum) to give a good representation of the noise level, and to truly measure the separation between the most positive and most negative excursions. There is an almost overwhelming temptation to be optimistic here, particularly when you have spent an hour setting up the instrument specially for the measurement. The widest excursions will occur with low probability, so there will seem to be a big gap between them and what looks like the main part of the noise (Figure 3.8); it is so appealing to dismiss peaks as 'spikes' rather than incoherent noise that an iron will is needed. It is fair enough to exclude peaks which really are

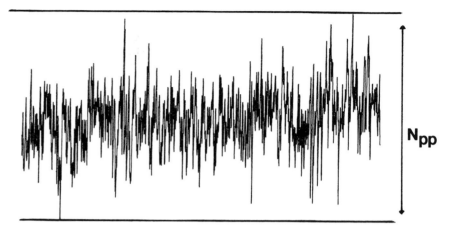

Figure 3.8 A fair peak-to-peak noise measurement is *really* peak-to-peak.

instrumental spikes though; these can be treated as a separate defect. Repeating the experiment a few times will show whether or not a contentious peak is random or always present.

The noise measurement is usually made on a plotted spectrum with a ruler, so it is important to make sure that this process is not introducing any excessive errors. A specific problem with plotters is that the faster they go the more they tend to cut the corners off things, so if the software or hardware allows variable speed plotting the slowest available should be used. It also should be obvious that you need to amplify the noise plot sufficiently so that an accurate measurement of it can be made. There is not much point trying to compare the ratio of noise a few millimetres high to a peak 20 cm high; multiply it up so that it is several centimetres high too. Obvious or not, one sees this point ignored rather often.

The factor which is used to convert the peak-to-peak noise to *rms* is 2·5. This number arises by choosing arbitrarily a point in the Gaussian distribution beyond which the probability of a noise component of that amplitude occurring is small enough to be 'negligible'. If the standard deviation of the distribution is σ, the point chosen is $2\cdot5\sigma$; approximately 99% of the noise samples will fall within $\pm2\cdot5\sigma$ of zero. Thus the definition of the signal-to-noise ratio $\mathscr{S}$ is:

$$\mathscr{S} = \frac{2\cdot5 \times A}{N_{pp}} \tag{3.1}$$

where A is the height of the chosen peak and N_{pp} the peak-to-peak noise. This should be directly comparable with the more meaningful signal-to-*rms* noise value provided this includes the factor of 2 mentioned above. In fact, since if a fairly large section of the noise has been taken to make the peak-to-peak measurement this may be several thousand sample points, the factor to convert to *rms* should probably be a little larger; 2·5 is universally used though. Differences in signal-to-noise ratio measurements of less than about 10% should be discounted anyway, so such fine distinctions are not important.

Now we need to turn to the 'standard' conditions under which the measurements are made. For the matters discussed in the preceding paragraphs, everyone at least agrees what *ought* to be done, even if it is not always followed in practice. When it comes to exactly how to set up the instrument to generate the test spectrum there is generally no specific convention beyond the idea that a single scan with a $\pi/2$ pulse (see next chapter) should be used. This leaves free such parameters as the spectral width, filter width and window function for processing the spectrum. The only thing to do if you are trying to compare instruments with a view to purchase is to decide on the conditions yourself and insist that the tests are performed in your presence under these conditions.

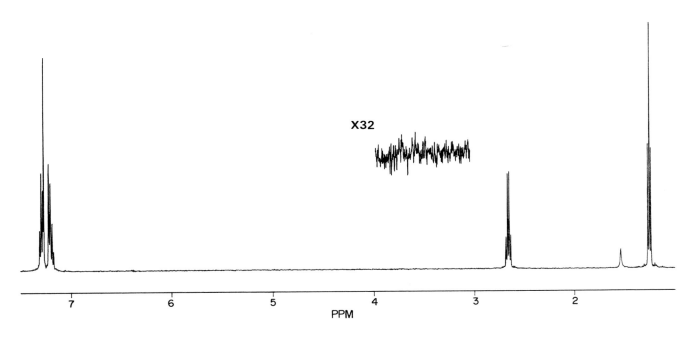

X32

Figure 3.9 The ethyl benzene test for proton sensitivity. The height of the central lines of the quartet at 2·7 p.p.m. is compared with a sample of noise; the result in this case is $\mathscr{S} = 270{:}1$.

Presuming that what we want to test is the performance of the instrument under conditions similar to those of its intended use, for proton sensitivity a 10 p.p.m. range with an acquisition time of 2 to 3 seconds, a filter bandwidth somewhat greater than the sweep width and matched exponential multiplication before the transform would be suitable. The standard proton test sample is 0·1% ethyl benzene in deuterochloroform, the measurement being made on the centre peaks of the ethyl quartet (Figure 3.9). These lines are uncommonly broad as proton lines go, due to unresolved coupling with protons in the aromatic ring, so matched exponential filtration requires 1 - 1·5 Hz of line broadening.

Choice of conditions for measuring carbon sensitivity is less straightforward. In routine ^{13}C spectroscopy spectra are obtained with low digital resolution (2 - 3 Hertz/point). Peak height measurements made on such a poorly digitised spectrum are very unreliable unless substantial line broadening is applied, but then this would be far from a matched filter and would also obscure variations due to resolution and decoupling effects. The latter in particular is a very important aspect of the heteronuclear performance of an instrument, as variations in broadband proton decoupling efficiency can cause large linewidth changes. One of the two commonly used test samples (the ASTM test, a mixture of 60% by volume deuterobenzene and dioxan) specifically excludes decoupling, and by virtue of the large natural linewidth of deuterobenzene reduces the dependence on lineshape. This tests the probe and preamplifier performance separately from other factors, but more informative from the chemist's point of view is the *total* performance. Thus I prefer the more old-fashioned 10% ethyl benzene in chloroform, the measurement to be made on the largest of the aromatic carbon resonances with proton decoupling.

A full spectral region (say 200 p.p.m.) should be measured, but with a long enough acquisition time to properly digitise the lines after matched filtration, perhaps 2-3 seconds. This is no problem with modern data systems which have plenty of memory. The achieved linewidth for this sample is rather variable, depending on the type of decoupling employed (see the discussion of decoupling techniques in Chapter 7); what is essential is that it is *broadband* decoupling. You can nearly always get a better result by using on-resonance coherent decoupling of the aromatic protons, but this is not a legitimate way to perform the test.

Other common heteronuclear test samples are: ^{31}P, 1% trimethylphosphite in deuterobenzene (measured *without* proton decoupling); ^{15}N, 90% dimethylformamide in deuterodimethylsulphoxide (may be measured with or without decoupling, make sure you know which!); ^{19}F, 0·1% trifluoroethanol in deuteroacetone; ^{2}H and ^{17}O, tap water (no lock). If you have an interest in a nucleus that is not usually specified, make your own test sample and take it round to the different manufacturers.

Supposing that we have a nice set of carefully measured signal-to-noise ratio values, it still remains to interpret properly what they mean. You will note that all of the test samples mentioned above contain a specific *concentration* of solute, not a specific amount. This means, for instance, that if we make a sensitivity measurement in a 10 mm diameter tube it will be much bigger than the same measurement using the same solution in a 5 mm diameter tube. Is the probe which takes 10 mm tubes more sensitive? Of course not, it just contains more material (in fact it is likely to be rather *less* sensitive, see below).

This is presumably an obvious point, but what is often forgotten is that there is a second potentially variable dimension to the active region of an NMR probe, the *length,* and this is not usually known. Improvements in magnet construction have allowed probes to be built in recent years with longer coils; this is part of the origin of increased sensitivity specifications. Increasing the length of the coil in this way increases the measured signal-to-noise ratio obtained with the standard samples, but may not improve sensitivity when the total quantity of material is limited (in fact it is probably detrimental in this case).

The quantity of most interest, then is the signal-to-noise ratio *per mole of solute* obtainable in a given probe. What we need to do is find the volume of the active region of the probe, work out the amount of, say, ethyl benzene present in that volume, and divide this into the signal-to-noise measurement to get a properly comparable figure. The volume of the active region means, ideally, the minimum sample depth which can be used without loss of lineshape multiplied by the internal cross-sectional area of the tube. The latter figure can be obtained from the specifications of the test sample manufacturer; typically 5 mm tubes have an internal diameter around 4·2mm and 10 mm around 9 mm. The former may be rather troublesome to measure during a short demonstration, so if you do not want to take the manufacturer's word for it the alternative is to measure the length of the receiver coil in the probe. The true minimum usable sample length will be considerably greater than this, but will be approximately proportional to it so the comparisons will be meaningful; this is the only way to get *real* data to compare spectrometer performance.

A final point which is rarely considered is that all the conventional test samples are moderate concentration, covalent molecules in solution in organic solvents. Probes optimised for operation with such samples may not be at all optimised for solutions in water which are 0·5M in inorganic salts; however many of the most interesting samples come dissolved in this kind of solvent. If you are going to be concerned with solutions of this kind, it is worthwile devising your own test sample which models the real problem, and getting measurements made on this.

3.4.3 Factors Affecting Sensitivity

The probe

Far and away the most critical component of any NMR spectrometer is the probe, which contains the apparatus which must detect the tiny NMR signals without adding excessive noise to them. Although the first stage of the receiver (the pre-amplifier) can also make a significant contribution to the noise level, design of this type of circuit has been extensively researched as it is used in radar and radio equipment. There *should* therefore be little difference between pre-amplifiers made by the various instrument manu-facturers, as they can all draw on the same literature information. The probe itself, however, is specific to NMR, a particularly unusual aspect of its design being the requirement that its components do not distort the static field in the region of the sample too much. Competition and progress in the area of probe design has been fairly continuous throughout the development of NMR. The user of a spectrometer naturally has to take the probes offered by its manufacturer, but there is still usually a wide range of choice. I am going to survey here a number of factors to consider when deciding what kind of probe to use for a particular experiment.

The first thing to pick is what size to use. This is expressed as the diameter of sample tube which can be accommodated; common sizes are 5, 10 and 15 mm. With the recent development of biological and medical applications of NMR some manufacturers now also offer big probes, for instance rat-sized, even for their high resolution machines, but those are outside the scope of the present discussion. Chemists often seem a little confused in the matter of probe diameters, a confusion which has a largely historical basis. Traditionally, insensitive nuclei such as ^{13}C have been run in larger diameter probes. Equating weak signals with large probes can lead to an incorrect approach to the problem, unless you recall *why* this is done.

The object of using a larger probe for a less sensitive nucleus is to enable you to get as much solute as possible into its active region. That is, it is advantageous only when the factor limiting the total availability of sample is its *solubility* in the chosen solvent. It is surprisingly rare that this is the case, at least for observation of the common nuclei ^{1}H, ^{13}C and ^{31}P, since saturated solutions of most covalent compounds are *strong* solutions, from an NMR point of view, for these nuclei on modern medium or high field instruments. Of course, there are many exceptions to this when the solu-bility is such that it is useful to have a large sample volume, and for observation of less sensitive nuclei such as certain metals this situation is still the norm.

When the limiting factor is not solubility, but rather the total quantity of material available, then the situation is reversed. It is advantageous to work with the smallest possible probe, as it will have *intrinsically* higher sensitivity (the reasons for this are quite technical; if you are interested see references 2,3). Aside from the intrinsically better electrical performance, all the practical problems of lineshape and sidebands, and for heteronuclei of broadband proton decoupling, are reduced as you change from larger to smaller probes. In an environment in which the majority of samples are of non-synthetic origin (for instance biological work, or analytical studies of natural products) I would be inclined to choose mainly small diameter probes when configuring an instrument. Synthetic chemists, in contrast, can usually provide sufficient sample that use of larger probes provides a significant timesaving.

A second problematical choice faced when choosing an instrument configuration is whether to deal with heteronuclear operation using broad-band or selectively tuned probes. The broadband option is very tempting, as it seems that you are getting a multiplicity of probes for the price of one (or more likely *two*, as broadband probes can be quite expensive). This is true enough, but of course something must be sacrificed, and that is sensitivity. I will make a distinction here between fully broadband and simply tunable probes. In the latter category, for example, would fall proton probes designed so that they can be retuned to the nearby frequ-encies of fluorine (slightly lower) or tritium (slightly higher). This type of tunability sacrifices nothing. Fully broadband probes can be tuned over a far wider frequency range, maybe as much as a factor of ten from the lowest to the highest frequency. In this case you must expect to lose perhaps a factor of 2 in signal-to-noise ratio around the frequency for which the probe is optimised (usually ^{13}C), and more far from this frequency.

The question that must be asked when deciding whether to buy a broadband probe instead of two selective ones is what is the real requirement for running all those extra nuclei. I think to justify degrading performance on the common nuclei you need a rather finely balanced expectation of the usefulness of the less common ones. If you hardly ever want to run them, then why worry at all? It is usually possible to persuade a colleague or instrument manufacturer's representative with access to a differently equipped spectrometer to run an occasional spectrum. On the other hand, if running, say, ^{15}N or ^{109}Ag spectra is going to be a major activity, then would it not pay to have a probe optimised for this job, rather than a broadband probe which is likely to be far from optimum at these frequencies? It seems to me that it is only in a vague area of intermediate level of use of a wide variety of nuclei that a broadband probe is the best choice. However, I am ignoring any *practical* advantage arising from not needing to change probes; if your main aim is to run ^{13}C, ^{31}P and ^{2}H spectra of strong samples quickly then a broadband system may be ideal.

Tuning it up

The central component of the probe is simply a piece of wire, formed into a shaped coil, in which the sample sits after descending into the magnet. This is, if you like, the aerial which receives the NMR signals; in most probes designed for use with superconducting magnets it also transmits the pulses. In order for the output of the transmitter to be properly dissipated into the sample, and for the NMR signal to be properly amplified by the receiver, it is necessary for the impedance of this wire to be *matched* with (i.e. equal to) those of the transmitter and receiver. This is a consequence of the *maximum power theorem*, which you might perhaps recall in its DC form from school physics. At radiofrequencies things are rather more complicated, as the impedance consists of reactive elements as well as resistance.

As a chemist operating a spectrometer it is hardly necessary to appreciate the technicalities of impedance transformation at radiofrequency, but it *is* necessary to be able to adjust a probe so it is properly matched. There are several reasons for this. Optimum sensitivity will only be achieved when this adjustment is made correctly. Pulse widths are also minimised by correct matching, which reduces off-resonance effects (see Chapter 4). Pulse widths are usually reproducible provided probe tuning is adjusted between samples, so in cases when pulse width calibration is impossible for reasons of sensitivity multi-pulse experiments can still be performed using predetermined values (see Chapter 7). Heteronuclear probes often have a second coil for broadband proton decoupling, and tuning will be required if this is to work properly (particularly if modern composite pulse decoupling is in use - Chapter 7 again). Broadband probes must necessarily be tuned from one frequency to another (this often involves a combination of exchangeable or switchable capacitors with the fine adjustment described here). Finally experiments involving use of low power irradiation (e.g. homonuclear decoupling or the nuclear Overhauser experiment) can only be performed in a reproducible way if tuning is optimised on the sample of interest. Different spectrometers have various schemes for determining correct tuning, but all have the same fundamental aim. I observe that beginners often find this adjustment tricky; this need not be so providing you know what you are trying to do. In the following paragraphs I survey two common schemes for tuning probes and try to give you some idea of what to aim for when carrying out the adjustment.

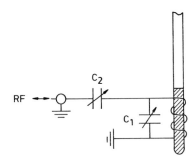

Figure 3.10 A typical probe resonant circuit (schematic; in a vertical bore superconducting magnet the transmit/receive coil would not in fact be solenoidal).

The circuit for converting the impedance of a wire (which is resistance plus inductance) into pure resistance of the correct value (50Ω) in almost universal use is shown in Figure 3.10. The two capacitors are usually mounted *inside* the probe near the coil, and they can be adjusted by long screwdrivers that pass up the probe body; they need to be adjustable because different samples have different effects on the inductance of the coil. We need not worry about *why* this circuit works, but it is useful to appreciate the different effects of the two tunable components. C_1 changes the *resonant frequency* of the circuit, which we want equal to the frequency of the nucleus we are observing, while C_2 changes its *impedance*. The tricky thing is that neither adjustment is completely independent of the other, so optimisation requires repeated adjustment of each (this is akin to the problem of adjusting interacting shims discussed previously).

To determine the correct setting for the two capacitors we need some indication of the quality of the match and whether the correct resonance frequency has been reached. There are two common approaches to this problem. The first, and generally best, is to actually measure the probe response using an *rf* bridge. This is a device analogous to the more familiar resistance (or *Wheatstone*) bridge, but designed for AC work. It has four terminals (these devices are available in handy, small boxes with four BNC connectors), two of which are used as input for a test signal and output to a measuring device (ideally an oscilloscope). On one of the remaining two is placed a known impedance (a 50Ω load), and on the other the probe under test (Figure 3.11). When the impedance of the probe matches that of the known load the bridge is balanced, and the output from the test connection reaches a minimum.

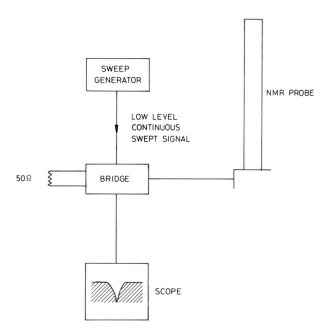

Figure 3.11 Tuning with a bridge.

The ideal procedure for using such a setup is to sweep the frequency of the test signal across the desired resonance of the probe repetitively. The output shows a dip at the resonant frequency, the depth of which indicates the quality of the match. Unravelling the interaction between the tuning and matching adjustments is now very easy, because you can see the two effects separately as variations in the position and depth of the dip. Some spectrometers are equipped to perform probe tuning in this way. If yours is not, and you are seriously interested in performing the highest quality experiments, it is worth trying to obtain the required equipment. Sweep generators and high frequency oscilloscopes are a little expensive to justify their purchase solely for this purpose, but there is an active second-hand market (in the U.K. at least) in these devices, or your general electronics workshop may possess them. The *rf* bridge itself is a cheap component; you can use a bridge without a swept frequency source (e.g. using the spectrometer's decoupler to generate a signal) but this loses most of its advantage.

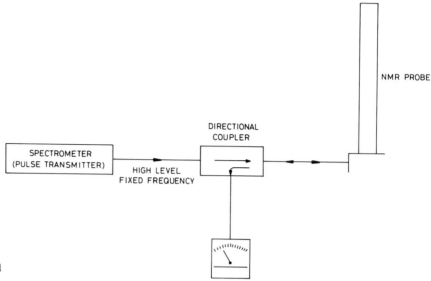

Figure 3.12 Tuning with a directional coupler.

The second approach is probably more common, and involves inserting a device known as a *directional coupler* between the output of the transmitter and the probe (Figure 3.12). This produces an output proportional to the power *reflected back from the probe*, which in turn depends on the quality of the tuning and matching. Usually the reflected power level is displayed on a meter; this may be built into the spectrometer, or separate *directional power meters* (sometimes known as VSWR meters) are available at moderate cost. In a sense this is a more direct indication of proper matching to the transmitter, because minimum reflected power will indicate the best match whatever the actual output impedance (the bridge method assumes that this is 50Ω).

There are, however, a number of problems. Most directional couplers are designed to operate with rather high power levels, which is fine if it is the pulse transmitter you are trying to match but may be a problem if it is a decoupler of some kind. Another big snag is that you no longer have a separate indication of resonance and matching, just a combined effect. It is therefore necessary to adopt a systematic procedure for adjusting the capacitors, for which see the section on shimming (3.3.4 - 'Adjusting the shims') and read 'tuning' and 'matching' for 'Z' and 'Z^2'. It becomes easy to tell when you are in the right place if you have experience of the probe you are working with, because the optimum setting of the matching control gives the 'sharpest' response to changing the tuning. In contrast, if everything is completely wrong the controls will feel unresponsive, and it can seem impossible to make progress. This should only arise if things have been accidentally misadjusted, or if through inexperience you turn the controls too much and get 'lost'.

Dynamic range and ADC resolution

The necessity to convert NMR signals into digital form prior to performing a numerical Fourier transform has a variety of effects on the experiment. The decomposition of the spectrum into discrete points in the frequency domain has already been discussed in the previous chapter, and will be further pursued (probably *ad nauseam*) in Chapter 8. This effect becomes quite obvious once you examine the spectrum closely, and so is usually well understood by anyone who has operated a spectrometer. However, there is a much more subtle consequence of the digitisation process, which relates to the *amplitude* rather than the frequency of the signals. It is important to be aware of this, because in some circumstances it can lead to the complete disappearance of peaks - the ultimate in poor sensitivity. I assume in the following discussion an elementary knowledge of computer terms such as 'bit' and 'word'.

When the electrical NMR signal arrives at the computer, it must be converted into a stream of numbers. We already know how to determine the speed at which this should be done, and for how long we should continue doing it. What remains to be considered is the manner in which these numbers are to be represented and stored. The *analogue-to-digital converter* (ADC) will sample a voltage, and output a binary number proportional to it, which is added to the computer memory as the FID is built up. The important factors influencing this process are the number of bits the ADC uses to represent the signal voltage, the maximum speed at which it can perform the conversion, and the number of bits in a single word of computer memory. In combination these three things determine how wide a range of *frequencies* can be characterised, how wide a range of *amplitudes* can be detected, and the maximum number of scans which may be accumulated.

Other things being equal, it is advantageous for the ADC to represent the signal with the largest possible number of bits (in practice other things are *not* equal; the problems arising from using a large number of bits are discussed later). The reason for this can be understood by considering the range of signal amplitudes which could be represented using, say, 12 bits (typical word lengths for ADC's used in NMR are 12-16 bits). We can suppose that the receiver of the spectrometer has been adjusted so that the maximum signal occurring during each scan just fills the ADC, and causes it to output the largest number it can. If one of the 12 bits is used for the *sign* of the value, this will be $2^{11} - 1$, or just over 2000. Smaller signals will cause the output of smaller numbers; evidently the smallest possible output is 1. The ratio of these values, about 2000:1, is the *dynamic range* available with the 12 bit digitiser.

The NMR signals must 'fit in' to the digitiser, or problems ensue. If the biggest signal is too big, then severe distortion of the spectrum arises (Figure 3.13). If the biggest signal just fills the ADC, then signals more than 2000 times smaller than this lie below its resolution, and will not be detected at all. No amount of time averaging will recover a signal which is too small for the digitiser, because it is simply never being measured. However, if the *noise* amplitude is greater than the the minimum digitisable value, small signals 'ride' on the noise and *can* be detected. This means that the ADC only limits the detection of signals in the presence of *both* a wide dynamic range and a very high signal-to-noise ratio. Typically these conditions only arise in ^{1}H NMR.

Figure 3.14 illustrates the effect of ADC word length on signal detection. The sample used for this demonstration also included a signal (not plotted in the figure) 100 times larger than the largest of the three lines; the smallest line (only visible in the upper two traces) is in turn another 100 times smaller, so the total dynamic range is 10,000:1. The spectrum was accumulated three times using ADC word lengths of 10, 12 and 16 bits, all other settings remaining the same; the largest peak is plotted to the same scale each time. Notice how the noise level appears greater with a shorter ADC word, to the extent that the small signal is invisible on the 10 bit trace; the extra noise is *quantisation noise* from the digitisation process. Dynamic range limitations of this kind are most common in biological applications, where proton spectra often must be acquired in undeuterated water. They may be tackled by selection of the maximum available ADC word length, careful adjustment of receiver gain so that the biggest signal really does fill the ADC, and by taking steps to attenuate the troublesome large signal, a topic discussed in Chapter 7.

There are at least two purely practical reasons why the maximum ADC length cannot always be used. Generating a longer word is slower, so the maximum rate of digitisation and hence maximum spectral width is reduced. Typical 12 bit ADC's need around 3 μs to perform each sampling operation, so they can run at a maximum speed of 300 kHz, and therefore

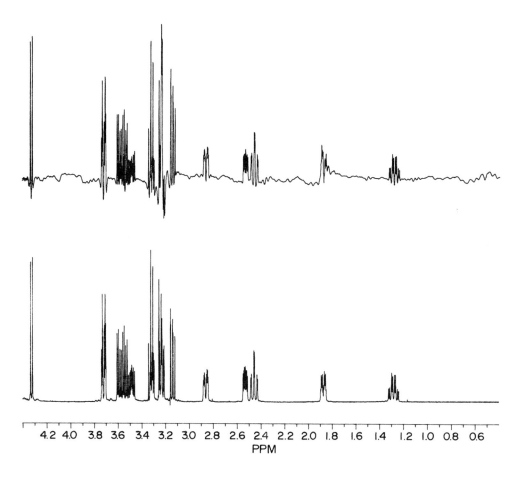

Figure 3.13 Severe baseline distortion resulting from ADC overflow; the lower trace shows the spectrum obtained with correct receiver gain and the upper trace the effect of increasing the gain too much.

can characterise a spectral range of 150 kHz according to the Nyquist criterion. For a 16 bit ADC this might be reduced to 30-50 kHz. However, even at 500 MHz this is perfectly adequate for proton spectra in solution, so this is not a major problem.

A more troublesome consequence of using the longer ADC arises over long term accumulation. Each time a scan is stored, a string of 16 bit numbers is added into the computer memory. If these continue to be represented as binary integers, they can only be accumulated so long as the total word length of the computer can accommodate the cumulative value of each data point. To take a ridiculous case, if one data point arises from an exactly filled 16 bit digitiser (i.e. its value is $2^{15} - 1$ plus sign), and the computer word length is also 16 bits, the second time a scan is taken the maximum value it is possible to store will be exceeded.

In fact this is *not* such a ridiculous case, because 16 bits is a common word length in small computers. Historically this has been quite a problem, because of the prohibitive expense of computer memory, and a number of schemes have been devised to permit signal averaging with words that are really too short[5]. Spectrometer data systems have also often been designed with exotic word lengths such as 20 or 24 bits, as a compromise between necessary storage capacity and cost. As this should not remain a problem much longer, owing to the dramatic fall in computer prices, I will not discuss the matter further, except to mention that for a 12 bit ADC a 24 bit word is adequate for most applications (permitting several thousand scans to be accumulated), while with a 16 bit ADC a 24 bit word is restrictive and 32 bits would be preferable. For a more extensive discussion of these matters, see references 4,5.

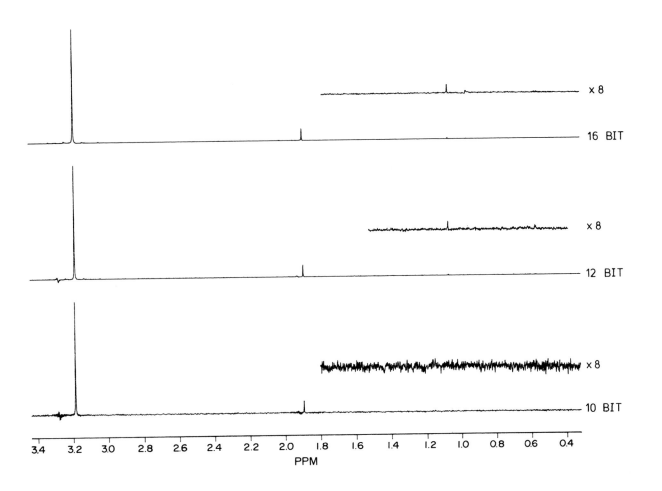

REFERENCES

1. W. W. Conover, in *Topics in Carbon-13 NMR Spectroscopy*, (ed. G. C. Levy), **4**, 38-51, Wiley, 1984.
2. D. I. Hoult, in *Topics in Carbon-13 NMR Spectroscopy*, (ed. G. C. Levy), **3**, 16-27, 1979.
3. J. N. Shoolery, *ibid.*, 28-38, 1979.
4. M. L. Martin, J-J. Delpuech and G. J. Martin, *Practical NMR Spectroscopy*, pp. 120-131, Heyden, 1980.
5. J. W. Cooper, in *Topics in Carbon-13 NMR Spectroscopy*, (ed. G. C. Levy), **2**, 411-419, Wiley, 1976.

Figure 3.14 The effect of ADC resolution on sensitivity (see text for details).

4

Describing Pulse NMR

4.1 INTRODUCTION

In this book we are looking at physical processes, and yet being Chemists we will not be inclined to model them in a mathematical way. This presents something of a dilemma, since, in trying to avoid getting bogged down in the theory of pulsed resonance, it is all too easy for the text to degenerate into a catalogue of experiments, providing no insight into what is happening inside the sample. On the other hand, any attempt at a rigorous approach to describing even a 'simple' two pulse experiment such as COSY (Chapter 8) is bound to entail more mathematics than most of us could be bothered to read. I believe it is possible to steer between these alternatives in a useful way, as I think a significant number of concepts important in pulse NMR can be appreciated through simple physical pictures.

Specifically, what *can* be described in a digestible form includes relaxation processes (but not all the mechanisms which give rise to them), the effects of ideal *rf* pulses on the bulk magnetisation, simple spin echoes and the concept of *frequency labelling* central to two-dimensional spectroscopy. Two other concepts of great importance are less amenable to this approach, these being the process of *coherence transfer* and the creation and properties of *multiple quantum coherence*; however the first of these can be appreciated in some cases through the use of population diagrams for simple systems.

Thus what I intend to do in this chapter is develop a vocabulary for describing pulse experiments in a pictorial way that we can use later. You might like to whet your appetite for this kind of thing by looking at some of the other chapters first if you find the discussions here a bit far removed from reality, but a basic understanding of this material will be required to appreciate the rest of the book fully. The best incentive for thinking about what happens during the experiments comes, in fact, from the desire to optimise them for the solution of real problems, so if you can use a spectrometer and try out some multi-pulse spectra that would be even better.

4.2 THE COMPONENTS OF THE EXPERIMENT

4.2.1 Introduction

NMR is concerned with the interaction between an oscillating magnetic field (the *rf* field) and the net magnetisation of a sample, which originates from some arrangement of its constituent nuclei brought about by applying a static magnetic field. It is a phenomenon which is invariably observed in bulk matter, so that a surprisingly large proportion of the theory of NMR can be expressed in *classical* terms. For instance, the Bloch theory of absorption is entirely classical and requires no knowledge of quantisation of energy levels. Ironically, this is often a source of confusion to chemists brought up with quantum theory, who feel quite happy about discrete transitions between various energy levels of a system, but become uneasy when the bulk manifestation of the existence of such transitions turns out to be a simple, classical magnetisation of the sample.

In reality, the ability to approach nuclear resonance from both quantum mechanical and classical viewpoints provides considerable advantages and is one of the beauties of the subject. For our purposes the classical picture is often going to be most productive, and so much of the description below relates to the *net* behaviour of the magnetisation of the sample. It is extremely important to keep this in mind throughout; places where the switch from microscopic to bulk descriptions is made are carefully high-lighted in the text. We shall consider in turn the behaviour of nuclei in a static field and the nature of electromagnetic radiation, and then find a convenient way of connecting the two together. Throughout the following sections the discussion will be about spin-$\frac{1}{2}$ nuclei. The symbol **B** is used for the magnetic induction or flux density, which is the appropriate measure of magnetisation in materials with non-zero permeability. Many other texts and publications use **H**, the magnetic intensity, or use **B** and **H** interchangeably; the distinction is not important for our empirical approach to the subject. I also switch fairly freely between angular velocity or speed (in radians/s), represented by a vector **ω** or scalar ω respectively, and the corresponding frequency represented v (in Hertz), assuming the reader will make the mental substitution $\omega = 2\pi v$ when necessary.

4.2.2   The Nuclei

Probably you are familiar with the image of a spin-$\frac{1}{2}$ nucleus as a small bar magnet, the magnetic moment arising since the nucleus has charge and spin. Placed in a static field, the nuclear magnet has an energy which varies with orientation with respect to that field. On the microscopic nuclear scale, the possible energies are quantised, and in our mind's eye we relate the two possible values of the quantum number m ($\pm \frac{1}{2}$) to parallel and antiparallel orientations of the magnet and the field. The NMR absorption is then a consequence of transitions between these two energy levels, stimulated by the applied *rf* field. To extend this picture to include the bulk description of the experiment, we need to consider the motions of the nuclear magnets more thoroughly.

The nuclear angular momentum **J** and the magnetic moment **μ** arising from it can both be represented as vectors, the constant of proportionality between them being called the *gyromagnetic ratio* γ. It is this constant which determines the resonant frequency of the nucleus. The interaction of the nuclear magnetic moment and angular momentum with the applied field **B**$_0$ (Figure 4.1) can be expressed in classical terms by the equation:

$$\frac{d\mathbf{J}}{dt} = \mathbf{\mu} \times \mathbf{B}_0 \qquad (4.1)$$

This differential equation is not particularly difficult to solve (see section 4.3.2), but we do not even need to trouble ourselves with that. Instead note that the equation of motion for a body with angular momentum in a *gravitational* field is analogous with 4.1, provided we equate the angular momentum vector with that of the magnetic moment and the gravitational with the magnetic field. Why do this? Well, you already know the solution to the second case: it is the motion of a gyroscope. A gyroscope in a gravitational field *precesses,* that is the axis of its rotation itself rotates about the field direction. Exactly this motion is performed by the nuclear spins.

A single nucleus, then, gives rise to a magnetic moment which rotates at some speed around the applied field; this is referred to as the Larmor frequency of the nucleus, and is just its NMR absorption frequency **ω**. You already know, of course, that this frequency depends on the strength of the applied field, and on intrinsic properties of the nucleus reflected in its

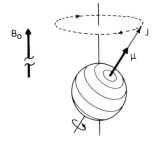

Figure 4.1 A spinning charge with angular momentum **J**, such as a proton, gives rise to a magnetic moment **μ** (= γ**J**). In a magnetic field its axis of rotation precesses around the direction of the field, like a gyroscope.

gyromagnetic ratio: $\boldsymbol{\omega} = -\gamma \mathbf{B}_0$. The rotation may be clockwise or anti-clockwise, depending on the sign of the gyromagnetic ratio, but is always the same way for any particular nucleus.

If we define the z axis of a set of Cartesian coordinates to point along the static field direction, then this single spin has a stationary component of its magnetic moment aligned along the z axis, and a component rotating with the Larmor frequency in the x-y plane. You can, if you like, equate the z component of the magnetic moment with the original bar-magnet picture; it is the orientation of this component which determines the energy of the system. Note in passing that since the frequency of Larmor precession depends on the applied field, it is a natural measure of the field strength from an NMR point of view. Such usage is very familiar ("We've got a 500 MHz magnet"); it will turn out also to be useful for describing the strength of the oscillating *rf* field. Beware, however, of forgetting that the precession frequency relates to a specific nucleus: the 500 MHz magnet (for protons) is equally well a 125 MHz magnet for carbon.

Now consider a large number of such spins, all possessing the same Larmor frequency (Figure 4.2). We know that the parallel orientation of the z component of each spin along the applied field direction is of lower energy than the antiparallel one. So, assuming a thermal equilibrium is somehow achieved, we would expect a difference in populations of the two states, with a surplus in the lower energy state according to the Boltzmann distribution; only the surplus spins are shown in the figure. Therefore, along the z axis, a net magnetisation of the sample should be present parallel with the applied field direction. Meanwhile, all the contributing spins have components precessing in the x-y plane. However there is no reason to prefer one direction in this plane over another, as all have equal energy, so that the *phase* of precession is random. Therefore for a very large collection of spins at equilibrium there will be no net magnetisation in the x-y plane, and so the total magnetisation of the sample is stationary and aligned along the z axis.

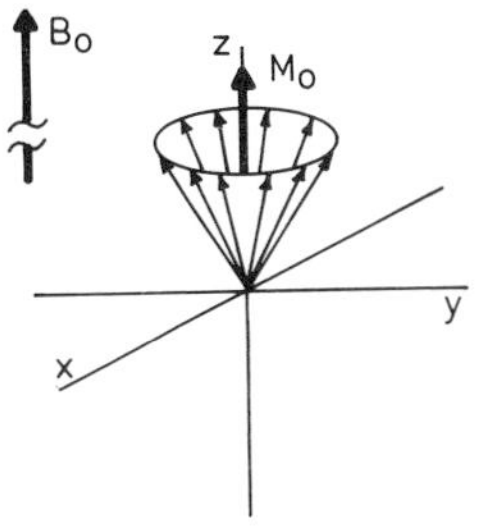

Figure 4.2 Out of a large collection of moments, a surplus have their z components aligned with the applied field, so the sample becomes magnetised in that direction.

4.2.3 The Radio-frequency Field

The alternating voltage applied across the ends of the coil in the NMR probe induces an alternating magnetic field throughout the sample. The geometry of the coil is arranged so that this field is perpendicular to the applied field (i.e. it is in the x-y plane). Ideally this $\boldsymbol{B}_1$ *field* (field 1 because the static field is $\mathbf{B}_0$) would be uniform everywhere in the sample and would fall abruptly to zero at the sample boundaries; naturally the true situation deviates from this ideal. The amplitude of the oscillating field is substantially smaller than that of the static field (thousands of times smaller). Its frequency is for the time being taken to be that of the NMR resonance to be observed, i.e. the Larmor frequency.

Look at Figure 4.3. In the upper part of the figure, a half-cycle of the oscillation of the magnetisation due to the *rf* field is represented. The lower part of the figure shows two magnetisation vectors of constant amplitude rotating about an axis (the z axis) in opposite directions. The angular frequency of the rotation is the same as the radio frequency. Convince yourself that the resultant of these two vectors behaves in just the same way as the oscillating vector above; in other words this pair of counter-rotating vectors is an equivalent representation of the *rf* signal. Decomposing the oscillating magnetic field into two rotating vectors in this way helps to clarify the way in which it interacts with the magnetisation of the sample.

4.2.4 The Rotating Frame

Now consider what happens when the *rf* field and the sample magnetisation interact. The conceptual problem here is that our *rf* field is in

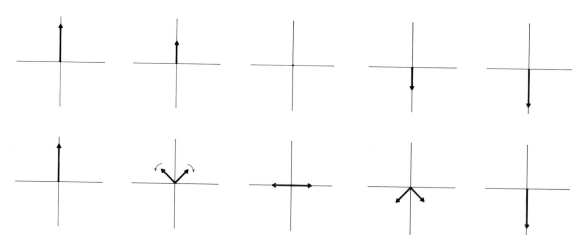

Figure 4.3 An oscillating magnetic field, as at the top of the figure, is equivalent to two counter-rotating magnetisation vectors.

motion, and while the net magnetisation of the sample is currently static along the z axis, if it becomes displaced from that axis (as indeed it will momentarily) it will evidently have precessional motion about the static field. There are too many different rotating motions here to comprehend without risking seasickness, but the problem is ingeniously solved by observing it from a slightly different perspective. We would like to eliminate the various rotations present in the experiment, and this can be achieved if we choose a set of coordinates that rotate along with the nuclear precession and watch what is happening from this point of view.

Reverting to the microscopic scale for a moment, if the rotating set of coordinates is chosen to rotate at the same speed and in the same direction as the nuclear precession, then each individual nuclear magnetic moment appears static in that frame. As there is apparently no precession, and the cause of the precession was the external field, evidently in this frame of reference the external field has disappeared. The net magnetisation of the sample looks just the same, though, along the z axis as before (Figure 4.4).

Figure 4.4 The typical arrangement of fields in an NMR experiment. In the lab. frame on the left, we have the static field, the sample magnetisation and the two counter-rotating vectors due to the *rf*. Switching to the rotating frame simplifies things by eliminating the static field and freezing one of the *rf* vectors (the other one we simply ignore).

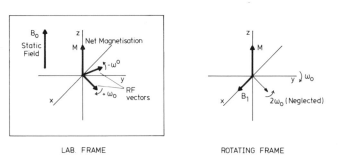

Of the two components into which the *rf* field can be decomposed, one is static in the x-y plane, since the frequency of the *rf* was chosen equal to that of the Larmor precession. The other, which was originally moving at equal speed in the opposite direction, is in this new frame rotating at *twice* the Larmor frequency, and has no significant effect on the experiment (justifying this statement unfortunately would require a considerable digression, so it must be taken as granted).

Originally, of course, this coordinate transformation was invented as an algebraic manouevre to simplify the solution of the equations of motion of the magnetisation, but we can steal the idea and use it just to help visualise what is going on. A similar trick is used in a more familiar context in the analysis of circular motion, where the change of coordinates causes a fictitious force (the centrifugal force) to *appear* rather in the same way in which this transformation has made the $\mathbf{B}_0$ field *disappear*. We will return and look at this a little more formally later. In representing static and rotating frames it is conventional to give the x and y axes different labels in each frame, for instance x, x' and y, y', to emphasise that the coordinate systems differ. However, in this book we will be working in the rotating frame nearly all the time, and only in an informal way, so I will not use this labelling convention. On the few occasions when static frames are used (mainly in the following sections), this is indicated explicitly in the figures.

4.2.5 A Pulse!

Now we are in a position to think about what happens during an *rf* pulse, that is to say what happens when we turn on the $\mathbf{B}_1$ field for a while and then turn it off again. In the rotating frame, both the sample magnetisation and the $\mathbf{B}_1$ field vector are static, one along the z axis and the other at right angles to it, say along the x axis (Figure 4.5). Two magnetisations at right angles like this exert a force on each other, and since the sample magnetisation is associated with some angular momentum (of the nuclei), the net result (remembering gyroscopes again) is a *torque* acting around $\mathbf{B}_1$. The sample magnetisation is driven around the $\mathbf{B}_1$ field vector (i.e. around the x axis) at a speed depending on the field strength, eventually passing through the x-y plane. Assuming that we somehow know how fast the sample magnetisation is moving, we could calculate when it would arrive along the y axis, and could turn off the $\mathbf{B}_1$ field at that moment. We would have brought about a rotation of the sample magnetisation through 90 degrees - a 90° or $\pi/2$ (radians) pulse. By waiting different lengths of time the magnetisation could in theory be rotated through any angle, returning every 360° to its starting position. In real life though numerous factors prevent this process being extended indefinitely, and usually only a few rotations are possible before the magnetisation disappears.

Figure 4.5 A pulse! While the *rf* is on, the sample magnetisation is driven round; we can turn it off whenever we like (in this case, just as it reaches the y axis).

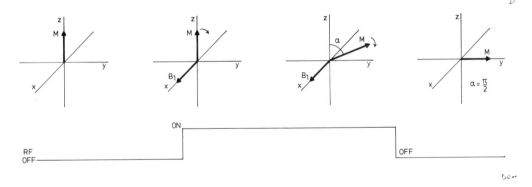

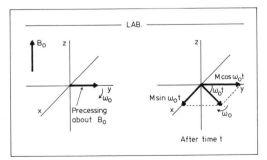

Figure 4.6 After the pulse is over, the sample magnetisation remains in the x-y plane. In the lab. frame we see it precessing about the static field, and thereby generating radio signals.

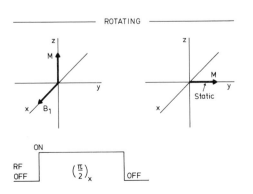

Consider now the consequence of turning off the **B**₁ field at the moment of $\pi/2$ rotation. We are left with a magnetisation vector pointing along the y axis of the rotating frame, the **B**₁ vector having been removed (Figure 4.6), and would like to know what this means. Having previously persuaded ourselves that the applied *rf* signal could be represented, in the rotating frame, as a static vector plus another one rotating at twice the Larmor frequency, we might imagine (correctly) that this new static vector we have created also has some of the essence of an *rf* signal. Jumping back into the static laboratory coordinates (Figure 4.6) clarifies what is going on. In the laboratory frame the sample magnetisation is rotating about the z axis at the Larmor frequency; projecting this rotating vector onto the x and y axes decomposes it into *two* radio signals which are in some sense orthogonal. These are the signals which are measured as the *Free Induction Decay* (FID) in an NMR experiment. Whether they are detected separately or mixed together in some way depends on the experimental set up; just how is explained later.

It is now easy to understand the effect of using a pulse length of other than $\pi/2$; some cases are shown in Figure 4.7. All other pulse lengths leave some of the sample magnetisation along the z axis, where it creates no signal. Only the component in the x-y plane is precessing in the manner required to generate voltages in the receiver coil. Thus, $\pi/2$ pulses (or in theory $3\pi/2$, $5\pi/2$ etc.) appear to generate the maximum signal (this is true, *for a single scan experiment*). In contrast, π (or 2π, 3π ⋯) pulses generate *no* signal, as the magnetisation is all along the z axis. In Chapter 7, this property is exploited to allow the determination of the duration of a π pulse, and hence of the **B**₁ field strength.

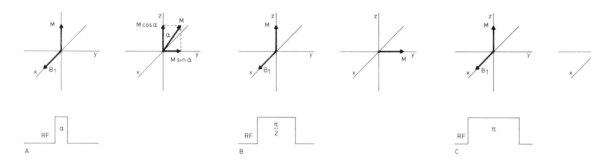

Figure 4.7 Some pulses with various flip angles. Many NMR experiments are built up from combinations of $\pi/2$ and π pulses only.

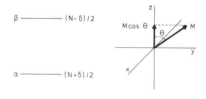

Figure 4.7a The initial populations of a two-level system (left), and the effect of a pulse of flip angle θ on the z magnetisation (right).

4.2.6   Vectors and energy levels

By concentrating on the net magnetisation as a classical quantity, as we shall throughout most of this book, we avoid getting into the deep waters of quantum mechanics and the density matrix. This is the limitation of our pictorial approach; we will not be able to understand in detail anything which hinges on quantum mechanical properties, such as *coherence transfer* and the creation of *multiple quantum coherences* (Chapter 8). This is a necessary simplification in a book (almost) free from formulae, and is, I hope, to your taste. However, it is counterproductive to ignore the fact that nuclear systems have quantised energy levels completely, and we can at least determine the effect of pulses on the *populations* of those levels quite easily. Thinking about populations can then give us a partial insight into experiments involving coherence transfer, and this approach is taken in later chapters.

Consider a system with only two energy levels α and β (Figure 4.7a), and suppose there are N nuclei which must be in one or other state. If the energies were equal, then there would be $N/2$ nuclei in each, but as the α state has slightly lower energy, it has a small population excess. If there are δ more nuclei in α than in β, then their populations are $(N+\delta)/2$ and $(N-\delta)/2$ respectively. To calculate what happens to the populations when a pulse of flip angle θ is applied, we concentrate on the z component of magnetisation. It is convenient to consider only the *excess* populations (i.e. the deviations from $N/2$) P_α and P_β of states α and β, which initially are $+\delta/2$ and $-\delta/2$.

The z component of magnetisation at any moment is proportional to the population difference between the levels:

$$\mathbf{M}_z \propto P_\alpha - P_\beta \qquad (4.2)$$

so initially $\mathbf{M}_0 \propto \delta$. We also know, come what may, that

$$P_\alpha + P_\beta = 0 \qquad (4.3)$$

Now, after the θ pulse, the z component is (Figure 4.7a):

$$\mathbf{M}_z = \mathbf{M}_0 \cos \theta \qquad (4.4)$$

so evidently

$$P_\alpha - P_\beta = \delta \cos \theta \qquad (4.5)$$

Combining this with 4.3, we calculate the new populations:

$$P_\alpha = \frac{\delta \cos \theta}{2}$$

$$P_\beta = \frac{-\delta \cos \theta}{2} \qquad (4.6)$$

These simple formulae allow us to relate what we already know of $\pi/2$ and π pulses to the behaviour of the quantised energy levels. If $\theta = \pi/2$, then $\cos \theta = 0$, and there is no excess in either state, i.e. the pulse *equalises* the populations. If $\theta = \pi$, then $\cos \theta = -1$, and the populations are *inverted*.

4.3 REALISTIC EXPERIMENTS

4.3.1 Introduction

Throughout section 4.2 we have been working with the simplifying assumption that there is only one frequency involved in the experiment - that all the nuclei have the same Larmor frequency, and that the applied *rf* signal also has that frequency (it is exactly *on resonance*). This, of course, is untrue in real spectroscopy, where the whole object of the experiment is to measure variations in the resonance frequencies of different nuclei in the sample. In extending the rotating frame picture to encompass several frequencies at once, in most of this book I am going to behave in quite cavalier fashion, selecting the frequency of the coordinate rotation entirely for visual convenience, and ignoring all consequences of non-ideality in the experimental set-up. However, before entering this fantasy world of infinitely strong and uniform $\mathbf{B}_1$ fields and relaxation times which are infinite during pulse sequences and conveniently short the rest of the time, I want to examine briefly how moving away from resonance can be treated in a relatively simple way. If you don't like this kind of thing, please feel free to ignore section 4.3.2. Parts 4.3.3 and 4.3.4, however, describe the conventions we will use later in talking about the experiments, and require attention.

4.3.2 When the Pulse is Off-resonance

In order to understand what happens when a finite $\mathbf{B}_1$ field is applied away from exact resonance, we need to take a more detailed look at the motion of the magnetisation in the rotating frame. Throughout this section I will refer to the magnetic moment $\boldsymbol{\mu}$ of a single nucleus, but the behaviour of the net magnetisation **M** is the same. Even if you find the algebra in this section disheartening, the *geometrical* results described at the end and in the figures should be comprehensible, and I would recommend paying most attention to these.

The equation of motion of an angular momentum **J** subject to a torque $\mu \times \mathbf{B}_0$ arising from the interaction of the nuclear moment and the static field $\mathbf{B}_0$ was given earlier:

$$\frac{d\mathbf{J}}{dt} = \mu \times \mathbf{B}_0 \qquad (4.1)$$

By the definition of the gyromagnetic ratio γ:

$$\mu = \gamma \mathbf{J} \qquad (4.7)$$

therefore eliminating **J**

$$\frac{d\mu}{dt} = \gamma \mu \times \mathbf{B}_0 \qquad (4.8)$$

in the *static* frame. The rate of change of μ in a frame rotating with angular velocity ω is then given by

$$\frac{\delta\mu}{\delta t} = \gamma \mu \times \left(\mathbf{B}_0 + \frac{\omega}{\gamma}\right) \qquad (4.9)$$

(For a justification of this, see Slichter[1], page 12). Suppose we pick a value ω_0 for ω such that $\omega_0 = -\gamma\mathbf{B}_0$. Then in the frame rotating with angular velocity ω_0:

$$\frac{\delta\mu}{\delta t} = 0 \qquad (4.10)$$

i.e. the magnetic moment is static. Therefore in the laboratory frame it is precessing with angular velocity $-\gamma\mathbf{B}_0$, which is the Larmor frequency; this exceptional ease of solution of the vector differential equation 4.8 is the reason for using the coordinate transformation.

Now we will add the $\mathbf{B}_1$ field, and choose ω such that one component of $\mathbf{B}_1$ is static in the rotating frame. In this case:

$$\frac{\delta\mu}{\delta t} = \gamma \mu \times \left(\mathbf{B}_0 + \mathbf{B}_1 + \frac{\omega}{\gamma}\right) \qquad (4.11)$$

Now, if $\mathbf{B}_1$ is exactly on resonance, $\omega = \omega_0$, so:

$$\begin{aligned}
\frac{\delta\mu}{\delta t} &= \gamma \mu \times \left(\mathbf{B}_0 + \mathbf{B}_1 + \frac{\omega_0}{\gamma}\right) \\
&= \gamma \mu \times (\mathbf{B}_0 + \mathbf{B}_1 - \mathbf{B}_0) \\
&= \gamma \mu \times \mathbf{B}_1 \qquad (4.12)
\end{aligned}$$

This equation has the same form as 4.8, which we already know describes a precession about **B**, so in this case the magnetisation rotates with velocity $-\gamma\mathbf{B}_1$ around the direction of $\mathbf{B}_1$, as stated informally before.

If $\omega \neq \omega_0$ (i.e. if $\mathbf{B}_1$ is off-resonance) then:

$$\begin{aligned}
\frac{\delta\mu}{\delta t} &= \gamma \mu \times \left(\mathbf{B}_0 + \frac{\omega}{\gamma} + \mathbf{B}_1\right) \\
&= \gamma \mu \times \left(\frac{\omega - \omega_0}{\gamma} + \mathbf{B}_1\right) \qquad (4.13)
\end{aligned}$$

We can regard $(\omega - \omega_0)/\gamma$ as representing a reduced static field $\mathbf{B}_r$ in the rotating frame; $\boldsymbol{\mu}$ now precesses about the resultant of $\mathbf{B}_1$ and $\mathbf{B}_r$ (Figure 4.8), i.e. about a *tilted effective field* $\mathbf{B}_{eff}$. This gives us a criterion for deciding how big $\mathbf{B}_1$ should be: we do not want to deviate too far from the ideal of rotating all the vectors around the same axis in the rotating frame, as generally experiments will be designed with the assumption that this is what is occurring. In other words, we want the angle θ in Figure 4.8 to be small. θ is given by

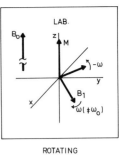

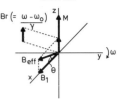

$$\tan \theta = \frac{\omega - \omega_0}{\gamma B_1} \qquad (4.14)$$

So γB_1 (which is the rate of precession about $\mathbf{B}_1$, remember, or if you like it is the field strength measured in frequency units in the same way as we did for the static field in section 4.2.2) must be large compared with the maximum resonance offset $\omega - \omega_0$ we are likely to encounter.

This can be clarified by looking at some concrete examples. Suppose we decide arbitrarily that θ should be no larger than $1°$. For a proton spectrum run over 10 p.p.m. at 200 MHz, the maximum offset is 1000 Hz (5 p.p.m.). Tan θ is about 0·017, so the required field strength is 1000/0·017 which is about 60 kHz. This corresponds to a $\pi/2$ pulse length of about 4 μs, which in a 5 mm probe at this frequency is perfectly practical. Arguing like this from a requirement to rotate the magnetisation of all the different nuclear species around approximately the same axis through to a calculated pulse length of a few μs is one way to justify the need for short, strong pulses in FT NMR.

If we do the same calculation for carbon observation on a 500 MHz spectrometer (carbon frequency 125 MHz), where the maximum offset is around 15 kHz, we come up with a desired $\pi/2$ pulse of 0·3 μs. In fact, on a high resolution liquids spectrometer this value is more likely to be 15 to 20 μs in a 10 mm probe, which would give a maximum tilted effective field angle of around $45°$. This is a major stumbling block for many multi-pulse experiments, and investigating ways round it is an active research area. The ingenious concept of composite pulses, which is one partial solution to this problem, is mentioned elsewhere in the text (Chapter 7).

The consequences of the off-resonance effect vary with pulse length. For a $\pi/2$ pulse, where the main object may be to leave behind no z magnetisation, tilting the rotation axis means the magnetisation vector has further to go to get to the x-y plane. On the other hand, the effective field $\mathbf{B}_{eff}$ is larger than $\mathbf{B}_1$, so the vector moves more quickly. Thus there is an element of 'self-compensation', and the $\pi/2$ pulse is not too bad at eliminating all the z magnetisation over a wide offset range. Vectors with different offsets do arrive in the transverse plane with *phase* errors however (Figure 4.9a).

Figure 4.8 If the pulse is not exactly on resonance, then the effect of the static field is not entirely removed in the rotating frame. The sample magnetisation then rotates about the resultant of the residual field and $\mathbf{B}_1$.

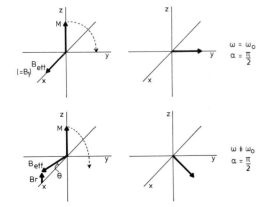

Figure 4.9a For a $\pi/2$ pulse the off-resonance effect is not too bad, if all we are interested in is how much z magnetisation is transferred to the x-y plane $\cdots$

In the case of a π pulse the major concern may be to create no x-y magnetisation; the increased rate of precession about $\mathbf{B}_{eff}$ is no help in this respect because it just tends to put the vector into the x-z plane (Figure 4.9b). The strong variation in inversion efficiency with offset can be a major source of error in experiments such as the inversion-recovery T_1 measurement described later.

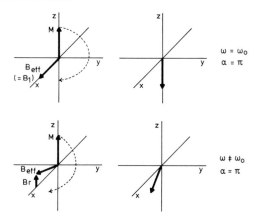

Figure 4.9b ⋯ however, for a π pulse we are interested in inverting the magnetisation, and tilting the rotation axis is highly detrimental.

4.3.3 Several Frequencies At Once

Now, whether you have been following the previous section or have taken the shortcut, we will drop any pretence at formality and take a look at how we can informally treat samples containing more than one line. This will involve completely neglecting the effect described in the previous section; thus when we apply a pulse to a sample we will assume that all the magnetisation vectors due to the different spin species present experience the same rotation about the same axis. The result for a $\pi/2$ pulse will therefore be a net magnetisation, comprised of the sum of the magnetisations of the various species, aligned along an axis in the x-y plane. Very often we will be concerned with how this magnetisation evolves during an interval before we apply another pulse. As the rotating coordinate system can only be chosen to move at one particular angular velocity, evidently this will involve more than just a static vector remaining on its original axis. It is easiest to get a feel for what happens by looking at some specific cases.

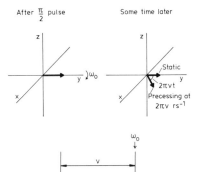

Figure 4.10 Two lines in the rotating frame. We choose the frame to suit the problem at hand; here one line has been made static, so the other precesses at its chemical shift frequency.

Consider just two lines, with some chemical shift difference v between them (Figure 4.10). If we choose the rotating reference frame so that it has the same angular velocity as the frequency of one of the two lines, then the magnetisation arising from those nuclei will remain along the y axis after the $\pi/2$ pulse. The contribution from the other line, although it arrives initially along the y axis, is precessing at a different speed (*slightly* different; remember that chemical shifts are usually measured in parts per million), so its precession will not be entirely cancelled out in the rotating frame. Rather, it still precesses about the z axis, at a frequency equal to its chemical shift difference from the other line. We can easily work out where it gets to after some time t, particularly if you remember that Hertz used to be called cycles-per-second; thus every $1/v$ seconds it has travelled once round the rotating coordinate system. The angle through which it moves in t seconds is $2\pi v t$ radians. This conversion of the absolute precession frequencies of the nuclei (in MHz) down to frequencies which represent the chemical shift differences between them (usually in kHz) might strike you as being reminiscent of the physical process in which an *rf* reference frequency is subtracted from the NMR signals prior to digitisation (Chapter 2), and indeed we can consider the angular frequency of the rotating frame to be this reference frequency.

Particular interest will arise later on in the case when the various components in the x-y plane are parts of a multiplet, in other words when the frequency separations are coupling constants rather than chemical shifts. Figure 4.11 shows the evolution of the components of a triplet; the reference frequency (speed of the coordinates) has been chosen on the centre line (at the chemical shift, in other words), so *that* one remains fixed.

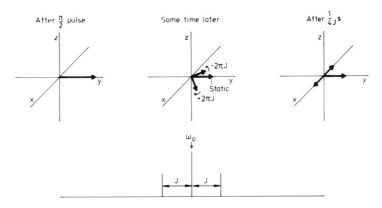

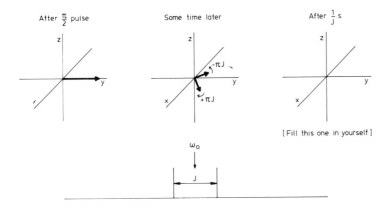

Figure 4.11 When looking at multiplets it is normally appropriate to set the reference frequency at the chemical shift; here is the result for a triplet.

The other two move in *opposite* directions in the rotating frame, at $+J$ and $-J$ cycles-per-second ($\pm 2\pi J$ radians/s). They all realign along the $+y$ axis every $1/J$ seconds. Try working through the same thing for a doublet with the same coupling constant J; set the reference frequency at the chemical shift again (i.e. halfway between the lines, see Figure 4.12). You should find that the components realign at the same time as, but in a different place from, those of the triplet.

Figure 4.12 Try working out where the doublet components get to after $1/J$ s.

4.3.4 Axes and Phases

So far we have been treating directions in the x-y plane of the rotating frame in an arbitrary way, assigning the $\mathbf{B}_1$ field as 'say along the x axis' and generating an initial sample magnetisation therefore along the y axis. Also, when we jumped back into the laboratory frame after a $\pi/2$ pulse, we were left with two *rf* signals 'in some sense orthogonal', each oscillating along different axes of the static coordinates. We need to attribute some physical significance to these statements, and the behaviour of the magnetisation in the static frame after a pulse gives us a first clue as to what this is. Consider Figure 4.13, where I have reverted to treating the simplest possible case of a single line with the pulse and reference frequency exactly on resonance. The sample magnetisation is rotated through 90° about the x axis and arrives down the y axis in the rotating frame; then we jump back into the laboratory frame and watch what happens as the magnetisation precesses at the Larmor frequency.

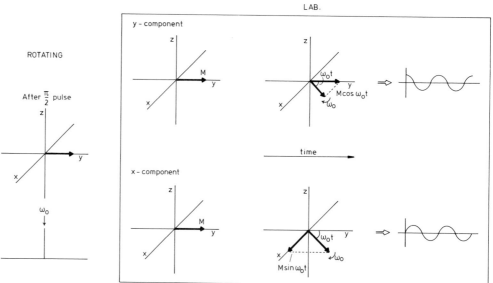

Figure 4.13 A closer look at the sample magnetisation in the lab. frame after a $(\pi/2)_x$ pulse reveals *two rf* signals, differing only in phase (by 90°).

Suppose there is some way to detect separately the projections of the magnetisation onto the x and y axes of the lab. frame, without worrying for the time being about how this might be done. Then evidently the only difference between the two components of the magnetisation is that the y component starts at a maximum and the x component at a minimum immediately after the pulse; both then oscillate at the same frequency. In other words, they differ only in *phase*, the x component oscillating as a sine function and the y component as a cosine. The voltages induced in the receiver coil are proportional not to the magnetisation, but to its *rate of change*, so the x component induces a maximum voltage at time zero and the y component a minimum (the derivative of sin being cos and that of cos $-$sin). The two signals 90° out of phase are called the absorption and dispersion parts of the magnetisation, and give rise to the two corresponding forms of the Lorentzian line which we have already encountered in Chapter 2. In CW NMR the spectrometer is adjusted to select the absorption component (although of course the signals are not excited quite as described above, the two components still exist); in FT NMR the real and imaginary parts of the transform correspond with the absorption and dispersion signals (or more generally with mixtures of the two, as we will see shortly).

Now let us try the same thing again, but with the $\mathbf{B}_1$ field along the y axis (Figure 4.14). This is easy - the result is identical, except that the sin

Figure 4.14 The same as 4.13, but after a $(\pi/2)_y$ pulse.

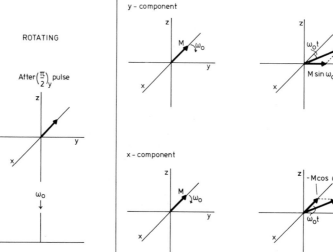

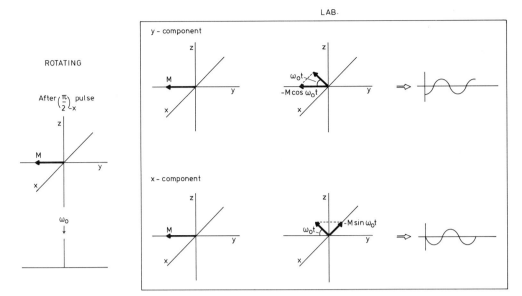

ROTATING

After $\left(\frac{\pi}{2}\right)_{-x}$ pulse

y - component

x - component

Figure 4.15 The same as 4.13, but after a $(\pi/2)_{-x}$ pulse.

and cos components of the resulting signal are interchanged (and the cos component is inverted). Moving the $\mathbf{B_1}$ field round another 90° so that it lies along the $-x$ axis (Figure 4.15) seems to restore the original situation, except that relative to the first case *both* components are inverted (their phases shifted by 180°). If signals from $+x$ and $-x$ pulses are separately transformed and the $+x$ absorption part designated 'positive', then the $-x$ experiment will appear to give a 'negative' peak. It is important to appreciate that this is due to a phase change in the signal, not to a change from absorption of energy to emission of energy; signals are being *emitted* by the sample in both cases.

The manner in which the separation of the absorption and dispersion signals is brought about experimentally need not concern us in intricate detail, but it helps to have some idea what is involved. As mentioned several times already, the NMR signals are detected by subtracting a constant reference frequency, leaving signals in the audio range to be digitised for transformation; we have equated the frequency of this reference with the angular velocity of our rotating coordinate system. The device which brings about this subtraction (commonly a component known as a double-balanced mixer, although there are other possibilities) responds also to the phase relation between the signal and the reference (it is a *phase sensitive detector*), and adjusting the phase of the reference frequency allows the selection of one or other component of the signal, or indeed of any mixture of the two.

If the phase of the reference corresponds exactly with the phase of the NMR signal, then it is detected as a pure absorption mode, while if it differs by exactly 90° the result is pure dispersion. More usually it is neither necessary nor practical to adjust an FT spectrometer to work in this fashion, and a mixture of components is detected. Just as we can take the frequency of the reference to define the speed of the rotating coordinate system, we could take its *phase* as labelling the x and y axes, but since this does not correspond exactly with experimental procedure on a pulse spectrometer we will use a different definition, discussed shortly. If you have worked with a low-field CW spectrometer, you will almost certainly have needed to adjust the receiver reference phase for pure absorption lineshape, and observed the result of small errors in the adjustment which give rise to an absorption line mixed with a small proportion of a dispersion line (Figure 4.16).

By now I hope you can guess the significance of 'setting the $\mathbf{B_1}$ field along the x axis' - this refers to some particular phase of the *rf* pulse. Only *relative* phase is important, so we can just call the first pulse we perform an

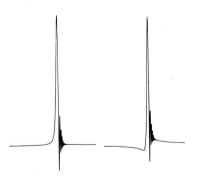

Figure 4.16 The effect of detector phase error is quite familiar from CW spectroscopy (one of only two 60 MHz spectra in this book!).

x pulse, then when we want a y pulse we step the phase 90°, 180° for a $-x$ pulse and 270° for a $-y$ pulse (increments other than 90° are also needed for certain experiments). It is these transmitter phases which we take to define the labels on the rotating coordinate axes; ideally we would then set the receiver reference phase so that it also corresponded with one of the axes, but as already mentioned this proves impractical on a pulse spectrometer. Instead the receiver phase has some arbitrary (but *constant*) relation to the transmitter, and a mixture of absorption and dispersion components is measured. The two are separated numerically after Fourier transformation; phase errors arising from other sources than misadjustment of the receiver reference are also corrected at this time as discussed later.

The point of all this will become much clearer when we come to discuss real experiments. Devising proper phase relations between sets of pulses in order to select particular components of the total magnetisation is an important element of pulse NMR. From now on I will specify the pulses in the form $(\pi/2)_\varphi$ where φ represents the phase of the *rf* and hence the axis *about which the magnetisation is rotated* in the rotating frame. Thus a $(\pi/2)_x$ pulse rotates around the x axis, leaving the magnetisation directed along the $+y$ axis; $(\pi/2)_y$ around the y axis and so on (Figure 4.17). Rotations are by convention taken to be clockwise (the actual direction depends on the sign of γ, but since this is fixed for a given nucleus it does not matter which way we choose to imagine the rotation). Receiver reference phases are also varied in many experiments, and these will be simply referred to as x, y etc. with the implicit assumption that appropriate numerical phase corrections are subsequently applied when necessary. There is an added complication regarding receiver phase which is discussed in the next section.

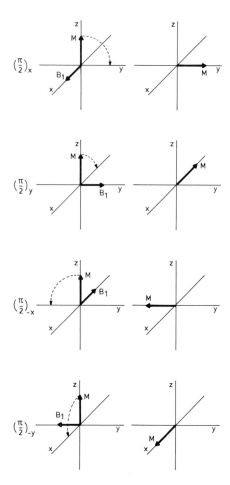

Figure 4.17 Virtually all experiments use only the four pulse phases shown here.

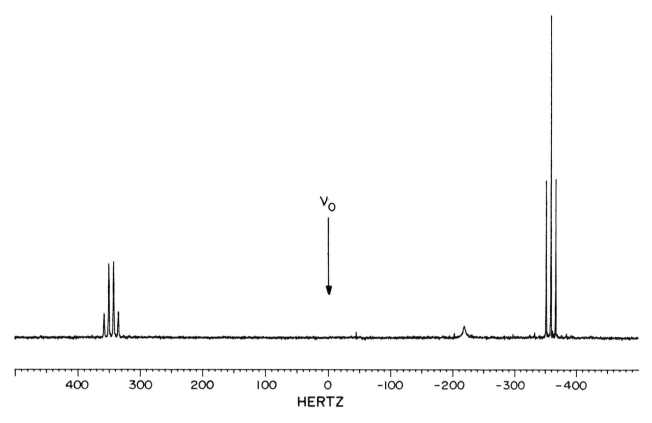

ν_0

HERTZ

Figure 4.18 The middle of the spectrum seems an obvious place to put the detector reference, but we then have to contend with both positive and negative frequencies. Two-phase (quadrature) detection is required to distinguish these.

4.3.5 Quadrature Detection

This is essentially a technical feature of spectrometers, aimed at improving sensitivity, and you may feel that this section is not relevant for the non-specialist. If we were only going to discuss one-dimensional spectra I might almost concede the point; however the problem to be considered here crops up again in a slightly different guise in 2D NMR, and we will be better equipped to understand that if we have previously tackled the 1D case. Also the artefacts known as quad images arising from imperfect quadrature (quad) detection may interfere even with 1D spectra when samples with very high dynamic range are examined. An ingenious method for dealing with quad images serves as an introduction to the idea of *phase cycles*, which are of paramount importance in multi-pulse NMR. If you are new to NMR I would recommend omitting this section on a first reading, and perhaps returning later when the need to understand these concepts has become more pressing.

I have remarked many times already that the signals in a pulse NMR experiment are detected by subtracting a reference frequency similar to the resonance frequencies of the nuclei, then digitising and transforming the result to give the frequency spectrum. A problem arises regarding the choice of this reference frequency, owing to a property of the Fourier transform. One might suppose that a nice place to put the reference would be in the centre of the spectral region of interest, so that half of the spectrum lies above and half below it (Figure 4.18). In particular, this would minimise the necessary speed of the analogue to digital converter (ADC), since if the spectral width is F we only need to characterise frequencies up to $F/2$ (so, by the Nyquist criterion, the ADC needs to run at F Hertz). Unfortunately, if only a single component of the magnetisation is detected and used as input to the Fourier transform, the resulting

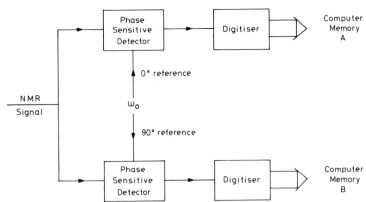

Figure 4.19 The experimental set up for quad detection.

frequency domain spectrum makes no distinction between positive and negative frequencies; a time domain signal with frequency $v_0 + \delta$ (where δ is the offset from the reference) gives rise to two peaks in the frequency domain spectrum at $\pm \delta$, so that frequencies above and below v_0 are not distinguishable.

We can get a feel for why this should be if we imagine ourselves acting as the NMR detector by watching the signals in the rotating frame. Distinguishing frequencies above and below v_0 corresponds with distinguishing the direction in which a magnetisation component precesses in the rotating frame (which is rotating with frequency v_0, of course). If we were somehow observing the vector, but were compelled to observe only its projection onto a single axis, we would just see a magnetisation oscillating up and down whichever way it was in fact rotating. In a sense, our perspective is too narrow to appreciate fully what is going on. We can expand the perspective by simultaneously observing the projections onto *two* axes; comparing the magnetisation along each axis from moment to moment would allow us to work out which way the vector was rotating.

This is the scheme known as *quad detection*. Experimentally it corresponds with using two phase-sensitive detectors whose reference frequencies are identical, but whose phases differ by 90° (Figure 4.19). For simplicity we can assume that one is set correctly to detect the cosine component of the magnetisation; the other therefore detects the sine component (actually each detects a mixture of both components). The two signals are digitised separately and treated as the real and imaginary parts of a *complex* spectrum; when this is subject to Fourier transformation positive and negative frequencies are properly distinguished. To understand fully why this is we would need to digress further into the mathematics of Fourier transforms than would be healthy, but we can get some idea of what is happening if we take as granted one property of the transform. This is that it preserves the symmetry of functions.

By symmetry in this context I mean the behaviour of functions with respect to change of sign of their variables. We can distinguish two cases: if $f(-x) = f(x)$ then we call the function f *even*, while if $f(-x) = -f(x)$ we call it *odd*. A moment's thought reveals that sine is an odd function and cosine an even one; this is illustrated in Figure 4.20. Oddness and evenness

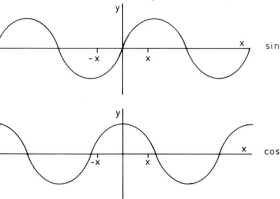

Figure 4.20 Clearly sin is odd and cos even (see text).

in the time domain (i.e. whether the sine or cosine component is involved) transforms into oddness and evenness in the frequency domain, which corresponds with whether the amplitudes of the two components of the absorption part of the complex transform at $\pm\delta$ have the same or opposite signs. Thus when we do a transform including *both* components (i.e. the complex transform for quad detection) one line in the frequency domain is reinforced while the other is cancelled; this brings about the distinction required. Figure 4.21 illustrating this is adapted from section II.A.3 of Fukushima and Roeder[2], where there is a more extensive discussion in this vein if you find it interesting.

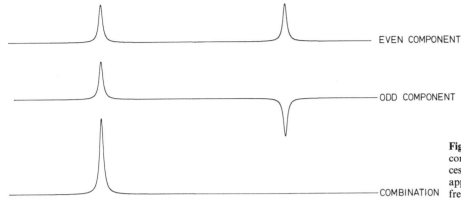

Figure 4.21 We can imagine the two components of a complex transform processed separately and then combined, to appreciate how positive and negative frequencies are distinguished.

Perhaps this seems an elaborate way to proceed, when a simple alternative would appear to be placing the reference frequency at one edge of the spectral region so that all detected frequencies have the same sign (Figure 4.22). The problem with this scheme is that, while we can arrange that there are no *signals* to one side of the reference frequency, there will inevitably still be *noise*. Without quad detection this extra noise is added into the noise present in the interesting region, resulting in a decrease in the signal-to-noise ratio by a factor of $\sqrt{2}$ (it is $\sqrt{2}$, not 2, because the noise is random; the *central limit theorem* of statistics applies).

Figure 4.22 This alternative placing of the detector reference circumvents the problem of negative frequencies, but with single-phase detection extra noise (to the right of the reference here) will be combined into the spectrum.

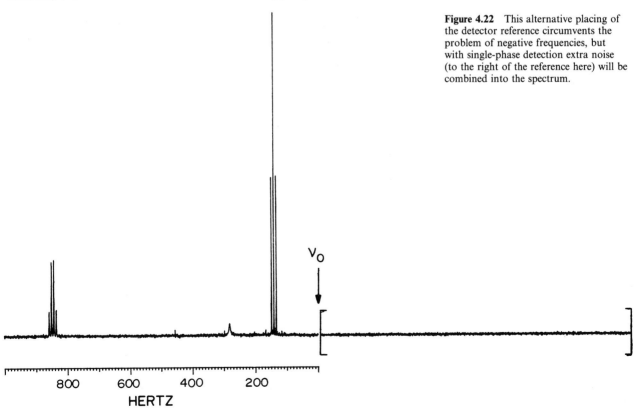

This improvement in signal-to-noise ratio is the principal advantage of using quad detection, and has led to its universal adoption on modern spectrometers. There are also several more or less subsidiary advantages. With the reference in the centre of the spectral region only half as large a frequency range needs to be characterised, so a slower analogue-to-digital converter will do. On the other hand, you need two ADC's, so in a way the total digitisation effort remains the same. Generally it is inconvenient to have to generate several radio frequencies which differ only slightly, so with most spectrometers it is necessary for the pulses to have exactly the same frequency as the reference. Without quad detection this means that the pulse must be at the edge of the spectrum, not in the centre, and so off-resonance effects (section 4.3.2) are more severe.

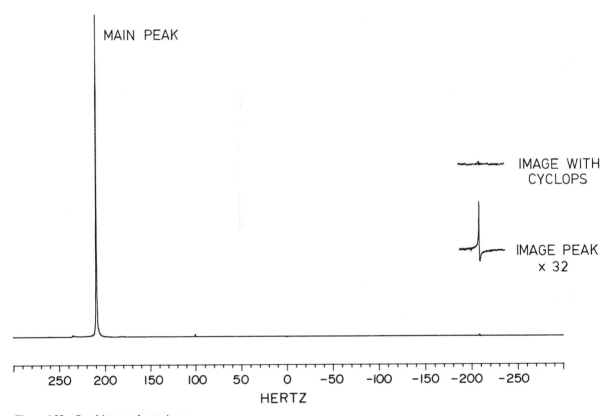

Figure 4.23 Quad images due to imperfect quad detection can be a nuisance in high dynamic range samples. They are much reduced by the CYCLOPS phase cycle.

Problems with quad detection

As with most things, the use of quad detection to improve sensitivity also introduces certain disadvantages. The chief problem is that we are relying on the cancellation of the unwanted component by adding two signals which have been through different parts of the hardware. This will only work properly if the amplitudes of the signals from the two channels are exactly equal, and their phases differ by exactly 90°. Naturally perfection in both these respects cannot be achieved in practice, and artefacts due to incomplete suppression of the so-called *image* peaks will be present to some extent. These take the form of reflections of peaks about the centre of the spectrum, which can be distinguished from small genuine peaks by the fact that they show different phase and move when the reference frequency is changed (Figure 4.23). The amplitude of such quad images is usually less than 1% of the parent line, so they only become troublesome when you are examining weak lines in the presence of strong ones.

Phase cycling

Quad images can be reduced below the level attainable by balancing the channels, using an ingenious idea known as *phase cycling*. The specific details given here are not so important to understand, but the concept is, as it is widely used in multi-pulse experiments. Referring back to Figure 4.19, the signal is split into two parts after it has been amplified to a sufficient level; these are detected against the reference frequencies in quadrature and each component separately digitised into two regions of computer memory **A** and **B**. It is not hard to see that any imbalance between the two channels of the receiver would be eliminated if we could arrange that, on average over a number of scans, each contributed equally to the data in **A** and the data in **B**. *However,* we must remember that only signals with 0° phase should be sent into **A**, and only those with 90° phase into **B**, so if we swap the receiver channels we also have to swap the phases of the signals passing through them.

How can the phases of the signals in the channels be swapped? Well, perhaps the obvious thing would be to exchange the reference phases. We must be careful here, though, because we are making a circular argument: the reference frequencies themselves are very likely to be a source of the imbalance we are trying to eliminate. We want to keep the receiver

Figure 4.24 Here is how the first two steps of CYCLOPS allow the rôles of the two receiver channels to be reversed, thus averaging out imbalance between them.

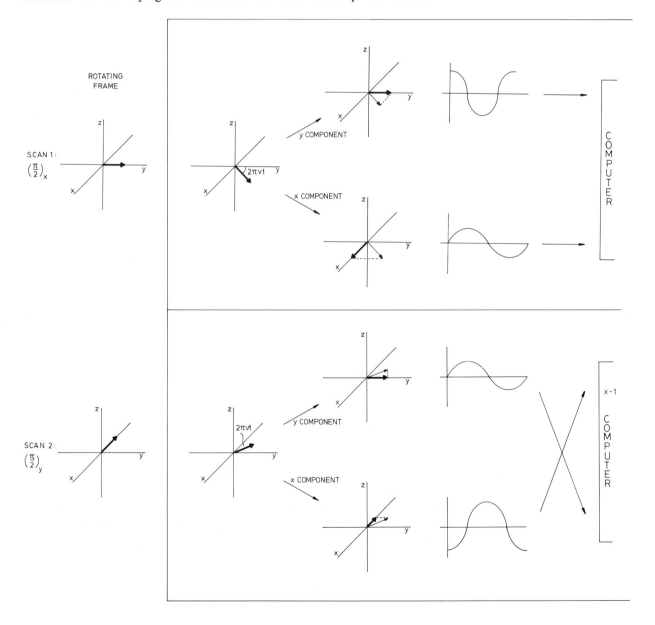

hardware just the same, swap the outputs between **A** and **B** (which is only a software operation) and at the same time shift the phase of the signals by 90°. In fact, we already saw how to do this in section 4.3.4: *shift the phase of the pulse by 90°*. This interchanges the absorption and dispersion signals as required (Figure 4.24; if you look at this carefully you should notice that one of the signals is also inverted, and so must be multiplied by -1 after digitisation), and the remaining operations are simply data routing which the spectrometer control software can do. This leads to a two-step phase cycle, alternating the phase of the transmitter between 0° and 90° on alternate scans and swapping the data destinations at the same time.

It also proves advantageous to shift the transmitter pulse by 180°, together with which we must *subtract* the signals from both **A** and **B**. This eliminates any spurious signals whose phases are independent of that of the pulse, such as might arise from electronic faults or outside interference.

Table 4.1   The CYCLOPS phase cycle for reducing quad images
The final two columns indicate the required data routing during each scan. The two receiver channels, differing in phase by 90°, are referred to as '1' and '2', and the two memory blocks as '**A**' and '**B**'.

Scan	Pulse Phase	Receiver Mode	A	B
1	x	x	$+1$	$+2$
2	y	y	-2	$+1$
3	$-x$	$-x$	-1	-2
4	$-y$	$-y$	$+2$	-1

Combining this with each of the 0° and 90° experiments leads to a four step cycle called CYCLOPS (Table 4.1), which is now standard for all spectrometers employing quad detection. Figure 4.23 also illustrates the difference in image size with and without CYCLOPS. It is conventional when specifying the phase cycle for an experiment to call the four different receiver modes x, y, $-x$ and $-y$ as though they were really phase changes of the reference signals. The actual details of what is happening are left implicit, and are dealt with by the spectrometer software.

Phase cycles are widely used in multi-pulse NMR, both for suppressing artefacts, as here, and to select particular components of the total magnetisation. This is discussed in much more detail in other chapters, particularly those about 2D NMR. For the cycle to have the desired effect the number of scans accumulated must obviously be a multiple of the number of steps in it, and for 2D experiments in many cases this becomes one of the factors which determine how long the experiment will take. Although *devising* the cycles can safely be left to the originators of NMR experiments, it is useful to develop some fluency in *interpreting* them for implementation on a particular spectrometer. Often converting what is written down in a publication into suitable language for your spectrometer's phase control routine can be the major effort in implementing new experiments. It is not that it is really difficult, but just that small slips or omissions are easy to make and usually lead to exotic and uninformative symptoms. For this reason it can be considered essential to try out new experiments on samples for which you know what the answer should be, and preferably ones which give this answer quickly.

The Redfield method

I mention for completeness that there is an alternative implementation of quad detection in quite widespread use. Again, the main interest to us is that analogous methods are used in 2D NMR, and you may find yourself with a spectrometer that works this way (for instance, those made by Bruker). The object of this experiment is to bring about quad detection using only one ADC, which is an expensive component.

This is achieved by digitising the signal as if it were being detected single-phase, so the rate of digitisation is twice the spectral width. The reference frequency, though, is at the centre of the spectrum as for normal quad detection. The receiver phase is incremented by 90° *after each point has been sampled*. The effect of this can be appreciated by imagining a single frequency component being digitised. At each sample point the phase of the signal has advanced 90° more than otherwise would have been expected, because of the phase change of the receiver. Thus, from the point of view of the digitiser, it looks like a signal of higher frequency. If the total spectral width is F, then we are digitising at $2F$, and since the phase shifts are 90° ($= 1/4$ cycle) the frequency appears to be increased by $2F/4 = F/2$. Now, relative to the reference frequency, what we started with were frequencies running from $-F/2$ to $+F/2$, so adding $F/2$ converts the range to 0 to F and the problem of negative frequencies has disappeared. The input to the Fourier transform now consists of real numbers, not complex pairs, so slightly different data processing is required. Because the phase is incremented for each sampled point, this experiment is often referred to as Time Proportional Phase Increment or TPPI.

In practice actually changing the receiver reference phase, potentially rather often if the spectral width is large, may be difficult. A convenient alternative is to use two detectors as for normal quad detection, but only one digitiser (Figure 4.25). Switching this from one detector to the other takes care of the 90° phase shift, and the 180° and 270° steps can be obtained by multiplying the 0° and 90° points by -1 before storing them. Images arise in this experiment similar to those in two-channel quad detection, and the same phase cycle is used to reduce them.

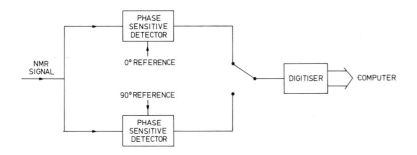

Figure 4.25 One possible implementation of Redfield's scheme (TPPI) for quad detection.

4.3.6 Phase Errors and Phase Correction

The first time you use a Fourier transform spectrometer you will encounter the need for *phase correction* of the frequency domain spectrum after transformation. The adjustment is fairly straightforward for 1D spectra, although it is important to realise that it cannot always be made exactly, as it rests on certain assumptions about the phase error which may not be true. When we come to phase-sensitive two-dimensional spectra, fully interactive phase adjustment is usually impractical, and we need to devise ways of determining the phase correction from parts of the data only, or from suitable model 1D experiments. For this reason, I want to examine the sources of phase errors in FT spectra and the numerical process by which we compensate for them. Once again, this discussion is not essential to understanding many parts of the book, but later on you may find you want to return here when the problems mentioned arise in a more interesting context.

One simple source of phase errors has already been mentioned several times, and that is the difficulty of adjusting the phase of the reference frequency so that the pure sine and cosine components are detected in the two channels. This causes equal error throughout the spectrum; that is, the phase change is independent of frequency. What we get if the reference is misadjusted is a spectrum in which the real and imaginary parts $\mathscr{R}$ and $\mathscr{I}$, which we would like to represent the absorption and dispersion components **A** and **D** respectively, are actually mixtures of the two. That is, we want:

$$\mathscr{R} = \mathbf{A}$$
$$\mathscr{I} = \mathbf{D} \tag{4.15}$$

but instead we get:

$$\mathscr{R} = \mathbf{A}\cos\theta + \mathbf{D}\sin\theta$$
$$\mathscr{I} = \mathbf{A}\sin\theta - \mathbf{D}\cos\theta \tag{4.16}$$

where θ represents the phase error. Evidently we can regenerate the separate **A** and **D** components by taking linear combinations of the $\mathscr{R}$ and $\mathscr{I}$ parts thus:

$$\mathbf{A} = \mathscr{R}\cos\theta + \mathscr{I}\sin\theta$$
$$\mathbf{D} = \mathscr{R}\sin\theta - \mathscr{I}\cos\theta \tag{4.17}$$

the value of θ being determined experimentally by observing the lineshape as θ varies.

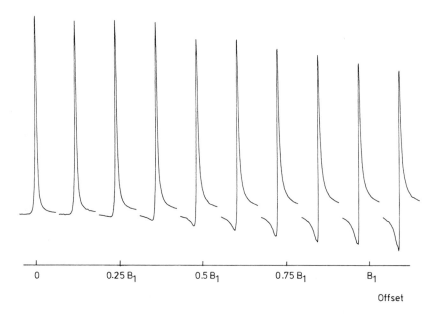

Figure 4.26 The off-resonance effect causes phase changes after a $\pi/2$ pulse. These are experimental lineshapes obtained using a very weak $\mathbf{B}_1$ field (about 80 Hz), so that other sources of phase error were negligible in comparison. Note how the amplitude of the signal changes only slowly as the offset is increased.

A variety of other experimental problems cause phase errors that are not independent of frequency. We have already seen the effect of the pulse being off resonance in section 4.3.2; Figure 4.26 demonstrates the progressive phase change which occurs as the pulse is moved further and further from the line to be excited. Another unavoidable cause of frequency dependent phase changes is the fact that sampling of the free induction decay cannot begin instantaneously after the pulse. At the very least, the pulse has finite length and the receiver cannot be switched on during it. In practice a delay of perhaps tens of μs will also be needed after the end of the pulse to allow the receiver electronics to recover. The manner in which this introduces phase changes is illustrated in Figure 4.27; during the delay before the sampling begins signals with different frequencies will have advanced in phase by different amounts. Finally, various elements of the receiver electronics may not transmit all frequencies with equal phase; this is particularly true of the audio filters used between detection and digitisation. All of these effects are lumped together and assumed to introduce extra phase changes that are a linear function of frequency v, that is θ is expressed as:

$$\theta(v) \;=\; \alpha + \beta v \qquad (4.18)$$

α and β are then the two parameters to be experimentally adjusted. The best way to do this is to select two peaks a long way apart and adjust α observing one and β observing the other, as then the effect of the frequency dependent term is maximised and small errors can easily be detected.

There are several circumstances in which this phase adjustment can be difficult or impossible. The assumption that all the phase errors can be modelled as a linear function of frequency may not be true. Even if it is, judging whether a line is in pure absorption mode is rather subjective, especially if the baseline of the spectrum is not completely flat. This is often a bad problem near the edges of the spectrum, so that if it is desirable to make accurate phase corrections (for instance to improve the accuracy of integration) it can be helpful to arrange that the interesting peaks are in the centre of the spectral range. Spectra acquired with short acquisition times and poor digitisation (as is typical for ^{13}C for instance) can show line shape distortions which look like phase errors, but are not[3].

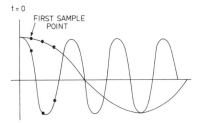

Figure 4.27 Delayed start of sampling is an important source of frequency-dependent phase error. By the time we get round to measuring the first data point, signals differing in frequency which were in phase directly after the pulse have diverged significantly.

4.4 RELAXATION

4.4.1 Introduction

In most forms of spectroscopy we put some energy into a sample, for instance by shining a light through it, and find that more is absorbed at some frequencies than at others. Where does the energy eventually go? If pressed, most people would suggest something like 'it is turned into heat'; however one is rarely concerned with how this might happen, at least in optical spectroscopy. In NMR, by contrast, where the energy goes, or more particularly *how fast it gets there* is a matter of prime importance. The reason for this is that the transition energies involved in NMR are so small that in many circumstances the timescale for attaining thermal equilibrium between levels can be very long, and this constrains the way we may carry out experiments.

When we observe an electronic transition of a molecule by shining ultraviolet light on it, excited electrons find their way back to the ground state so quickly (e.g. in a few tenths of a picosecond) that measuring the excited state lifetime may be a difficult matter. When we excite the magnetic resonance of a nucleus, however, the excited state may live for *minutes*. In pulse NMR, where the idea is to perform experiments repetitively in order to improve the signal-to-noise ratio, this may prove troublesome. In fact, the success of FT NMR owes a great deal to good fortune regarding the speed with which certain interesting magnetic nuclei return from excited to ground states, as had this been either much slower or much faster than it happens to be the experiment would not have been so advantageous.

As chemists rather than physicists, there are two reasons why we might be interested in *relaxation* (the general term for movement towards equilibrium). The first is that the relaxation parameters described below do correlate with structural features of molecules, and particularly with their motions. Unfortunately, these correlations are not understood in such a general sense as are those of chemical shifts and coupling constants, and although relaxation measurements are sometimes used to good effect in structure elucidation, it still seems to be difficult to lay down guidelines for their application. One problem that contributes here is the difficulty of making accurate and reproducible measurements of relaxation times, since many extraneous (from a chemical point of view) factors contribute to them, as we shall see. Because of this, and because of the overall theme of not discussing empirical correlations in NMR in this book, this aspect of relaxation will not be pursued[4].

The second aspect, and one that will concern us greatly, is that consideration of relaxation processes permeates the design of NMR experiments; this is evident even in the simplest single-pulse case, where it is precisely the decay of the magnetisation through relaxation that we are observing during the 'free induction decay'. From a practical point of view, we will need some idea of relaxation times to achieve optimum sensitivity in any experiment, something we always have to struggle for even on the best of modern instruments. The concept of *transverse* relaxation discussed later leads to the idea of *spin echoes*, which form a building block of many interesting experiments. The rotating frame model which has just been introduced makes it rather easy to appreciate the distinction between the different kinds of relaxation of importance in NMR.

4.4.2 Towards Equilibrium

If we have a (diamagnetic) sample which is not in a magnetic field, it will not be magnetised. Presumably when we put it in the magnet, the induced field which we expect to arise does not spring instantaneously into existence, but will appear over some interval. Calculating from fundamental principles how long this should take, and what form the approach to equilibrium adopts, is a whole area of physics in itself, and cannot be achieved without extensive approximation. In the Bloch theory of NMR, the problem is vastly simplified by *assuming* that the equilibrium will be approached exponentially, and treating the time constant of this exponential as a parameter which one might set out to measure. It is rather too easy to forget, in the midst of arguments about relaxation times, that exponential relaxation is a hypothesis, not a universal fact; fortunately it often turns out to be close to the truth for the molecules in solution with which we are concerned. Putting this idea formally, the hypothesis is that the induced field will build up according to the equation (Figure 4.28):

$$\frac{d\mathbf{M}_z}{dt} = \frac{\mathbf{M}_0 - \mathbf{M}_z}{T_1} \tag{4.19}$$

where $\mathbf{M}_0$ is the magnetisation at thermal equilibrium, so that if the magnetisation is initially 0:

$$\mathbf{M}_z = \mathbf{M}_0(1 - e^{-t/T_1}) \tag{4.20}$$

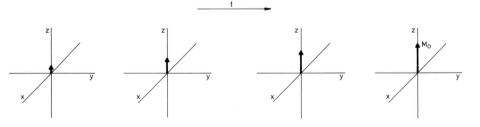

Figure 4.28 Longitudinal relaxation; according to the assumption of exponential growth of the induced field, we should have to wait infinitely long to reach $\mathbf{M}_0$.

$$\frac{dM_z}{dt} = \frac{(M_0 - M_z)}{T_1}$$

$$M_z = M_0(1 - e^{-t/T_1})$$

The time constant T_1 is known as the *longitudinal* or spin-lattice relaxation time. The latter term arose because much of the original theory of relaxation was worked out for solid samples, where the interaction of the magnetic moments with the rigid lattice was of interest. For solutions it does not convey the character of what is happening, and I will avoid using it; you will encounter it widely throughout NMR literature however. *Longitudinal* helps us remember that it is the behaviour of the z component of the magnetisation that is involved: any time the magnetisation is moved away from the z axis, we assume that in the absence of external influences it will tend to return there exponentially with time constant T_1.

We can now return to the rotating frame picture of a pulse experiment and add a little more detail (Figure 4.29). In the previous discussion, after the $\pi/2$ pulse the net magnetisation has been left precessing around the z axis, apparently indefinitely. Adding the model for exponential relaxation, we have the z axis magnetisation reappearing with time constant T_1, and therefore the magnetisation in the x-y plane disappearing at least this quickly (we will see shortly that it may sometimes disappear faster than the z magnetisation reappears, but obviously the opposite cannot be true). In a multiline spectrum there is no reason why all the nuclei should experience relaxation at the same rate, so a set of T_1 values may exist for the different nuclear environments in a molecule. Giving the excited states a finite lifetime in this way corresponds in the frequency domain to giving the resonances finite linewidth; as relaxation times in solution NMR are often quite long, NMR resonances are correspondingly very sharp compared with, for instance, the lines in an ultra-violet absorption spectrum.

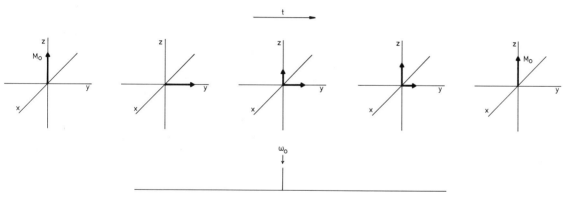

Figure 4.29   What happens after a pulse with relaxation included.

Why are T_1 values often long? The answer to this would fill several books; however we can get at least a qualitative idea of the problem. The first point to note is that there should be absolutely no problem dissipating the tiny energies of NMR transitions into the general thermal energy of a sample, compared with which they are negligible. Therefore, it is not lack of somewhere to send the energy, but lack of a means to move it there that slows down relaxation. We know that for closely spaced energy levels such as we have here the probability of spontaneous emission is extremely low (about 10^{-25} per second) and can be neglected. So relaxation must occur by stimulated emission, and the observation that T_1's tend to be rather long implies that there do not happen to be suitable stimuli available.

In other words, since the resonances are stimulated by oscillating magnetic fields, such fields oscillating at the correct frequency are not abundantly available under the conditions we normally use for observing NMR, so the nuclear energy levels are not connected very well with the surrounding environment. In developing theories of relaxation we try to think of sources for suitable fields to stimulate the transitions, and then calculate how effective they will be. For spin-$\frac{1}{2}$ nuclei in solution the dominant source is the magnetic (dipolar) interaction between nuclei, modulated by molecular motions; thus the prediction is that relaxation rates depend on such things as temperature, solution viscosity, molecular size and structure and in some circumstances the applied magnetic field. This complex area is extensively treated in classic NMR texts such as Abragam[5] or Slichter[1].

It is not difficult to devise experiments to measure T_1; any sequence leading to a signal whose intensity depends on the relaxation time will do.

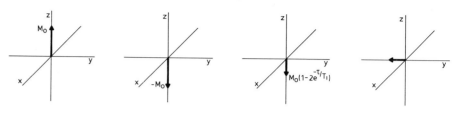

Figure 4.30 The inversion-recovery method for measuring T_1.

One popular method is illustrated in Figure 4.30 (and here is our first multi-pulse NMR experiment!). The idea is to invert the z magnetisation with a π pulse, then wait a while for it to relax back towards the $+z$ axis, and finally apply a $\pi/2$ pulse and measure the signal. Note that when the interval τ is less than $T_1/\ln 2$ the sampled signal arrives along the $-y$ axis; we already know that this corresponds to an apparent negative peak if we phase-correct so that $+y$ is positive. The result of such an experiment with varying τ is shown in Figure 4.31; obviously one can extract the T_1 graphically from the varying peak heights. While the concept of this experiment is simple, obtaining accurate results in a practical length of time may not be, but this is not a problem which need concern us. More often we will be interested in quick and approximate determination of T_1's to allow the sensitivity of an experiment to be optimised, for which see Chapter 7. Discussion of the advantages and disadvantages of various approaches to T_1 measurement can be found, for example, in the book by Martin, Martin and DelPuech[6].

Finally, note that we can control T_1 values to some extent by controlling the availability of relaxation pathways. A very common source of accelerated relaxation is the presence of paramagnetic substances, whose unpaired electrons provide an effective stimulus for NMR transitions. If we

Figure 4.31 Experimental results from an inversion-recovery sequence. The group of peaks at lower field (i.e. to the left) all have slightly longer T_1's than those of the high-field group.

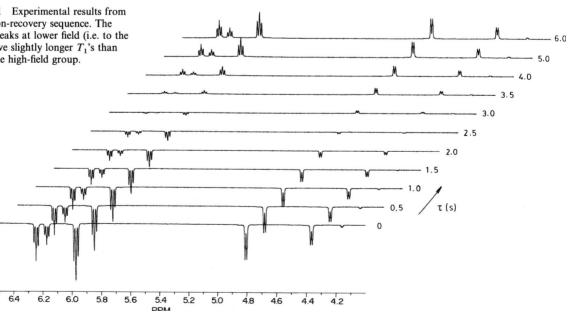

need to shorten T_1's, for instance to speed up an experiment or to increase the accuracy of quantitative measurements, paramagnetic material can be introduced into the sample deliberately. Chromium (III) acetylacetonate is a common reagent used for this purpose. On the other hand, samples as normally prepared already contain some paramagnetic material - dissolved oxygen - and this must be removed by degassing if the sharpest possible lines or measurements of T_1's or other relaxation phenomena such as the nuclear Overhauser effect are needed.

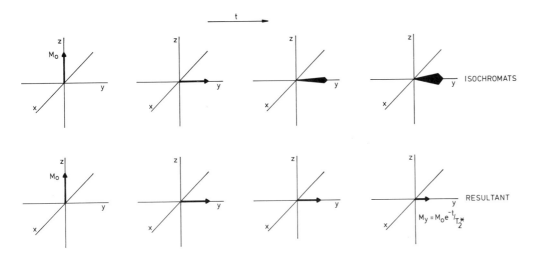

4.4.3 Relaxation in the *x-y* Plane

Returning yet again to the simple single pulse experiment on a sample with a single line exactly on resonance with the reference frequency, forget for a moment that longitudinal relaxation exists, and consider the net magnetisation, static in the *x-y* plane of the rotating frame (Figure 4.32). Without longitudinal relaxation, is there any reason why the magnetisation should not remain there forever? Remembering that the total magnetisation originates from the a large number of nuclei spread throughout the sample volume, we can at once think of a rather prosaic reason why it should not: the static field cannot be perfectly uniform.

Think of the sample being subdivided down into small regions, small enough that the field *is* perfectly uniform within each region. The total magnetisation is the sum of the contributions from these regions; each region contributes a vector precessing in the static frame with perfectly well defined frequency (these are often called *isochromats*), but the frequency may vary from one region to another. In the rotating frame, this corresponds with the vector initially aligned along the *y* axis gradually becoming 'blurred' as some of the contributing isochromats precess a little faster than the frame, some a little slower (Figure 4.32). Thus the total magnetisation *does* decay, even without longitudinal relaxation. The process is not changing the energy of the sample, as no transitions between levels are occurring, but the amount of order present is decreasing. In other words, the *enthalpy* remains constant but the *entropy* increases.

This too is a form of relaxation, referred to as transverse or spin-spin; again the second term is sometimes a little misleading and I will not use it, however we will see how it arises momentarily. *Transverse* reminds us that the process occurs in the *x-y* plane and does not necessarily involve the *z* magnetisation. We assume once more that the process takes place exponentially, and call the time constant T_2. It turns out to be possible to separate experimentally the loss of order due to magnetic field inhomogeneity (which is really a kind of experimental artefact) from loss due to other causes (see below), and the symbol T_2 is usually reserved for the latter; the time constant of the directly observed decay, which may be dominated by the inhomogeneity, is generally written T_2^*.

Figure 4.32 Transverse relaxation. The various isochromats contributing to the net magnetisation gradually 'fan out' for various reasons (above); the result is assumed to be exponential decay (below).

Any process that causes loss of transverse magnetisation, including return of magnetisation to the z axis, contributes to T_2. Thus, in the absence of any other mechanism for transverse relaxation, T_2 must equal T_1, because obviously magnetisation which is along the z axis is not in the x-y plane. Were it not for the problem of the static field inhomogeneity, this would quite often be the case in liquids, and the envelope of the free induction decay would be an exponential with time constant T_1. In the solid state, in contrast, however uniform the static field, local magnetic fields due to varying environments in the material cause T_2 to be very short, while lack of motion-induced relaxation may give rise to very long T_1's; both facts make observation of NMR in solids more challenging than in liquids.

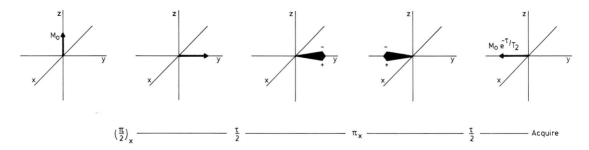

Figure 4.33 The formation of a spin echo by refocusing of magnetisation which has been dephased through static field inhomogeneity.

4.4.4 Spin Echoes

Introduction

Something extremely interesting happens if we apply two *rf* pulses in quick succession (quick relative to T_1, that is) to a sample. The experiment is easiest to visualise for the sequence $\pi/2$ - τ - π (Figure 4.33), although other pulse angles generate similar effects. Consider yet again the simplest case of a single line exactly on resonance. After the $\pi/2$ pulse, the inhomogeneity of the static field causes the contributing isochromats to 'fan out' gradually, leading to a blurring of the vector in the x-y plane. In Figure 4.33, the front edge of the blurred vector has been labelled '+' to indicate that the isochromats on that side are precessing a little faster than the frame, while the back edge where they are precessing a little slower has been labelled '−'. A π_x pulse then rotates all the isochromats together around the x axis, and leaves the blurred vector centred on the $-y$ axis. The important feature to note about this situation is that the '+' side of the vector now *lags behind* the average direction of the isochromats, while the '−' side is ahead. Thus the fast vectors are catching up with the average position and the slow ones are falling back towards it. Assuming they are all just as fast or slow as they were during the 'blurring' phase (about which more in a moment), after a further time interval τ they will all realign along the $-y$ axis; they have been *refocused* by the π pulse.

If we carried out this experiment and observed the xy magnetisation (i.e. the NMR signal) from immediately after the $\pi/2$ pulse, we would expect therefore to see a signal which died away until the π pulse, then built up again during the second τ interval, and after that faded as before (Figure 4.34). The regeneration of the signal at time 2τ is called a *spin echo*, and it is a ubiquitous component of multipulse NMR experiments.

The spin echo has a number of exceptionally useful properties as we shall see shortly, but first let us think a little more carefully about what is required in order that echoes should be generated. The key requirement is that isochromats have the same precession frequency before and after the π pulse. If we assume that only static field inhomogeneity contributes to this,

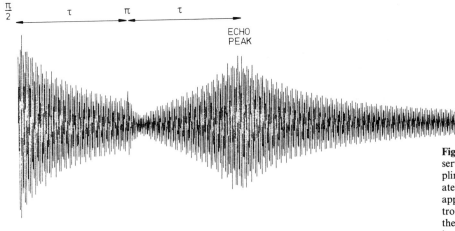

Figure 4.34 Experimental signals observed during an echo sequence. Sampling of the data commenced immediately after the $\pi/2$ pulse; a π pulse was applied at time τ as indicated (the spectrometer receiver does not like this, so the signals are distorted in the period immediately following the second pulse). The signal rebuilds to a peak at time 2τ then decays in normal fashion.

then this is equivalent to saying that the nuclei that make up the isochromat remain in the same part of the field throughout the echo. In other words, diffusion of molecules within the solution must be slow relative to τ, and also, since the sample is usually spinning, so must be the spin speed. Neither requirement is fully met in practice, but provided τ is not too long echo generation is effective.

It is of some interest to ask whether, since the amplitude of the echo is evidently not dependent on the homogeneity of the field, the rate of decay of the amplitudes of a series of echoes with variable τ would simply equal T_1. The answer is sometimes 'no' (but not too often for small molecules in solution), because it is possible to lose x-y magnetisation for reasons other than inhomogeneity, but still without concomitant generation of z magnetisation. One mechanism that would cause this is linked relaxation of two nuclei, where one goes from ground to excited state simultaneously with another going the other way; this allows loss of phase coherence in the x-y plane but does not affect the net z magnetisation. This is the origin of the term spin-spin relaxation often applied rather sweepingly to all transverse relaxation processes, but it is a mechanism that is of most importance in the solid state. The time constant derived from the decay envelope of a series of spin echoes is regarded as the 'true' T_2 value mentioned earlier, as it is independent of homogeneity; factors which make this differ from T_1 in liquids include coupling to a quadrupolar nucleus (i.e. one with spin $> \frac{1}{2}$), or chemical exchange.

Spin echoes for measuring T_2

T_2 measurements are fraught with experimental difficulties. The obvious approach of creating echoes with variable τ and measuring the decay of the amplitudes is no good, because effects such as diffusion will become more pronounced as τ increases. As an alternative, consider the sequence:

$$(\pi/2)_x - \tau - \pi_x - 2\tau - \pi_x - 2\tau - \pi_x - \cdots$$

(the Carr-Purcell experiment) where we might try to measure the amplitude of the signal at the centre of the 2τ intervals, or more practically perform the experiment repetitively with an increasing number of delays and measure the last echo each time. If τ is chosen to be short (relative to the rate of diffusion), this eliminates the problem.

Unfortunately, as we already know, pulses never work exactly as advertised, and π pulses are particularly prone to problems. The most rudimentary of pulse defects - incorrect length - will sabotage the Carr-Purcell experiment, since the magnetisation is constantly being driven round in the same direction and any errors will accumulate. As a good exercise in

working with the rotating frame vector model, compare the Carr-Purcell
experiment with the Carr-Purcell-Meiboom-Gill (CPMG) variation, which
is:

$$(\pi/2)_x - \tau - \pi_y - 2\tau - \pi_y - 2\tau - \pi_y - \cdots$$

and for the latter convince yourself that a) using π_y pulses instead of π_x still
generates echoes (but along the $+y$ axis only) and b) cumulative errors due
to incorrect pulse length do not arise. To solve part (b) it is easiest to
consider the fate of, say, the fastest moving isochromat only, and look at
the consequence of the pulse angle being not π, but some slightly different
value $\pi + \varepsilon$. You should find that the effect of the error cancels on the *even*
numbered echoes. If you get stuck, see Martin, Martin and Delpuech[6],
pages 281-283, for a nice discussion of these experiments.

The CPMG experiment, incidentally, illustrates one of the two accepted
modes for naming NMR methods, that is stringing together the names of
the inventors. CPMG is quite an old experiment, as we can tell from the
lack of vowels in the condensed form of the name, which would not be
allowed to occur nowadays. More recent examples of this naming method
are usually easier to pronounce, for instance: WAHUHA (from Waugh,
Huber and Haeberlen). A peak of sophistication in inventor-derived names
is reached with the MLEV decoupling sequence, which not only immortal-
ises the name of Malcolm Levitt, but also is a pun on a solid state
experiment (MREV) which was itself named from its inventors (Mansfield,
Rhim, Elleman and Vaughan). The alternative method, which does not
need so much care in choice of co-workers, is making up some pertinent
acronym; for instance DANTE (Delays Alternating with Nutations for
Tailored Excitation) and WALTZ (Wonderful ALternating phase Technique for Zero residual splittings) are two of my favourites, as they both
combine a semi-plausible set of initials into a word which conveys something of the essence of the experiment (see Chapter 7).

Fortunately we do not need to be concerned at all with the measurement
of T_2. Of much more consequence usually is the overall transverse relaxation time including the effect of inhomogeneity T_2^*. This may be available
by inspecting the FID envelope, which will have decayed to $1/e$'th (about
0·4) of its initial amplitude after T_2^* seconds (Figure 4.35); however this can
be deceptive as the visible FID is often dominated by the solvent line,
which will generally have a longer T_2^* than lines from larger molecules.
More reliable is inspection of half-height linewidths (δv) in the transformed
spectrum, which are related to T_2^* by $\delta v = 1/\pi T_2^*$. Some idea of T_2^* values
is often needed to aid in choosing acquisition times, particularly in two-
dimensional experiments where one wants the shortest possible acquisition
time without loss of sensitivity. It is also important in the optimisation of
repetition rates and pulse angles (Chapter 7), where different criteria apply
depending on whether the repetition rate is short relative to T_1 or T_2 or
both, and in the choice of parameters for window functions (although
spectrometer software often formulates these in terms of linewidths rather
than relaxation times anyway).

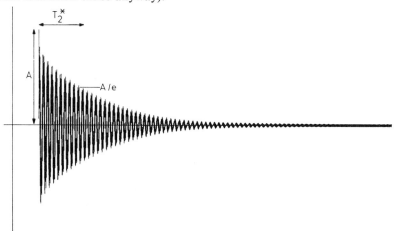

Figure 4.35 An FID; after one time
constant its amplitude has fallen by a
factor of $1/e$.

Useful properties of spin echoes

The importance of understanding spin echo formation is that the echo provides a building block for multi-pulse experiments with many convenient properties. The first of these is the effect of refocusing magnet homogeneity; we will see an experiment in Chapter 10 which uses echoes to measure spectra with *natural* linewidths, thus freeing the experimenter from the limitations of practical magnets to some extent. Two other useful features become evident when we consider the effect of the echo sequence on chemically shifted and *J*-coupled systems.

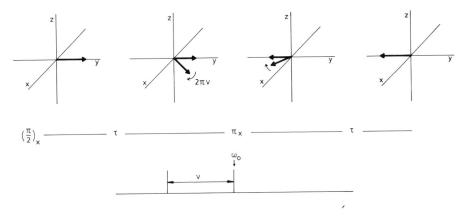

Figure 4.36 The effect of spin echoes on chemical shifts.

Consider the case of different chemical shifts first, as it is easier. Figure 4.36 shows an echo sequence performed on a sample with two lines only; for convenience the frame of reference is chosen so that one of them is static and the other therefore precesses with its relative chemical shift *v*. We already know that effects due to inhomogeneity of the field are eliminated at the peak of the echo, so the fanning out and refocusing of each line which occurs is omitted from the diagram for clarity. We can suppose that τ is rather short compared with $1/v$ so that the moving vector stays in the first quarter of the *x-y* plane; this will not affect our conclusion and is just to make the sequence easy to draw. The π pulse rotates the static vector onto the $-y$ axis and the moving vector into the second quarter of the *x-y* plane, where it keeps on precessing towards the $-y$ axis, realigning with the static vector at time 2τ. Thus *chemical shifts are refocused at the peak of the echo*. Of course we should hardly be surprised at this, since from the perspective we have taken there is no difference between a chemical shift and a field inhomogeneity - they are just different sources of varying local fields.

The importance of this property of the spin echo cannot be overemphasised. It permits us to construct manipulations of the nuclear magnetism *that are independent of chemical shifts*, at least in so far as the non-ideality of real pulses allows. This often means that selective experiments which might be tedious or impractical to apply can be generalised, leading to greatly increased efficiency. See, for example, the way in which the SPI experiment can be made into INEPT using a spin echo (Chapter 6).

Having disposed of inhomogeneity and chemical shifts in this way, we do not need to include them in the picture when we look at coupled systems. Consider a first order two-spin system which gives rise to two 1:1 doublets, the lines being separated by the coupling constant *J*. Let us look at one of the doublets (Figure 4.37). Superficially this seems to be the same as the previous two line case, but in fact it will prove helpful to choose the reference frequency in the middle of the doublet so that both components are precessing in the rotating frame, one at $+J/2$ and the other at $-J/2$ cycles per second. Up to the π pulse things proceed just as before, and the two vectors arrive in the second and third quarters of the *x-y* plane as

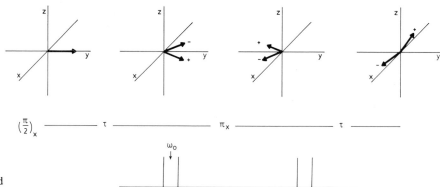

Figure 4.37 Spin echo on a coupled system (see text).

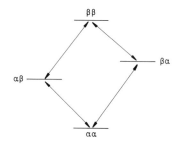

Figure 4.38 Energy levels of an AX system.

indicated in the figure. Now, before rashly proposing that they will then continue towards the $-y$ axis, we must bear in mind that the other nucleus which is causing the splitting of the line has also experienced the π pulse. The effect of this is very important, as it makes homonuclear coupling react differently to other types of line splitting during the spin echo.

To understand what happens here, we must revert briefly to thinking about the quantum mechanical aspect of the system. If we call the two states of the spin-$\frac{1}{2}$ nucleus in the field α and β, the two components of the doublet arise from the transitions of nuclei whose neighbours are in one or other state (Figure 4.38), i.e they are $(\alpha\alpha)$ to $(\beta\alpha)$ and $(\alpha\beta)$ to $(\beta\beta)$ (I note in passing that there is some variation between - and sometimes within - existing texts in choice of α or β to represent the lower energy state; here and throughout the book I am using the former, which seems to correspond with popular usage). The α and β states of the neighbouring nucleus occur with nearly equal probability, so the two lines are apparently of equal intensity. The π pulse *inverts the populations* (section 4.2.6), that is it turns every α state into a β and vice-versa. The result of this is that all the nuclei which had α neighbours and were contributing to, say, the line which was precessing at $+J/2$ (which line it in fact is depends on the sign of J), now find themselves with β neighbours and are therefore precessing at $-J/2$. Likewise for the other half of the nuclei, and so the π pulse *interchanges the labels of the two lines.*

Returning to the vector picture armed with this information, we see that at time 2τ the components of the doublet are *not* refocused, because the direction of their precession is reversed during the second half of the sequence. Indeed, they are not left aligned with anything in particular unless τ coincidentally bore some relationship to J; Figure 4.37 is drawn assuming that this was not so. More often experiments require that multiplet components are arranged in some specific way at the peak of the echo. For instance, if we took τ to be $1/4J$ seconds then for the case of a doublet we would generate antiphase components aligned along the $\pm x$ axes at the echo (Figure 4.39).

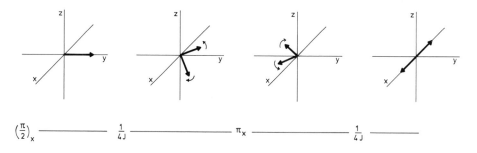

Figure 4.39 Echo sequence on a doublet with $\tau = 1/4J$.

It is very important to realise that the lack of refocusing of homonuclear coupling by a spin echo is *because it is homonuclear*, not just because it is a coupling. If the doublet we were looking at arose in a heteronuclear system, for instance if it was the proton-coupled carbon signal of a methine group, then the two components would refocus along with the chemical shifts and inhomogeneity, because the protons would not experience the π pulse and the labels of the lines would therefore not be exchanged. In this case we have the option of whether or not the components refocus in fact, because we could choose to apply a π pulse to the protons as well if the spectrometer hardware allowed this. We will see an experiment in Chapter 10 which exploits this choice to distinguish patterns of protonation.

4.5   DESCRIBING PULSE NMR

These are the ingredients of pulse NMR experiments. We have *rf* pulses, which can be used to rotate the magnetisation through any angle, typically $\pi/2$ or π radians. The rotation will be clockwise, about any of the four directions x, y, $-x$ or $-y$, or sometimes, in a few experiments, about other directions in between these. These rotation axes are selected by setting the *phase* of the pulse; we take this to define the labels on the axes of a rotating coordinate system. NMR signals are detected by subtraction of a fixed reference; we take its frequency to define the speed of rotation of the rotating frame. The phase of the reference has an arbitrary (but constant) relation to the phase of the pulses, but this does not matter, because the apparent detector phase will be adjusted numerically after Fourier transformation of the signal.

Once rotated from the z axis to the x-y plane, magnetisation components precess (in the rotating frame) according to their offsets from the detector reference frequency. At the same time, magnetisation reappears along the z axis exponentially with time constant T_1, and disappears from the x-y plane exponentially with a different time constant T_2^*. If we like, we can eliminate parts of the precession which do not arise from homonuclear coupling, and concomitantly the contribution to the decay of the transverse magnetisation from the static field inhomogeneity, by application of π pulses to generate spin echoes. The time constant of the (exponential) decay of the amplitude of a series of spin echoes, generated at increasing times after the initial pulse, is called T_2.

This vector model, which concentrates on the behaviour of the *net* magnetisation of the sample, will get us a long way in describing pulse NMR. However, we should not be lulled into a false sense of security, because it is far from revealing the whole truth. In coupled spin systems, the effect of the pulses on the relative populations of (and, in general, the *phase coherences* between) energy levels is not described. We saw how to calculate the effect of a pulse on populations in section 4.2.6, but this is only part of the story; the phase relation between states cannot be modelled in this way. However, this is as far as we can go without introducing excessive abstraction, and will be enough to let us appreciate the general features of many experiments.

For the less faint-hearted, I hope that the questions left unanswered in this text will prove to be a stimulus to explore elsewhere. The most explicit description of nuclear systems is given in terms of *density matrix theory*[7]. In this picture, we take the usual quantum mechanical model of the wavefunction of a system as a sum over its eigenstates. The coefficients of each term in the sum, being complex numbers, can be thought of as containing both amplitude and phase information. To model a real sample, we have to take the *average* values of the coefficients for a large number of similar systems in different environments throughout the sample. These ensemble averages are the elements of the *density matrix*, which can be thought of as a map of the average pairwise relationships between levels in the system at a given moment. Pulses are then represented as operators which change the density matrix, which evolves in between them according to the Hamiltonian for the system, allowing a quite explicit calculation of the result of any sequence.

The problem with the density matrix method is that it is hard to relate to physical pictures, and for systems with more than two or three spins the calculations rapidly become cumbersome. Alternative approaches which are often more fruitful include the *product operator formalism*[8], and *tracing coherence pathways*[9,10]. The former method combines elements of both the vector model and the more formal density matrix approach, in a way which nicely links the quantum mechanical description with physical pictures. The latter is particularly well suited to the practical problem of devising phase cycles for experiments, and is relatively easy to understand. For us, though, these matters must remain unspoken, as we have to press on and look at some experiments.

REFERENCES

1. C. P. Slichter, *Principles of Magnetic Resonance,* Springer-Verlag, 1980.
2. E. Fukushima and S. B. Roeder, *Experimental Pulse NMR - a Nuts and Bolts Approach,* Addison-Wesley, 1981.
3. M. B. Comisarow, *J. Mag. Res.,* **58**, 209-18 (1984).
4. For examples of the application of T_1 measurements to ^{13}C and ^{31}P structural problems, see: J. R. Lyerla and G. C. Levy, in *Topics in Carbon-13 NMR Spectroscopy,* (ed. G. C. Levy), **1**, 79-148, Wiley, 1974; P. A. Hart, in *Phosphorus-31 NMR - Principles and Applications*, (ed. D. G. Gorenstein), 317-347, Academic Press, 1984; T. L. James, *ibid.,* 349-400. Applications in ^1H NMR are less common.
5. A. Abragam, *Principles of Nuclear Magnetism*, Oxford University Press, 1983.
6. M. L. Martin, J-J. Delpuech and G. J. Martin, *Practical NMR Spectroscopy*, Heyden, 1980.
7. See reference 1, Chapter 5.
8. O. W. Sørenson, G. W. Eich, M. H. Levitt, G. Bodenhausen and R. R. Ernst, *Prog. Mag. Res.,* **16**, 163-192, (1983); Although quite mathematical, this is an extremely lucid paper with many illustrations, and is highly recommended further reading.
9. G. Bodenhausen, H. Kogler and R. R. Ernst, *J. Mag. Res.,* **58**, 370-388, (1984).
10. A. D. Bain, *J. Mag. Res.,* **56**, 418-427, (1984).

5

The Nuclear Overhauser Effect

5.1 INTRODUCTION

Experiments based on the nuclear Overhauser effect (nOe) occupy a very special place in the repertoire of modern NMR methods. If you have already made some use of ^{1}H NMR in structure elucidation, it is quite likely that you will have encountered equilibrium nOe difference spectroscopy, with which we will mainly be concerned in this chapter. This experiment has the unique distinction of being *fundamentally* different from the rest of the methods we will be discussing; it is the only technique that does not depend on the presence of scalar coupling for its operation. Instead, the interaction involved is the direct magnetic coupling between nuclei (the dipolar coupling), which does not usually have any observable effect on spectra recorded in solution. The nOe provides an indirect way to extract information about this dipolar coupling, which in turn can be related to internuclear distances and molecular motion. Measurements of these parameters for molecules in solution are hard to obtain by other means, making the nOe an extremely important phenomenon.

In trying to understand the nOe, and to see how we can get structural information from it, we are faced with several problems. The nOe is an aspect of nuclear relaxation; it is a change in intensity of one resonance when the transitions of another are perturbed in some way. To get to the bottom of this process, we need to analyse the possible relaxation pathways available in multispin systems, discover mechanisms which might cause these pathways to operate, and calculate the relative effectiveness of these mechanisms. The *concepts* behind such an analysis are not really very difficult, but the *detail* of it certainly is. For instance, we will see shortly that even a two-spin system has up to six relaxation paths available to it, each of which may be stimulated by various different mechanisms; it is easy to get lost in a welter of subscripts and superscripts when trying to model such a situation. The mechanisms which drive relaxation are related to molecular motion, which is clearly random and, for a decent size molecule, may be extremely complex. It is out of the question for us to attempt to analyse these matters properly; a glance at, say, Chapter 8 of Abragam[1] will rapidly convince you of this.

However, it is also unsatisfactory to make no effort at all to understand the physical processes behind the nOe. This would lead to the approach summarised by the statement: 'the nOe is proportional to the inverse of the distance between nuclei raised to the sixth power', which unfortunately is often the basis on which this experiment is interpreted. Much confusion can be caused by the uncritical application of this statement, which *might* be true, but very often is not. In the following sections I will attempt to show you enough about how the nOe works for you to be able to use it reliably. We will look at how it originates in the simplest possible case (a two-spin homonuclear system), how it relates to relaxation and dipolar coupling, and how it can be measured. We will see some formulae that allow quantitative extraction of internuclear distances in certain simple cases, but as a matter of practical necessity most realistic experiments will not be subject to quantitative analysis. In molecules of reasonable complexity, we will see how qualitative *comparison* of nOe's provides the best means to solve problems. Section 5.2 provides the theoretical background to the nOe, while sections 5.3 and 5.4 deal with its measurement and exemplify its interpretation respectively.

5.2 THE ORIGIN OF THE NUCLEAR OVERHAUSER EFFECT

5.2.1 Introduction

The nOe is a change in the intensity of an NMR resonance when the transitions of another one are perturbed. Throughout most of this chapter the 'perturbation' of interest will be *saturation,* which is to say that we eliminate the population difference across some transitions (by irradiating them with a weak *rf* field), while observing the signals from others. The nOe is then a manifestation of the attempt of the total system to stay at thermal equilibrium; we have forceably changed the population differences of part of it, so other parts change in compensation. Exactly how will be revealed by considering a very simple case: two nuclei, both spin-$\frac{1}{2}$. First, though, let's define what is meant by the nOe. Suppose the normal intensity of a resonance (i.e. that observed at thermal equilibrium and without perturbing the system) is I_0. Then if the intensity observed while saturating some other related resonance (and waiting for the new equilibrium to be established) is I, we define the nOe as:

$$\eta_i(s) = (I-I_0)/I_0 \qquad (5.1)$$

This expression is also often multiplied by 100 to make the figure a percentage. Note that, from this definition, the nOe will be positive if the new intensity is greater than the unperturbed intensity and negative if it is less (both cases are observed in practice). $\eta_i(s)$ indicates that this is the nOe at nucleus i when nucleus s is saturated (the notation used throughout this chapter is that of Noggle and Schirmer[2], except for the substitution of the more common symbol η for their 'f'). Methods of measuring the change in intensity of i, which can be rather small, are discussed in section 5.3; here we wish to discover how and why it might arise.

5.2.2 Pathways for Relaxation

In the previous chapter we looked at longitudinal relaxation from a macroscopic point of view. When we put a sample into the magnet, it is assumed that its induced magnetisation will build up by a first order (i.e. exponential) process with time constant T_1. This is only an assumption, but it is often confirmed by experiment. Underlying the macroscopic variation in magnetisation must be some quite complex process of equilibration across the various transitions of the nuclear system; it is the details of this process in which we are now interested.

Consider two spin-$\frac{1}{2}$ nuclei i and s with the same γ but different chemical shifts. We will assume they are in the same molecule, but not J-coupled. This system has four energy levels, corresponding with the nuclei being in the states $\alpha\alpha$, $\alpha\beta$, $\beta\alpha$ and $\beta\beta$ (Figure 5.1). Chemical shifts are generally very small in comparison with Larmor frequencies (parts per million), so the transitions of each nucleus are *nearly* equal in energy, making the states $\alpha\beta$ and $\beta\alpha$ nearly degenerate; for clarity the energy difference between them has been greatly exaggerated in the figure. Because we are assuming no J-coupling, the two transitions of nucleus i have *exactly* equal energy, as do those of nucleus s; the unperturbed spectrum consists of two singlets of equal intensity.

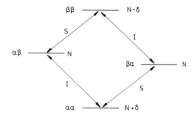

Figure 5.1 Energy levels and populations of a homonuclear AX system.

Now suppose the total number of nuclei in the system is $4N$. If the four levels were of equal energy, they would all be equally populated and each would contain N nuclei. However, as this is not the case, at thermal equilibrium a Boltzmann distribution will be set up. For simplicity we will assume that the difference in energy between states $\alpha\beta$ and $\beta\alpha$ is negligible, so that these states have equal populations. The state $\alpha\alpha$, being of lower energy, will then contain an excess of nuclei, while $\beta\beta$ will be deficient by an equal amount. If we call the excess or deficiency δ, we get the populations shown in Figure 5.1. The quantities of most interest here are the population *differences* between states, which can be summarised as follows:

i transitions:

$$\left.\begin{array}{l} \alpha\alpha - \alpha\beta \\ \beta\alpha - \beta\beta \end{array}\right\} \cdots \delta$$

s transitions:

$$\left.\begin{array}{l} \alpha\alpha - \beta\alpha \\ \alpha\beta - \beta\beta \end{array}\right\} \cdots \delta$$

$\Delta M = 0$ transition:

$$\beta\alpha - \alpha\beta\} \cdots 0$$

$\Delta M = 2$ transition:

$$\alpha\alpha - \beta\beta\} \cdots 2\delta \qquad (5.2)$$

The first four differences are across the normal transitions which give rise to the NMR lines, while the other two involve changes in the quantum number M (i.e. the total of the individual quantum numbers m for the two nuclei) of 0 ($\beta\alpha - \alpha\beta$) or 2 ($\alpha\alpha - \beta\beta$). While transitions of this type are not observed under normal conditions (because of quantum mechanical selection rules), we cannot safely assume that they will not be involved in the relaxation of the system, because as yet we have no theory for how this might occur. Instead, we should consider that if these population differences are disturbed the system will try to restore them by any available means. This leads to several possible pathways by which the system could relax (Figure 5.2), the operation of which we might try to confirm or disprove experimentally.

To proceed further we must make some assumptions. We will take it that the relaxation *across a single transition* is a first order process. That is, it is proportional to the extent to which the population difference across the transition differs from its equilibrium value. The rate constants for the various processes will be designated W, with a subscript to indicate the change in M involved (e.g. W_0 for $\beta\alpha - \alpha\beta$, W_1 for $\alpha\alpha - \alpha\beta$ etc.), and where necessary a superscript to indicate which nucleus the transition belongs to (W_1^i or W_1^s). In the most general case (where there might be J-coupling) there is no reason to assume that the two transitions of a given nucleus relax at equal rates (so we would need six rate constants), but here for simplicity we will do so.

Personally I find that this is the point in the argument at which confusion can set in, because a lot of new symbols have been introduced, so it is probably worth taking a few moments to reflect on what we are doing (by reference to Figure 5.2). We wish to dissect the relaxation of a system of interacting nuclei down to the level of individual transitions, and relate what we measure (for instance T_1) to what is actually happening. In a T_1 measurement, for example, we watch the approach of the longitudinal magnetisation to equilibrium. If we did this experiment for nucleus i of this two spin system, what we observe is the sum of the intensities of its two degenerate transitions. Thus T_1 would be related in some way to a

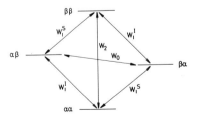

Figure 5.2 Connections between the energy levels of an AX system, which may be involved in relaxation.

combination of the rates $W_1{}^i$, W_2 and W_0. If these latter two rates were 0 (i.e. if the $\Delta M = 0$ and $\Delta M = 2$ transitions did not occur), then it is easy to see that $T_1{}^i$ would be given by:

$$T_1{}^i = 1/2W_1{}^i \qquad (5.3)$$

(the W's are *rate* constants, whereas T_1 is a *time* constant, and the total relaxation rate is the combination of the rates across the two transitions). Intriguingly, if W_2 and W_0 are non-zero, the total relaxation rate for nucleus i apparently must involve the population differences of nucleus s as well as its own. This contradicts the assumption made in defining T_1, which states that the rate of change of the macroscopic magnetisation due to a nucleus depends only on its own deviation from equilibrium, and says nothing about the magnetisation due to other nuclei. Thus, we can't necessarily expect a simple T_1 measurement to work in a multi-spin system. This is an important matter, but is not directly relevant to the present discussion; for more information see Noggle and Schirmer[2], Chapter 1 sections D, E and G.

In performing the nOe experiment, we force the population differences across some transitions to change, and observe the signals from others. Invariably we aim to affect all the transitions of a nucleus equally (unequal perturbation causes extra complications, discussed in section 5.3), so for instance we might saturate both transitions of s and then observe the signals due to i after the new equilibrium has been established. To analyse this experiment in the light of the preceeding remarks about relaxation, we need to write down a set of simultaneous rate equations for the various relaxation pathways, and solve them subject to the boundary conditions imposed by our saturation of s. This is straightforward, but slightly laborious, so I will not work through it here - the derivation can be found in reference 2. What we can do instead is estimate the possible effect of each pathway, by considering the circumstances immediately after the saturation of s has been achieved (Figure 5.3). The new population differences (to be compared with the equilibrium conditions, 5.2) are:

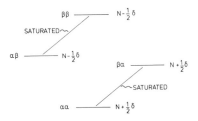

Figure 5.3 Populations of the levels immediately after the saturation of the s transitions.

i transitions:

$$\left.\begin{array}{c}\alpha\alpha - \alpha\beta \\ \beta\alpha - \beta\beta\end{array}\right\} \cdots \delta$$

s transitions:

$$\left.\begin{array}{c}\alpha\alpha - \beta\alpha \\ \alpha\beta - \beta\beta\end{array}\right\} \cdots 0$$

$\Delta M = 0$ transition:

$$\beta\alpha - \alpha\beta\} \cdots \delta$$

$\Delta M = 2$ transition:

$$\alpha\alpha - \beta\beta\} \cdots \delta \qquad (5.4)$$

The system as a whole is clearly no longer at equilibrium, and our hypothesis is that it will try to adjust itself closer to that state. We will consider this from the perspective of each possible relaxation pathway in turn. $W_1{}^s$ is, of course, irrelevant, because the population differences across those transitions are fixed by our saturation of the resonance. What of $W_1{}^i$? The population difference across each i transition at thermal equilibrium was δ, and we find that this is still the case, so from the point of view of $W_1{}^i$ no change is necessary. *If only single quantum transitions are active as relaxation pathways, saturating s does not affect the intensity of i,* or in other words there is *no* nOe at i due to s.

The real interest begins when we consider W_2 and W_0. The population difference between $\alpha\beta$ and $\beta\alpha$ is now δ, whereas at equilibrium it was 0. Thus W_0 acts so as to transfer population from the state $\beta\alpha$ to the state $\alpha\beta$, to try to restore a population difference of 0. This in turn is *increasing* the population of the *top* of one i transition, and *decreasing* the population of the *bottom* of the other one, thereby *decreasing* the total intensity of signals due to i (look at Figure 5.4). This tendency is counteracted by $W_1{}^i$, since from its point of view the i transitions were already at equilibrium, so the net result will depend on the balance of $W_1{}^i$ and W_0. *If W_0 is the dominant relaxation pathway, saturating s decreases the intensity of signals due to i*, or in other words there is a *negative* nOe at i due to s.

We can argue likewise for W_2. The population difference $\alpha\alpha - \beta\beta$ is now δ, whereas at equilibrium it was 2δ. Thus W_2 acts so as to transfer population from the state $\beta\beta$ to the state $\alpha\alpha$, to try to restore a population difference of 2δ. This in turn is *decreasing* the population of the *top* of one of the i transitions and *increasing* the population of the *bottom* of the other one, thereby *increasing* the intensity of signals due to i. This tendency is counteracted by $W_1{}^i$, etc. etc. *If W_2 is the dominant relaxation pathway, then saturating s increases the intensity of signals due to i*, or in other words there is a *positive* nOe at i due to s.

Solving the differential equations for this system properly lets us work out the balance achieved between $W_1{}^i$, W_2 and W_0 once the system has reached its new equilibrium. It is given by:

$$\eta_i(s) \;=\; \frac{W_2 - W_0}{2W_1{}^i + W_2 + W_0} \tag{5.5}$$

We have got to this point without thinking at all about how relaxation could be brought about - we have only considered abstractly the routes it might take. However, we are already in a position to test experimentally which pathways are followed. If nOe's occur in practice, this indicates that W_2 and/or W_0 processes (referred to collectively as *cross-relaxation*) are involved, and from the sign of the nOe we can discover which is dominant. What is found is that nOe's are *positive* for small molecules in non-viscous solution (implying that W_2 is important under these conditions), but *negative* for macromolecules or in very viscous solutions (implying a change to W_0 dominance). In between these cases is a region where W_2 and W_0 balance, and the nOe disappears. These very interesting observations tell us that relaxation is related to molecular motion, as we guessed in the previous chapter. We must now investigate in more detail how relaxation arises.

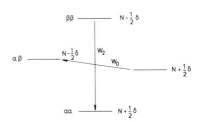

Figure 5.4 The initial direction of cross-relaxation after saturation of the s transitions.

5.2.3 Motives for Relaxation

Introduction

None of the transitions described in the previous section occur spontaneously at a significant rate. In seeking to explain nuclear relaxation, we have to think of mechanisms that might stimulate the transitions, and then calculate their effectiveness. The required stimulus is a magnetic field oscillating at the frequency of the relevant transition, so we examine the environment of the nuclei for such fields. This can be quite a complex problem, because it is possible to think of several potential sources of suitable fluctuating fields, but fortunately in most cases only one of these proves to be related to the internuclear distances we would ultimately like to measure. The quantitative treatment of relaxation is further complicated by the need to model the random motion of molecules in solution, so we will not be able to pursue the derivation of results in this area in any detail. However, it is not difficult to develop a feel for how relaxation mechanisms work on physical grounds; if you are then inclined to investigate the subject in a more quantitative way, any theoretical NMR text will include lengthy discussions of these questions.

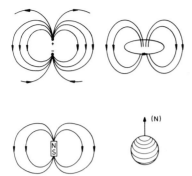

Figure 5.5 An assortment of dipoles: electric (top left), current loop (top right), bar magnet (bottom left) and nucleus (bottom right).

Magnetic dipoles and dipolar coupling

As I have already indicated, the interaction which gives rise to the nOe is the 'dipolar coupling between nuclei'. A basic obstacle to understanding this idea is not knowing what a magnetic dipole, let alone a dipolar coupling, actually is. Consider first an *electric* dipole. This consists of a pair of point charges, one positive and one negative (Figure 5.5). Surrounding them we imagine an electric field described by *lines of force*, which indicate the direction of the force on a positive test charge placed in the field. A *magnetic* dipole, in contrast, is a convenient fiction (because magnetic 'monopoles' do not seem to exist in the same way as electric charges), based on the observation that the lines of force describing the magnetic field around a loop of current (Figure 5.5) appear similar to those around an electric dipole, provided we do not go too near the loop. Noting this relationship saves physicists some effort in calculating the forces between current carrying wire loops, because they can translate results already obtained for the electrostatic case.

A nucleus (with charge and angular momentum) and a bar magnet are two more sources of magnetic fields of this type, and can also conveniently be described as magnetic dipoles (Figure 5.5). The vector μ, used in the previous chapter to represent the nuclear magnetism, is drawn along the direction of the dipole with the arrow head at the imaginary North pole. For our purposes the interaction between nuclei can then simply be assessed by noting whether the force lines from one nucleus oppose or reinforce the B_0 field at the other (Figure 5.6). There is a *local field* at each nucleus due to the other. Now, the nuclear dipoles are orientated by the external field, whereas the relative positions of the nuclei depend on the position adopted by the whole molecule, so the local fields may vary from one molecule to another. In amorphous solids, or polycrystalline powders, the positions of single molecules are essentially fixed, but they vary from one molecule to another, leading to a range of resonance frequencies and characteristic broad lineshapes. In single crystals on the other hand, there may only be a few, or one, relative orientations of the dipoles, leading to the direct manifestation of the dipolar coupling as a line splitting (the magnitude of which would depend on the orientation of the crystal relative to the static field). Note that this direct magnetic interaction is much bigger than the more familiar scalar coupling, and indeed frequently exceeds chemical shift differences, so the resulting variation in resonance frequency may be many kHz.

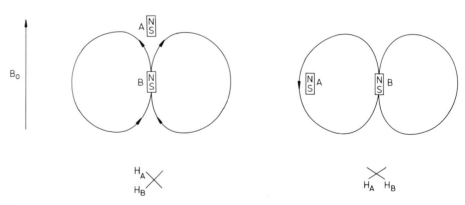

Figure 5.6 The interaction between two dipoles varies with their orientation.

In solution, however, we well know that linewidths may be a fraction of a Hz, which in the light of the existence of the dipolar coupling now seems surprising. This comes about because rapid molecular motion averages the dipolar interaction over all possible orientations of the molecule; calculations indicate that its average value is 0. This rapid reorientation of the dipolar interaction also proves to be a suitable source of fluctuating fields to stimulate longitudinal relaxation. Evidently the *strength* of the dipolar interaction will depend on the internuclear distance, so here at last we are making the connection between these physical ideas and the chemical information we require. It also seems reasonable that the dipolar coupling may be able to cause cross-relaxation (i.e. W_0 and W_2 processes) as required by the nOe, because it is a link between two nuclei (we will see some other relaxation mechanisms later which do not link nuclei and hence do not cause cross-relaxation). The remaining step is to look at how the motion of molecules in solution causes relaxation through the dipolar coupling, and how this relates to the W_0, W_1 and W_2 relaxation pathways.

Relaxation through dipolar coupling

To connect the dipolar coupling with longitudinal relaxation and the nOe requires a model of the motion of molecules in solution. For a large, complex molecule this is obviously not a problem that can be tackled in detail; the whole molecule will have several translations and rotations, and various bits of it may move in a different way to others, most of this motion being random. It is usual to make the gross approximation that the random motions can be summarised by a single parameter - the molecular correlation time τ_c.

Various ways of defining τ_c are encountered, depending on the particular problem to be solved; we might think of it like this. Suppose that the molecule moves from time to time from one orientation to another, and that each move occurs instantaneously. τ_c then characterises the range of times between moves. There will be a random distribution of 'waiting times' between one move and the next, and the correlation time is chosen such that waiting times shorter than τ_c seldom occur. This slightly odd definition (it might seem more straightforward to choose, say, the mean waiting time to characterise the motion) is advantageous, because the lower limit of the waiting time corresponds with the upper limit of the frequency distribution of the resulting fluctuating fields.

τ_c embodies the complex of influences on molecular motion into a single parameter, and as such will be a function of molecular weight, solution viscosity, temperature and maybe other more specific factors such as hydrogen bonding or pH. Assessment of whether one molecule will have greater τ_c than another must be made on the grounds of common sense and chemical insight; for instance, molecules with high molecular weight tend to move more slowly in solution than those with low molecular weight, and hence have longer τ_c. We then need some understanding of how variation in τ_c will affect relaxation times and the relative importance of the different relaxation pathways. Shortly I will give formulae which relate W_0, W_1 and W_2 to τ_c (for a two spin system), but first let's try to guess the relationship on physical grounds. To do this we have to think about how the distribution of the frequencies of oscillating fields due to the dipolar interaction varies with τ_c, and then compare this with the frequencies required to stimulate the W_0, W_1 and W_2 processes.

Noting that the *size* of the dipolar interaction does not depend on τ_c (because it only depends on the internuclear distance, which in this model we assume to be fixed), whereas its *rate of change* does, we can predict that the *total* amount of oscillating fields will be constant, while the upper limit of their frequencies will vary with the correlation time. Thus, if we plot graphs of the strength of the fluctuating fields (the *spectral density* due to the dipolar interaction, usually represented J) against their frequency ω for several values of τ_c, they enclose constant area but their upper limits vary (Figure 5.7). We will also assume, without attempting any particular justification, that the spectral density is essentially constant for $\omega \ll 1/\tau_c$, as indicated in the figure (more detailed theory can justify this assumption,

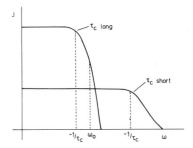

Figure 5.7 Spectral density fuctions for two different values of τ_c. A nucleus with Larmor frequency ω_0 would experience extreme narrowing with the shorter τ_c, but not with the longer one.

which is also confirmed by experiment). This then allows us to make some qualitative predictions about the variation of T_1, and even the W's, with τ_c (and hence with temperature, viscosity etc.).

The rate of relaxation will be determined by the intensity of the fluctuating fields with the right frequency to stimulate the transitions of interest; for instance, with the Larmor frequency ω_0 for the W_1 transitions. Consider first that $1/\tau_c$ is much greater than ω_0. In this case we are under the flat part of the spectral density curve; as τ_c *decreases* (say with increasing temperature), the height of the curve around ω_0 also *decreases* (because the area must stay constant, but the upper limit is moving further away), and therefore T_1 increases (slower relaxation). If we changed $\mathbf{B}_0$, and hence ω_0, we would expect no change in T_1 in this regime, because of the assumed flatness of the curve (refer to Figure 5.7). Now, if τ_c is gradually *increased*, by the same argument T_1 will *decrease*, but only up to the point where $1/\tau_c \approx \omega_0$. At this stage we leave the flat region of the spectral density and start moving off the end of it, so further reduction in τ_c causes an *increase* in T_1. In this vicinity we would also expect T_1 to become field dependent, because the spectral density is no longer flat. A plot of T_1 against τ_c is therefore predicted to show a minimum; this behaviour is confirmed in practice.

What of the relative amounts of W_0, W_1 and W_2 relaxation? The relevant frequency is clearly greatest for W_2 (being the sum of two ordinary transitions) and lowest for W_0, with W_1 in the middle. Thus, in the region $1/\tau_c \approx \omega_0$, we expect the W_2 relaxation to tail off first, while W_0 remains relatively effective. This, gratifyingly, is what we already know must happen from the variation in sign of the nOe with molecular size discussed in section 5.2.2. So it seems that this model of relaxation through dipolar coupling can explain, qualitatively at least, the observed dependence of T_1 and the nOe on molecular motion.

A quantitative treatment along these lines leads to the following expressions for the relaxation rates via dipolar coupling in a system of two spins separated by a distance r:

$$W_1{}^i \propto \frac{3\tau_c}{r^6(1+\omega_i{}^2\tau_c{}^2)}$$

$$W_0 \propto \frac{2\tau_c}{r^6(1+(\omega_i-\omega_s)^2\tau_c{}^2)}$$

$$W_2 \propto \frac{12\tau_c}{r^6(1+(\omega_i+\omega_s)^2\tau_c{}^2)} \tag{5.6}$$

where the constant of proportionality is the same in each case, and ω_i and ω_s are the Larmor frequencies of the two nuclei. Note that $W_1{}^i$ simply depends on the frequency ω_i of the i transitions, while W_2 and W_0 depend on the sum and difference frequencies respectively, as we might expect from the preceeding discussion.

The condition $1/\tau_c \gg \omega_0$ (i.e. the regime in which T_1 increases with decreasing τ_c) might alternatively be expressed as $\omega^2\tau_c{}^2 \ll 1$, in which case all the frequency dependent terms in the denominators of these expressions become negligible, and we can rewrite them:

$$W_1 \propto \frac{3\tau_c}{r^6}$$

$$W_0 \propto \frac{2\tau_c}{r^6}$$

$$W_2 \propto \frac{12\tau_c}{r^6} \tag{5.7}$$

The condition $\omega^2\tau_c{}^2 \ll 1$ is known as the *extreme narrowing limit* (the narrowing is extreme because under these conditions the dipolar broadening of the lines is completely averaged out). It is usual to assume that this applies when working with ordinary organic molecules in non-viscous solution, though it must be said that with the increasingly common use of high-field spectrometers to study complex molecules deviation from this condition is becoming more frequent. When working with proteins and other biological macromolecules extreme narrowing cannot be assumed, and the more complex expressions must be used instead. In section 5.2.4 we will put these expressions together with the formula for the nOe and see whether that helps us to measure internuclear distances.

Relaxation by other means

The dipolar interaction, modulated by molecular motion, is the most important relaxation mechanism in solution for protons and some other spin-$\frac{1}{2}$ nuclei such as ^{13}C. Here I mention briefly some of the other mechanisms known to exist; our main interest in these is to find and eliminate them during nOe experiments, because generally they do not involve cross-relaxation and hence reduce the nOe. A first point worth noting is that, while we have been discussing dipolar interactions between nuclei in the same molecule, there is no absolute requirement that dipolar relaxation be intramolecular. *Inter*molecular interactions will naturally involve greater distances, but this does not necessarily mean that they can be neglected (indeed, the first demonstrated nOe for protons was an intermolecular effect between chloroform and cyclohexane). From a practical point of view, since deuterated solvents will invariably be used during nOe measurements, the only relevant intermolecular interaction is likely to be between solute molecules; solutions should be kept dilute so as to minimise this.

Another dipolar mechanism which can be extremely effective is interaction with unpaired electrons. The magnetic moment of the electron is about 1000 times greater than that of the proton, so in the presence of paramagnetic materials even intermolecular interactions are strong and lead to very short relaxation times, completely short-circuiting cross-relaxation and the homonuclear nOe in the process. We can exploit this if, for some reason, we want to *eliminate* nOe's and shorten T_1's, by deliberately adding paramagnetic material to the solution (see Chapter 7). On the other hand, all solutions normally contain significant amounts of paramagnetic oxygen, which must be removed during nOe measurements (see section 5.3).

The next most commonly encountered relaxation mechanism only works for nuclei with spin greater than $\frac{1}{2}$ (quadrupolar nuclei). These are able to relax by interaction with *electric* field gradients as well as magnetic fields, a process which is usually very efficient. Quadrupolar nuclei (for instance ^{17}O or ^{14}N) are thus found to have very short T_1's and T_2's, with correspondingly broad lines. The only exceptions to this are nuclei in symmetrical environments (e.g. ^{14}N in compounds of the type $R_4N^+X^-$) or environments with low electric field gradients (the classic example is ^{14}N in isocyanides). Deuterium (spin-1), perhaps the most important quadrupolar nucleus from the chemist's point of view, can give relatively sharp lines in small molecules because of its low quadrupole moment; however it still relaxes quickly by the standards of spin-$\frac{1}{2}$ nuclei. Rapid relaxation independent of cross-relaxation means that the nOe is not a significant phenomenon for quadrupolar nuclei.

Several other possible relaxation mechanisms are known, and may be important in appropriate circumstances, but I will not burden you with them. For protons in medium sized molecules (I mean to exclude very small ones as well as big ones) in non-viscous solution, we can safely assume that intramolecular dipole-dipole relaxation is the principle mechanism, probably followed in undegassed solution by relaxation due to dissolved oxygen, then by other intermolecular dipolar interactions. For a survey of all other mechanisms with explanations at a comprehensible level, see Shaw[3], Chapter 9.

5.2.4 The NOe and Internuclear Distance

Maximum nOe's

We can now combine the expressions for the nOe and the various W's, in an attempt to extract information about internuclear distances. Recall that:

$$\eta_i(s) \;=\; \frac{W_2 - W_0}{2W_1{}^i + W_2 + W_0} \tag{5.5}$$

for a two-spin system which relaxes exclusively via dipolar coupling. If we assume extreme narrowing and substitute the values for the W's from 5.7 into this, we get:

$$\eta_i(s) \;=\; \frac{(12-2)\tau_c/r^6}{(6+12+2)\tau_c/r^6}$$

$$=\; \frac{1}{2} \tag{5.8}$$

This is perhaps not the expected result! The nOe is predicted to be *independent* of r, and equal to 0·5 (or 50%, as it would more commonly be written in a chemical context). We have not made a mistake here - the prediction is quite true; however it applies only for the completely artificial circumstance of two dipoles totally isolated from all other sources of relaxation. If the Universe consisted of two protons experiencing extreme narrowing, saturating the transitions of one would generate a 50% nOe at the other whether they were separated by Ångströms or light years. There is an important moral in this, which is that if our real chemical system approaches the situation of two isolated dipoles at all closely, then the nOe may cease to contain any information about internuclear distance. Understanding this point is crucial to making proper use of nOe measurements.

Unrealistic though this calculation has been, it still gives a useful result. The *maximum* positive homonuclear nOe we can ever obtain is 50%. In the *hetero*nuclear case the maximum nOe is given by $0·5(\gamma_s/\gamma_i)$, which when observing heteronuclei under conditions of broadband proton decoupling can lead to major sensitivity improvements. For instance γ_H/γ_C is about 4, so the maximum nOe is 200%, which leads to signals 3 times as strong ($I = I_0(1 + \eta)$) as they would be without nOe. On the other hand, some nuclei have negative gyromagnetic ratios (common ones are ^{29}Si and ^{15}N), so irradiation of protons may decrease or even invert their signals (e.g. γ_H/γ_{Si} is about -5, so the maximum nOe is about -250%, but if the full nOe is not realised enhancements close to -100% may arise leading to loss of signal). This point is discussed further in the next chapter, where the nOe is compared with another method for enhancing the intensity of signals from heteronuclei coupled to protons.

When the narrowing is not extreme

As spectrometer field strengths and our ambition to work with complex molecules both increase, it becomes quite common to encounter circumstances in which extreme narrowing cannot be assumed. The organic chemist, whose largest molecules are likely to have molecular weights of a few thousand at most, tends to fall into the unfortunate trap of balanced W_2 and W_0, which leads to negligible nOe. The only way out of this is to change τ_c by changing solvent or temperature, trying either to get back into the extreme narrowing region, or to move definitely into the W_0 dominated regime enjoyed by the biochemist (see below). There are also good future prospects for a new experimental solution to this problem ('rotating frame' nOe's[4]), but at the time of writing these experiments were not sufficiently well developed to include.

With very long τ_c, such as may be experienced by proteins or nucleic acids with molecular weights over 10,000, W_0 becomes completely dominant. Examining equation 5.5, and setting W_1 and W_2 to 0, we discover that the nOe reappears in this case, with a maximum value of -1. NOe's in macromolecules are therefore expected to be negative, and may be twice as big as those measured under extreme narrowing conditions; this has proved to be a very useful technique for assigning NMR spectra of proteins, particularly when applied as a two-dimensional experiment (see Chapter 8).

NOe's in realistic systems

We seem to have reached a pessimistic conclusion about the application of nOe's so far, but fortunately the nOe becomes related to internuclear distance when more interactions are included in the picture. There are two lines of attack we can adopt here. First, two-spin systems may show distance dependent nOe's when other competing sources of relaxation are present (as they invariably will be in practice); provided certain assumptions about the 'other' relaxation pathways and the molecular motion can be made, the relationship between the nOe and the internuclear distance can be calibrated experimentally. Second, in systems of more than two spins which are otherwise isolated from cross-relaxation sources, it may be possible to derive relationships between the *ratios* of internuclear distances (but not generally their absolute values) and the various nOe's. A particularly useful equation of this kind will be presented later.

Comparing two-spin nOe's

The first approach originates in a paper of Bell and Saunders[5] which is the foundation for much of the use made of the nOe in organic chemistry. Examining equation 5.5, we can identify the numerator as representing the net cross relaxation, often written σ_{is}:

$$\sigma_{is} = W_2 - W_0 \qquad (5.9)$$

while the denominator represents the total relaxation for nucleus i, for which the conventional symbol is ρ_i:

$$\rho_i = 2W_1{}^i + W_2 + W_0 \qquad (5.10)$$

For the special case of pure homonuclear dipolar relaxation between spin-$\frac{1}{2}$ nuclei, our previous calculations have told us that:

$$\rho_i = 2\sigma_{is} \qquad (5.11)$$

which is why the nOe, being the balance between these two, is 0·5 in this case. Now, in a more realistic situation there will be other relaxation mechanisms operating. If we can assume that these do not involve cross-relaxation, they can simply be included in the denominator of 5.5 as an extra contribution $\rho*$:

$$\eta_i(s) = \frac{\sigma_{is}}{\rho_i + \rho*} \qquad (5.12)$$

or, using 5.11:

$$\eta_i(s) = \frac{\rho_i}{2(\rho_i + \rho^*)} \tag{5.13}$$

Turning this upside down, we get:

$$\frac{1}{\eta_i(s)} = 2 + \frac{2\rho^*}{\rho_i} \tag{5.14}$$

From our model of dipolar relaxation we concluded that, in the extreme narrowing limit:

$$\rho_i \propto \frac{\tau_c}{r^6} \tag{5.15}$$

So equation 5.14 is of the form:

$$\frac{1}{\eta_i(s)} \propto \left(\frac{\rho^*}{\tau_c}\right) r^6 \tag{5.16}$$

Bell and Saunders' argument is then that in many cases it is acceptable to assume that ρ^*/τ_c is a constant for similar molecules, so that measured nOe's should be directly proportional to $1/r^6$. They tested this experimentally with a range of molecules of known geometry and found a good (indeed, quite remarkable) correlation between $1/r^6$ and η. This is the basis for using the nOe to compare internuclear distances in different molecules; however *extreme care must be taken to ensure that the assumptions involved are justified*. The following points in particular require attention:

1) For τ_c to be roughly constant, we must only compare molecules of similar molecular weight, using the same conditions of solvent and temperature. Specific interactions with the solvent, which might change τ_c, must also be absent, as must internal motion in the molecule which might cause nuclei to have an effective correlation time different from that of the whole molecule. This means, in effect, that the experiment is restricted to rigid molecules.

2) Extreme narrowing has been assumed, and care must be taken to work with small enough molecules so that this is justified at the field strength in use.

3) ρ^* is the extra relaxation arising from other dipolar interactions. For these to be constant, solvent and sample concentration must be the same each time, and most importantly there must be no variation in intramolecular interactions (other than the one being measured) between one compound and another. This latter point is almost impossible to realise in practice, and is the most probable cause of erroneous interpretation of nOe's, a question which will be examined in more detail later.

4) Since ρ^* is hopefully due to intermolecular interactions, there must be no structural features that might alter these between the compounds to be compared. For instance, all nuclei involved in the measurements should be on the periphery of the molecule, not buried within some cavity. There should be no specific interactions like hydrogen bonding which might cause the solvent to associate more strongly with one part of the molecule than another.

5) Great care must be taken to ensure consistent sample preparation, particularly with regard to removal of dissolved oxygen. Accidental introduction of oxygen can cause a considerable reduction in the measured nOe, leading to extremely misleading results. This point is pursued further in section 5.3.

The nOe for several spins

It will be evident from the preceeding remarks that using the nOe to make comparisons between *different* molecules may be less than straight-forward. Fortunately, many problems can be tackled by comparing nOe's within the same molecule, when some of the assumptions about things being equal are guaranteed to be true. In section 5.4 we will examine this approach qualitatively through examples; however for the restricted case of an AMX system which experiences no cross-relaxation from other sources (Figure 5.8) a quantitative expression can be derived[2]:

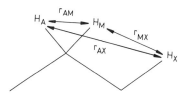

Figure 5.8 Internuclear distances in a three-spin system.

$$\left(\frac{r_{AX}}{r_{AM}}\right)^6 = \frac{\eta_A(M) + \eta_A(X)\eta_X(M)}{\eta_A(X) + \eta_A(M)\eta_M(X)} \quad (5.17)$$

The implication of this is that, provided all the relevant nOe's can be measured (or are negligible), knowledge of some distances within a molecule can be used to derive others. For instance, the distance between geminal protons at sp^3 carbon is relatively constant from one molecule to another, so if such protons form the AM part of an AMX system it may be possible to locate X by measuring nOe's. Once again we require a rigid relationship between the three nuclei (equal τ_c for both internuclear vectors), and no other significant sources of cross-relaxation (in other words, no other spin-$\frac{1}{2}$ nuclei nearby). While the latter requirement is unlikely to be satisfied properly other than in very simple molecules, it might be possible to arrange that there are no other cross-relaxation sources by making isotopic substitutions (e.g. deuterium for hydrogen). Alternatively, we may judge that, although there are other potential sources of cross-relaxation available, they are sufficiently far away to pose no threat to the accuracy of the calculation. This can be checked by investigating whether any nuclei other than those of the three-spin system of interest generate significant nOe's within it.

Two basic principles for multi-spin systems

It helps to clarify the rather abstract discussions of this section if we look at some particular cases (Figure 5.9). Consider first two protons in a molecule with no other spin-$\frac{1}{2}$ nuclei. If the experimental set up is optimised carefully, there will be few other sources of relaxation (i.e. $\rho^* \approx 0$), and on saturating either proton we should observe close to a 50% enhancement at the other, whatever their separation. This illustrates the first general principle:

Observation of an nOe between two protons does not, on its own, provide sufficient evidence that they are 'close'.

The additional knowledge needed to interpret this measurement is *either* a comparison experiment, as in the Bell and Saunders approach, *or* reasonable confidence that the proton at which the nOe is observed is *not* isolated from other sources of cross-relaxation. This latter condition corresponds with at least a three-spin case (Figure 5.9, diagram B). Suppose the protons are roughly in a line, with one end proton (H_A) nearer the centre (H_B) than the other (H_C). Irradiating H_A will still give a large enhancement at H_B, because, being closer than H_C, it will dominate H_B's cross-relaxation. By the same argument, irradiating H_C will only generate a small enhancement at H_B, so comparison of these two experiments now gives evidence about the relative internuclear distance. However, irradiation of H_B will give similar enhancements at both H_A and H_C, because from their points of view it is the only nearby proton, so this is not such a useful measurement to make. This illustrates the fact that, in general, the nOe's between a pair of nuclei are different in the two directions, because each nucleus may have different interactions with other sources of relaxation.

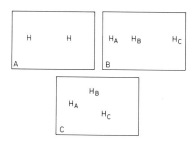

Figure 5.9 A two-spin system and two different arrangements of a three-spin system.

It is also quite interesting to consider what happens at H_C when H_A is irradiated. If we take it that, with the geometry we have here, H_A exerts no significant direct effect on H_C, then it experiences only its neighbour H_B's population differences *increasing* (as a result of the nOe from H_A). By the same argument by which *decreasing* a neighbour's population differences (by saturation) led to a *positive* nOe, we would expect this to generate a *negative* nOe at H_C. This is the famous 'three-spin' effect, and it is most important to take it in to account when interpreting nOe experiments. The detection of a negative homonuclear nOe (under extreme narrowing conditions) is very characteristic of a roughly linear arrangement of nuclei such as we have here. We might also imagine the effect propagating further through 'shells' of alternating positive and negative nOe's in long lines of atoms, but in practice indirect nOe's more than two nuclei away are almost always too small to measure.

Another aspect of the three-spin effect can be understood by imagining an alternative geometry such as Figure 5.9 diagram C. Here the H_A-H_C distance has been made much smaller, so that the direct positive nOe between these nuclei is no longer negligible. The observed nOe at H_C when H_A is irradiated now depends on the balance between the direct positive and indirect (through H_B) negative effects, and clearly for some appropriate value of the ratios of the internuclear distances these will cancel, leading to no nOe at all. So sometimes no effect is observed, even though the nuclei are 'close', in the sense that they are experiencing significant mutual cross-relaxation. Figure 5.10 indicates the range over which the three-spin effect leads to negative or zero nOe. This leads to our second general principle:

The absence of an nOe between two protons does not, on its own, provide sufficient evidence that they are 'far apart'.

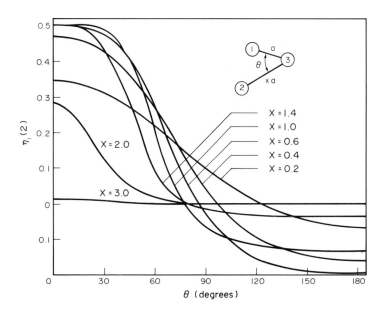

Figure 5.10 The variation of $\eta_1(2)$ with θ illustrates the three-spin effect.

This is to be applied even more strongly than the first principle, because nOe's can be small or undetectable for a wide variety of reasons. The overall conclusion here is that the interpretation of nOe's requires the assessment of *all* the interactions experienced by the nuclei involved. The steady state measurements may also need to be supplemented by measurements of transient nOe's (section 5.2.5).

The importance of these considerations can be graphically illustrated by the nOe's of a four spin system relaxing entirely by dipolar coupling, which have been calculated[6] (Figure 5.11). In the figure the arrows indicate nOe's, with the tail of the arrow at the saturated proton and the head at that which is enhanced; possible interactions which have not been marked are negligible. Note in particular the large nOe at H_A when H_B is irradiated, and compare it with the effect at H_C in the same experiment. An uncritical interpreter of this result would conclude that H_B is closer to H_A than to H_C, whereas in fact the opposite is true, and the large nOe at H_A arises because it has no other source of relaxation than H_B. Examining *all* the results together we would be less likely to make this mistake, and with more realistic relative distances and ρ^* non-zero such spectacularly misleading effects will not arise, but the conclusion is clear: *careful thought is essential.*

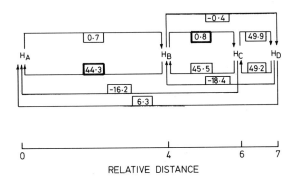

Figure 5.11 NOe's in a linear four-spin system. The values referred to in the text are in bold boxes.

5.2.5 Transient NOe's

Important extra information about molecular geometry can be obtained by measuring not the *value* of η, but the rate at which it arises[7]. The nOe itself depends on the competing balance between various relaxation pathways, which we have seen can be complex, but the initial rate at which it grows depends only on the rate of cross-relaxation between the relevant nuclei. For constant τ_c this will simply depend on the internuclear distance, so in a case such as the four spin system just mentioned (section 5.2.4) the initial rate of growth of $\eta_A(B)$ should be much less than that of $\eta_B(C)$, even though their final values are similar. Although nOe growth rates are less straightforward to measure than equilibrium nOe's, this simplicity of interpretation more than compensates for the experimental difficulty involved. In addition, the two-dimensional nOe experiment NOESY (Chapter 8) is necessarily based on the transient nOe technique described here.

In principle the rate of growth of η can be measured in exactly the same way as its equilibrium value, using the difference method described in section 5.3 together with variable periods of presaturation. However, since saturation cannot be brought about instantaneously, analysis of this experiment would require a rather complex model of the build up of both the nOe and the saturation. To avoid this difficulty it is more convenient to employ selective *inversion* of resonances as the initial perturbation, followed by a variable waiting period τ_m before acquisition of a spectrum (Figure 5.12). The difference method can still be used to assist in the measurement of intensities. A resonance which shows a positive nOe in an equilibrium experiment will, during a growth rate measurment, appear positive for short values of τ_m, with the initially small response growing to some maximum, then as τ_m continues to increase it will fade away as straightforward longitudinal relaxation returns the system to thermal equilibrium (Figure 5.13). The initial slope of the growth curve is the quantity of interest.

The practical problem with this experiment is that several time points will be required to characterise the nOe growth, so it is likely to take longer to acquire than an equivalent equilibrium nOe (but of course this will depend on how much presaturation time is used in the equilibrium experiment, as compared with the values of τ_m and the number of points in

Figure 5.12 The experimental scheme for a transient nOe experiment.

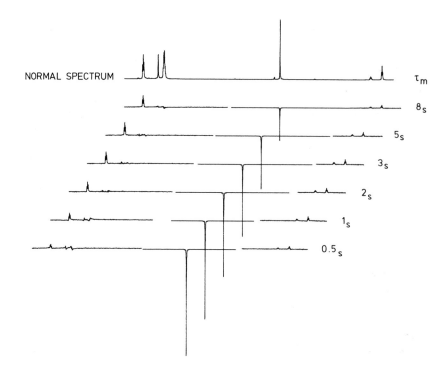

NORMAL SPECTRUM

τ_m

8_s

5_s

3_s

2_s

1_s

0.5_s

Figure 5.13 Transient nOe's (normal spectrum in the top trace, and nOe difference spectra with varying τ_m below). Following selective inversion of the singlet in the centre of the spectrum, responses build up in three other resonances. With even longer values of τ_m both the inverted peak and the resulting nOe's will decay to zero.

the growth rate measurement). In addition, since the early points on the curve obviously do not correspond with the maximum nOe, the signal-to-noise ratio required is much higher. Selectivity in the inversion pulse presents similar problems to those encountered with pre-saturation; however in this case very good results can be achieved using special shaped pulses (Chapter 7, section 7.7.3).

5.3 MEASURING NOE'S

5.3.1 Introduction

By way of respite from the mental strain of the previous section, I turn now to examine how we can measure nOe's in practice. The detection of the relevant intensity changes, which may be only a few percent, requires more care than most NMR experiments, and places considerable demands on both operator and spectrometer. Apart from describing the nOe difference method, and steps which can be taken to maximise nOe's and minimise distracting experimental artefacts, I also take the opportunity here to examine some aspects of instrumental performance, such as the deuterium lock and frequency stability, which are particularly important to difference spectroscopy. A good deal of this discussion is also relevant to other experiments requiring long-term spectrometer stability, such as two-dimensional NMR.

5.3.2 Sample Preparation

The main precaution that needs to be taken beyond those already discussed in Chapter 3 is the rigorous exclusion of paramagnetic materials, principally oxygen. An effective means for degassing a sample is the freeze-pump-thaw technique, preferably carried out directly in the NMR tube (tubes with tiny taps can be obtained from Wilmad to facilitate this operation). This requires the following sequence of events:

1) Cool the sample under an anhydrous atmosphere in liquid nitrogen or dry ice/acetone.

2) Evacuate the space above the solution by connecting it to a good vacuum line.

3) Isolate from the line, and warm back to room temperature.

4) Admit nitrogen and repeat from (1).

This sequence needs to be repeated many times (at least five). It is remarkably difficult to eliminate oxygen completely; this can readily be demonstrated by taking a solution of a small molecule like benzene, subjecting it to freeze-pump-thaw cycles and checking its proton T_1 before and after each cycle (see Chapter 7 for a quick method of T_1 measurement). See how long it takes to degass to 'constant T_1'; this will be a good test of your technique. Try to perform the degassing immediately before the nOe experiment, because even in a tube with a tap oxygen diffuses back in quite quickly; a vacuum line for this purpose is a useful accessory in an NMR lab.

A problem arises with aqueous solutions, which cannot be frozen in the NMR tube without risk of cracking it. In this case degassing must be carried out separately in a flask, and the solution transferred under an inert atmosphere. Beware, however, of using conventional syringe techniques for this, because metal ions are often dissolved from the needles. Another practice to be avoided is the 'degassing' of solutions by blowing nitrogen through them from a needle; this is ineffective and liable to introduce more new paramagnetic material than was originally present.

An alternative to the freeze-pump-thaw technique for aqueous solutions is simply pumping on the sample without freezing it (assuming of course the solute is involatile). With care (!) it is possible to perform this in the NMR tube, and after brief evacuation admit an inert atmosphere. Many repetitions of this cycle lead to relatively efficient degassing.

5.3.3   The Difference Method

Introduction

In measuring an nOe we are required to saturate one resonance and then compare the intensities of others with their equilibrium values. This could in principle be done simply by integrating resonances with and without a period of pre-saturation, but in practice this is not accurate enough to detect small nOe's. Instead the scheme known as *nOe difference spectroscopy* is employed (Figure 5.14). There are two elements to this experiment: the use of gated decoupling, and the difference method itself.

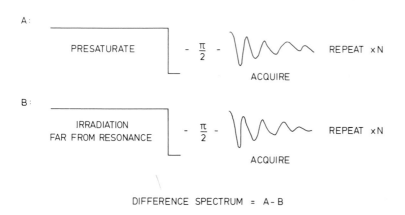

Figure 5.14 The experimental scheme for nOe difference spectroscopy.

The nOe is generated using *gated decoupling* to facilitate comparison of perturbed and unperturbed spectra. If we use the proton decoupler to irradiate a resonance for some time, new population differences will build up according to any nOe's which may be present. It is then not necessary to maintain the irradiation during the acquisition of an FID, because the populations will only return to their equilibrium values in a time of the order of the various T_1's. Since it is also desirable *not* to irradiate during the acquisition, so as to avoid any decoupling or Bloch-Siegert shifts, it is natural to turn off the decoupler just before the scan. This distinction between decoupling, which requires that the irradiation be present throughout the experiment, and the nOe, which can be generated and then persists, is sometimes confusing when first encountered, but should present no problem when you have thought about it for a few moments.

Having accumulated some scans with the system perturbed, an equal number are obtained without the perturbation. To make the two experiments as similar as possible, it is usual to simply shift the irradiation frequency well away from the resonances, rather than just turning it off. This minimises any peculiar instrumental effects which may arise; hopefully differences between the two spectra will then only reflect changes due to the nOe (but see the discussion of SPT effects later). Subtraction of the unperturbed from the perturbed FID, followed by Fourier transformation, generates a spectrum in which only signals which differ between the two are evident (Figure 5.15). One signal which will certainly differ is that which was saturated; this will appear inverted in the difference spectrum.

Figure 5.15 An nOe difference experiment. At the left, the lower trace is the 'off-resonance' experiment while the upper trace has been acquired with saturation of the peak marked with an arrow. The difference between the two spectra reveals several common features of such experiments: an enhanced peak (1); a peak due to subtraction error, which has dispersive character (2); the saturated resonance, which appears inverted in the difference spectrum (3); and failure to achieve perfect selectivity in the saturation (4).

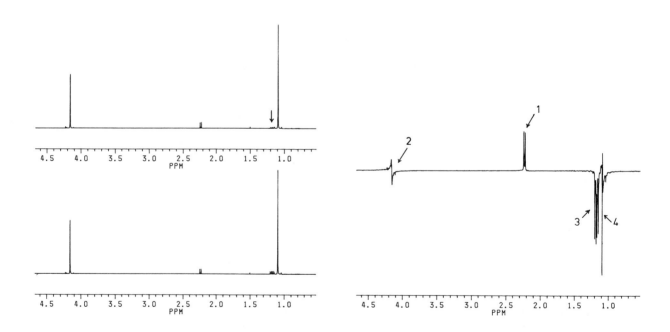

This technique *should* allow nOe's to be identified entirely unambiguously; their magnitudes can then be measured in a manner to be discussed later. In practice, however, if the spectral subtraction is to work properly high performance is required from the spectrometer. Many factors influence spectrometer stability, some of which are beyond the control of the operator; indeed difference spectroscopy is a good test of the quality of spectrometers (Chapter 7). The following sections survey the problems of difference spectroscopy, and what steps can be taken to ensure good results.

Subtracting Lorentzian lines

If we record an FID twice, under nominally identical conditions, we hope to obtain identical data. In practice, of course, no spectrometer is perfect, and the signals we actually obtain may differ in amplitude, phase and frequency between the two experiments. Thus when we attempt to subtract the FID's, which should cancel unperturbed peaks, some residual signals will remain depending on the type and extent of the instrumentally induced changes. How much residual signal we can tolerate will depend on the experiment being attempted; during an nOe measurement it is desirable to be able to detect 1% enhancements reliably, which ideally requires that residual signals due to imperfect cancellation are less than, say, 0·1%. In the following paragraphs I will give some indication of how stringent a requirement this is when related to aspects of spectrometer performance such as frequency and phase stability, but first let us consider how to make life as easy as possible by the manner in which the experiment is performed.

The basic scheme proposed for the difference experiment was the acquisition of a few scans with saturation of a resonance, followed by the acquisition of an equal number without the saturation. We might choose the number of scans according to the signal-to-noise ratio requirements of the experiment, but for proton NMR this is often not the limiting factor. Signal averaging also improves the cancellation of unwanted peaks, provided the instrumental instability that causes them is random. In this case, the basic cancellation we obtain from a single pair of scans (see below) will be improved as the square root of the total number of pairs, just as signal-to-noise ratio improves as the square root of the number of scans. In fact, since it is quite likely that instrumental instabilities will be composed of a mixture of random elements and systematic long-term drift (e.g. due to gradual temperature changes), the best procedure is to *interleave* the acquisition of saturated and non-saturated experiments over a period of time. For instance, we might acquire sixteen scans with the irradiation on resonance, sixteen off resonance, then repeat the whole thing ten times, making a total of 160 pairs of scans.

So in estimating the spectrometer performance required to give 0·1% residual signals, we can allow for some help from signal averaging, perhaps a ten- to twenty-fold reduction in a typical experiment. Our target for basic subtraction error in a single pair of scans would then be 1-2% residual signal. Experimentally, it is easy to reproduce the amplitude of signals accurately enough to give cancellation below this level, but errors due to phase and frequency instability may prove more troublesome (although these are really two aspects of the same thing, phase noise and frequency modulation being related, I will treat them separately here because the two types of results are both useful). The following treatment relates to what we actually see in the difference spectrum after transformation; that is, it is in the frequency domain.

Consider first the result of a small phase change φ from one scan to the next. If we take it that the real part of the first scan is obtained in pure absorption mode, then (in the notation of Chapter 4), the second scan, which we plan to subtract from the first, is represented by:

$$\mathscr{R} = \mathbf{A}\cos\varphi - \mathbf{D}\sin\varphi \qquad (5.18)$$

Now, since φ will be a small angle (unless the spectrometer is broken), we can simplify this considerably. The gradient of $\cos\varphi$ is a minimum at $\varphi = 0$, while that of $\sin\varphi$ is a maximum. So for a small change of φ away from 0, the amplitude of the absorption part of the signal hardly changes, while that of the dispersion part increases rapidly. Thus, to a good approximation, we can take it that in the difference spectrum in the presence of a small phase error φ the absorption part still cancels perfectly, and the residual signal has purely dispersive character with amplitude proportional to $\sin\varphi$. Residual signals of this type can clearly be seen in most of the difference spectra in this chapter. To take a concrete example, if φ is 1°, $\cos\varphi = 0\cdot99985$, but $\sin\varphi = 0\cdot017$; thus even an error as small as this will introduce residual signals as strong as we can tolerate. I know from careful measurement that the 500 MHz spectrometer used for most of the

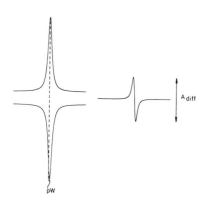

Figure 5.16 Subtracting two Lorentzian lines leaves a significant residual peak even when they differ in frequency only slightly (scarcely discernible at the left of the figure).

examples in this book (a 'state-of-the-art' instrument with digital generation of frequencies and phases) has medium-term phase instability of at least $\pm 2°$, which readily accounts for the residual signals obtained when it is used for difference experiments; this should be an area for future improvement of commercial spectrometers.

Another way of looking at the origin of residual signals is to consider subtracting two Lorentzian lines of equal amplitude and phase but different frequency (Figure 5.16). The results obtained here are also of use in the analysis of phase-sensitive two-dimensional spectra (Chapter 8), where overlapping antiphase lines are sometimes encountered. We saw one way of writing the equation of a Lorentzian line of amplitude A and half-height width W in Chapter 3 (Figure 3.3):

$$y = \frac{AW^2}{W^2 + 4(v_0 - v)^2} \tag{5.19}$$

Consider two lines, with $v_0 = +\frac{1}{2}pW$ for one and $-\frac{1}{2}pW$ for the other (i.e. they are separated by p times the linewidth; for cases of interest here p will be less than 1). If we let $x = 2v/W$, the *difference* lineshape can then be written:

$$y_{diff} = A\left(\frac{1}{1+(p-x)^2} - \frac{1}{1+(p+x)^2}\right) \tag{5.20}$$

This is not exactly the same as a dispersive Lorentzian, but it is possible to see that it shares the characteristic of passing through zero in the middle of the curve (at the point $x=0$, or halfway between the two lines) (Figure 5.16). We would like to know the amplitude of this line, so we have to find the x coordinates of its maximum and minimum. Differentiating 5.20, setting it equal to 0 and solving for x gives, after some exceptionally tedious algebra:

$$x = \pm \frac{1}{\sqrt{3}}(p^2 + 2\sqrt{p^4 + p^2 + 1} - 1)^{\frac{1}{2}} \tag{5.21}$$

This is quite an interesting equation. It tells us that, even for relatively large p, the measured frequency difference between the peaks of antiphase Lorentzians is significantly different from the true line separation; for instance with $p = 4$ the error is 17%. This is of some significance in two-dimensional NMR, where it may be desirable to measure proton couplings of a few Hz from antiphase patterns with linewidths of this order; this point is discussed in Chapter 8, where there also appears a graph of equation 5.21.

The amplitude of the residual peak is given by an inconveniently complex expression (plotted numerically in Figure 5.17), but from the point of view of difference spectroscopy we are only interested in what happens for very small p. Equation 5.21 tends to the limiting value of $x = \pm 1/\sqrt{3}$ as p tends to 0, so intriguingly the separation of the peak maxima also tends to a limit (of $W/\sqrt{3}$) whatever the actual separation. In this area, the amplitude of the line (measured from the negative to the positive peak, and expressed as a fraction of the amplitude of the original lines) can be approximated as:

$$A_{diff} = \frac{3\sqrt{3}p}{2} \tag{5.22}$$

So to get residual peaks as small as 2%, the frequency of our detected NMR line must change by no more than about 1% of its width from one scan to the next! For a typical proton line 1 Hz wide, this would mean that short term fluctuations have to be less than about 10 mHz, or 5 parts in 10^{10} on a 500 MHz spectrometer. This is just about the limit of the stability of temperature controlled crystal oscillators, so it might be

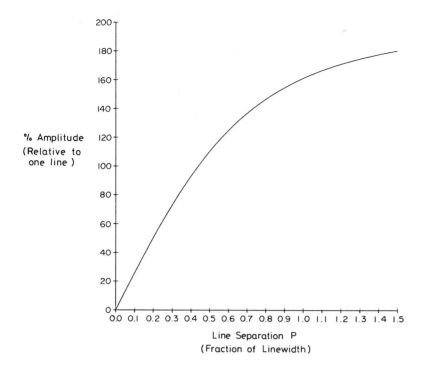

Figure 5.17 The residual amplitude in a difference spectrum as a function of the separation of Lorentzian lines.

achieved, but would require highly optimised instrument design and carefully controlled experimental conditions. We can also conclude that application of exponential multiplication during data processing will reduce residual signals of this kind, and for this reason it may prove desirable to use more line broadening than the matched filtration condition would indicate.

Optimum nOe difference spectroscopy

It is evident by now that measurement of small nOe's is likely to push a spectrometer to its limits. Thus, anything we can do to increase the nOe and improve the suppression of residual signals is likely to be worthwhile. To take the latter question first, the aspects of long term spectrometer stability which are capable of being influenced by experimental procedure are those related to the deuterium lock and the sample environment. Section 5.3.4 therefore provides a closer look at the lock system than we took in Chapter 3, while here I will briefly mention a few things that can be done to avoid environmental factors disturbing the experiment.

The most likely environmental cause of poor difference spectroscopy is varying temperature. The chemical shifts of many species show a significant temperature dependence, which may affect resonances of interest directly, or may move them indirectly via a shift in the frequency of the lock (this is a particular problem in aqueous solutions). Thermal regulation is thus highly desirable during nOe difference experiments, and is usually achieved by heating slightly above room temperature. This may also have beneficial side effects, since warming the sample can move it more definitely into the extreme narrowing region; in difficult cases with large molecules heating as much as they can stand may be necessary. From the point of view of careful regulation in the vicinity of room temperature, it is worth bearing in mind that the average variable temperature unit encountered on NMR spectrometers is likely to be something of a brute-force device; it may be better to obtain, or build, an alternative[8]. In any case, trials should be carried out both with and without temperature control, to determine whether an improvement is in fact obtained.

Another factor which, according to NMR folk-lore, is supposed to influence the outcome of difference experiments is human activity near the spectroscopy lab. Running nOe experiments in the small hours of the morning or on Sunday is therefore recommended, though personally I think that improved results under these conditions have more to do with long acquisition times than lack of footsteps above the ceiling. Nevertheless, it is certainly advisable to avoid traffic in the actual spectrometer room during difference spectroscopy, particularly for very high field spectrometers.

The second area over which we have some control is the potential size of the nOe's themselves. By minimising ρ^* and ensuring extreme narrowing, we give ourselves the best chance of detecting cross-relaxation. This means in practice eliminating other sources of relaxation by working with dilute, rigorously degassed samples in deuterated solvents of high isotopic purity. Choice of the least viscous solvent in which the sample is soluble may also help. Care should be taken to allow sufficient pre-saturation for the new equilibrium populations to build up, which means irradiation for at least three times the longest T_1 present, and preferably much longer. Indirect three-spin nOe's in particular are very slow to arise, because the direct nOe, which must build up first, itself requires a time scale of the order of the relevant T_1's to appear. A quick pre-check on T_1's (Chapter 7, section 7.5.3) is a very worthwhile investment to make, as for protons they may vary from less than a second to five seconds or more within the same molecule. Performing such measurements also gives the added bonus of additional information to aid in the interpretaion of the nOe data, since observation of an unusually long T_1 for a proton is symptomatic of its relative isolation from sources of dipolar relaxation. This in turn indicates the need for caution when drawing conclusions from nOe's it may experience (*cf* H_A of the four-spin system discussed at the end of section 5.2.4).

Selectivity and the SPT problem

We have assumed so far that it is possible to saturate all the lines of any multiplet at will. In reality, in a complex spectrum, perturbing one resonance without affecting others may present problems, and in fact this will often be what limits the applicability of the nOe difference method. The only way to proceed here is by trial and error, experimenting with different field strengths and examining the resulting degrees of wanted and unwanted saturation (see Chapter 7, section 7.7.2, for a general discussion of selective irradiation with low field strength). A likely outcome of this procedure is that we will be forced to use less than perfect saturation of multiplets, in order to avoid fatal interference with other nearby resonances. The resulting reduction in nOe's must be tolerated, but in coupled systems our problems may be further compounded by the effect known as *selective population transfer* (SPT).

SPT is likely to arise in a *J*-coupled system any time the transitions of one resonance are subject to *unequal* perturbation. Thus, in the ideal case in which all transitions are fully saturated there will be no SPT, but in struggling to make the irradiation selective we are likely to introduce it. I will not go into the mechanism of population transfer here, as it is discussed in some detail in Chapter 6, but simply point out the nature of the effect and how to eliminate it. In contrast with the nOe, which causes a net increase in the total integrated intensity of a multiplet, SPT only generates differential polarisation, or in other words antiphase lines which should have zero integral (Figure 5.18). This is helpful, because when we integrate resonances to measure nOe's the SPT signals should make no contribution, but the antiphase lines, which may be intense, are distracting nonetheless and worth eliminating.

Fortunately this is readily achieved, simply by acquiring the spectrum using $\pi/2$ pulses[9]. Since we should be leaving many times T_1 between scans, this makes sense from a sensitivity point of view as well, and can be

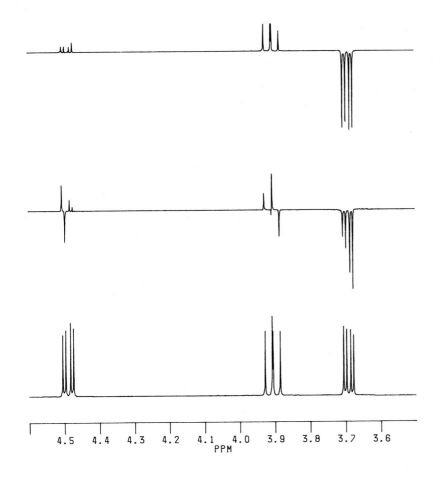

Figure 5.18 Uneven saturation of multiplets generates large SPT effects if the spectrum is acquired with a small flip angle (centre trace); these completely dominate the much smaller nOe. Acquiring the spectrum with a composite $\pi/2$ pulse (top trace) virtually eliminates the SPT and reveals the Overhauser effects.

considered mandatory. For really excellent suppression of SPT signals, the use of a composite $\pi/2$ pulse is recommended (Chapter 7, section 7.3); all the spectra in this chapter were acquired in this way (except Figure 5.18!).

Quantitative measurements

Once a satisfactory nOe difference spectrum has been obtained, it still remains to integrate the resonances and determine their various η's. There are two possible approaches to this. The most obvious method, which follows from the definition of η, is to integrate a multiplet in the off-resonance component of the difference experiment (i.e. transform part B of Figure 5.14 on its own), and assign its area as 100%; this corresponds with I_0. Integration of the corresponding multiplet in the difference spectrum then yields η directly, provided the transform is carried out with the same scaling factor (it may be necessary to set some flag in the spectrometer software to achieve this).

While this procedure certainly yields the actual enhancement of any multiplet, it may not necessarily be the most relevant measurement to make. In reducing the strength of the saturating field so as to achieve selectivity, we are quite likely to cause incomplete saturation. It is then reasonable to scale the measured η according to the degree of saturation, so that we estimate the value that would have been obtained had saturation been complete. This can be done simply by using the intensity of the saturated peak in the difference spectrum as a reference (e.g. assigning its area as -100% if it is a 1-proton multiplet), and integrating the enhanced peaks relative to this scale. Percentage enhancements quoted in section 5.4 were measured in this way.

Whatever method of measurement is chosen, it is important not to be over-optimistic regarding the accuracy of the results. The signal-to-noise ratio of a difference peak due to an nOe of a few percent will usually be small, so there is no point quoting its magnitude to several decimal places. Fortunately we are most often interested in determining whether η is, say, 5% or 20%, which should be straightforward. Even when attempting quantitative calculations using equation 5.17 tremendous accuracy is not essential, because the sixth-root dependence of the distance ratio makes it quite insensitive to errors in the individual η's.

5.3.4 Some Remarks about the Deuterium Lock

The use of a field-frequency lock during difference spectroscopy is essential to compensate for long term changes in the static field. However, on the time scale of individual scans the lock is not capable of providing significant stabilisation, for reasons which will become apparent shortly, but it *is* capable of degrading total performance. Thus it is important to ensure optimum lock operation, and while we have already discussed this in Chapter 3, it seems appropriate to re-examine it now we know enough to appreciate how the lock actually works. These considerations also apply in their entirety to two-dimensional NMR, where short term instability in the field-frequency ratio will be manifested as 't_1 noise' (Chapter 8).

The aim in using a lock system is to stabilise the spectrometer against variation in its static field or radio frequency, by keeping the *ratio* of the two constant. This is achieved by monitoring an NMR line, typically the deuterium resonance of the solvent, and arranging to control the static field such that this resonance is always in the same place. It is particularly convenient to work with the dispersion mode line for this application, because it has zero amplitude at resonance and positive or negative amplitude to each side (Figure 5.19); the detected signal can therefore be used directly in a feedback loop to generate corrections to the magnetic field. We can thus see immediately why the *lock phase* (Chapter 3) is such an important adjustment; completely incorrect phase (i.e. a mainly absorptive lock lineshape) will mean that there is no longer zero amplitude at resonance, and hence will disrupt the ability of the control loop to stabilise the field. Even small errors in the phase will make the gradient of the line around its centre less, hence reducing the effectiveness of the lock.

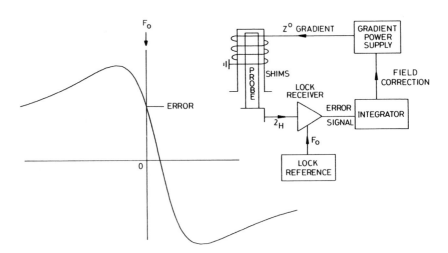

Figure 5.19 The lock feedback loop, which monitors the position of a dispersion mode line.

In an ideal, noise free environment, we might expect this system to be able to eliminate field/frequency variations on a fairly short timescale, limited perhaps by how quickly the correcting field could be adjusted, and by the need to avoid overshoot and oscillation in the feedback circuit.

However, NMR signals are not free from noise, and to avoid this noise propagating onto the static field through the lock (a disastrous situation!) we have to lengthen the time over which samples are taken. The lock signal is *averaged* (by an electronic integrator), and the time constant of this averaging may need to be quite long, from tens to hundreds of seconds (this is a usually a built-in characteristic of the spectrometer without any possibility of operator adjustment). There is no point expecting correction of changes in the field/frequency ratio which occur over shorter time periods than this, which is a shame because this is precisely the interesting range of times for difference spectroscopy.

So from one scan to the next the lock is unlikely to help. However, it is still essential in the medium to long term, and we can now see how its operation might be optimised. The correction signal derived from the dispersion mode resonance will be most sensitive to changes in the field/frequency ratio if the lock line is sharp, correctly phased and has a high signal-to-noise ratio. We can influence these factors, both by the manner in which the spectrometer is operated (good shimming and correct adjustment of lock phase and power), and by the choice of solvent (the deuterium linewidths of solvents vary greatly). Acetone is one of the best NMR solvents, because its low viscosity causes its deuterium resonance to be very sharp. More viscous solvents (e.g. dimethylsulphoxide, pyridine) are likely to give inferior lock performance. Water, though unavoidable in many experiments, is one of the worst, because its protons (deuterons) are exchangeable, and under certain conditions of pH, temperature and type of solute may give extremely broad signals. Experimental planning should take these factors into account, since minor changes in the conditions may have a large affect on the lock linewidth in aqueous solution.

5.4 USING NOE'S

5.4.1 A Very Simple Case

Consider an apparently straightforward problem: distinguishing between citraconic acid (**1**) and mesaconic acid (**2**). Here is a case where the Bell and Saunders approach will be easy to apply, as the two molecules are clearly very similar, and there is no danger of interference from other dipolar interactions. By performing the experiment in D_2O the only protons left in the molecule are those in which we are interested: the olefin and the methyl group. A useful rule of thumb to apply when investigating olefin isomers is that the *trans* : *cis* internuclear distance ratio is about 1·3 : 1. In this rather simple case, a more accurate calculation using standard bond lengths (and assuming averaging of the positions of the methyl protons) indicates that the methyl-olefinic proton distance in the *trans* isomer should be 1·32 times greater than that in the *cis* isomer.

The natural nOe to attempt to measure here is that at the olefinic proton when the methyl is saturated. The inverse experiment, although possible, is less attractive, because the relaxation of protons in a CH_3 group is dominated by their mutual dipolar interactions, making nOe's at methyls generally rather small. Figure 5.20 shows the proton spectra of the two acids, and the corresponding nOe difference spectra obtained while saturating the methyl groups. Great care was taken to ensure the establishment of true equilibrium nOe's (not a trivial matter in such small molecules - the T_1 of the olefinic proton of the *trans* isomer was around 18 s). Clearly the nOe obtained with citraconic acid is substantially larger than that for mesaconic acid, as expected; integrating the resonances leads to values for η of 0·29 and 0·06 for the two compounds. Assuming a $1/r^6$ dependence of η, the calculated distance ratio is therefore 1·30 : 1, and if the η's are accurate to $\pm 0·01$ (probably optimistic!), the error in this measurement is $\pm 0·04$. Thus, the observations are consistent with the known internuclear distances within experimental error.

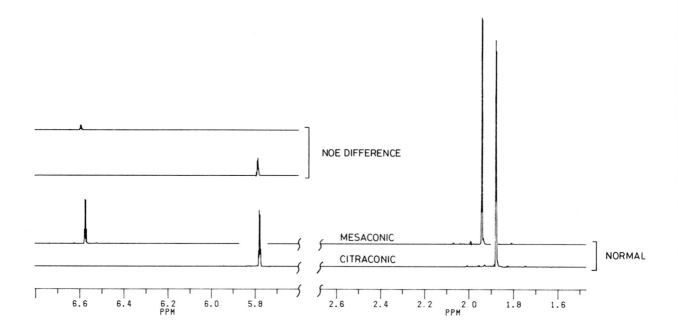

Figure 5.20 Difference spectra referred to in the text.

So, a perfectly straightforward and satisfying result. However, I would like you to take a moment to reflect upon the problems that would have arisen if *only one* isomer had been available. We would have obtained a result, say a 6% nOe, but what meaning could be attributed to it? In the absence of knowledge of related compounds, we have no way of telling whether this is a 'large' or 'small' nOe. Indeed, it would not be difficult to obtain a measurement as small as this from the *cis* compound, by a combination of poor sample preparation technique and careless choice of experimental parameters. For instance, if you neglected to check the T_1's and used only a few seconds pre-saturation, the nOe's in these compounds would scarcely have time to build up. The essence of the nOe is that it is a comparative method, and the only truly reliable comparison is amongst stereoisomers of a compound. The further you move away from this situation, the greater must be your caution in interpreting the results.

Another aspect of this experiment of some interest becomes apparent by considering the range of nOe's which can be measured. The maximum possible nOe is 50%, of course, and enhancements less than, say, 1% are hard to measure reliably. Thus, the maximum ratio we can obtain is 50 : 1, corresponding with a distance ratio of just less than 2 : 1. This is a kind of 'dynamic range' limit for the experiment; distances differing by more than this will not be comparable.

5.4.2 A Test of the Three-Spin Equation

Equation 5.17 can be very useful if isolated three-spin systems can be identified. As an example of this, consider acrylic acid. Once again, the carboxylic acid proton can be eliminated by deuteration, leaving a pure three-spin system, which is first-order at 500 MHz. Measuring nOe's in this compound presents some interesting technical problems, as the resonances are separated from each other by only 120 Hz or so, making selective irradiation tricky. A procedure which works well in compounds like this with simple first-order multiplets is to build up the saturation *line by line*[10]. Thus, the decoupler is set on resonance on a single line, and a very weak *rf* field (a few Hz) applied briefly. The irradiation frequency is then moved to each line of the multiplet in turn, and the whole sequence performed repetitively throughout the pre-irradiation period.

For acrylic acid in degassed methanol, the longest T_1 (that of H_X in formula **3**) was found to be around 10 s. During the nOe experiments, each line of a particular multiplet was subject to 0·5 s bursts of irradiation, giving a cycle for the four lines taking 2 s. This was repeated 45 times, to give a total pre-irradiation period of $9T_1$. Since an extremely weak *rf* field can be used by this means, the selectivity is very high, as can be seen from the difference spectrum in Figure 5.21 (for clarity, only one of the three experiments is illustrated here). The resulting values of η are collected in Table 5.1, using the labelling of the protons indicated in formula **3**.

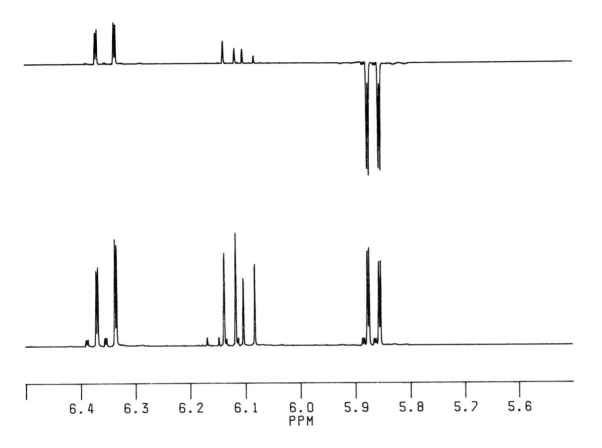

In applying equation 5.17 to data of this kind, care must be taken to choose the best distance ratio to calculate. Obviously there is only one geometrical relationship here, but it could be expressed in various ways. For instance, we might try to calculate the *cis* to *trans* distance ratio, or the ratio of either of these distances to the geminal distance (r_{AM} in **3**). These alternatives correspond with permuting the labels A M and X round the three protons before applying equation 5.17. It can readily be appreciated, by examining the equation and the typical data we have here, that the most accurate result is obtained by choosing the labelling so as to make the AM and AX interactions large in comparison with that between M and X. Alternative arrangements of the labels, such as in **4**, make the distance ratio depend on very small differences between inaccurate numbers, and therefore introduce large errors.

So on this basis the best distance ratio to calculate is that between *cis* and *gem* protons, which from bond lengths is expected to be 1·30 : 1. The result of inserting the values from Table 5.1 into equation 5.17 is 1·32 : 1, which is entirely satisfactory. To illustrate the importance of selecting the right ratio to calculate, try repeating the arithmetic with the labels permuted as in **4**. This amounts to comparing the *cis* and *trans* distances, for which the predicted ratio is 0·79 : 1. You should find that the result in this

Figure 5.21 One of three nOe experiments on acrylic acid.

case is much further from the true value. The problem here is that the M-X interaction dominates the experiment, but it is the difference between the A-M and A-X interactions that contains the information.

Table 5.1   NOE enhancements for acrylic acid (see formula **3** for labelling).

Irradiate	A	M	X
η_A	-	0·340	0·068
η_M	0·360	-	−0·015
η_X	0·140	−0·035	-

5.4.3   A Real Problem

In extending the use of nOe measurements to more interesting molecules than the contrived examples of the previous two sections, we have some difficulties to contend with. More than likely, the numbers of interacting protons will be such that the quantitative calculation of distance ratios becomes impractical. In a larger molecule the assumption of equal correlation times for all internuclear vectors, on which such a calculation rests, is less likely to be valid anyway, so we should not be too disappointed at this. In order to make progress in identifying the structure of a complex molecule, we have to partially abandon the 'Two General Principles' of section 5.2.4. Thus we *will* take the observation of an nOe to indicate the relative proximity of the nuclei involved, *but* we will always be on the alert for exceptional circumstances that make our conclusions questionable. With this in mind, let's examinine a realistic problem.

Compound **6** was isolated from the incubation of tripeptide **5** with the enzyme isopenicillin N synthetase[11] (IPNS, Scheme 1). The gross structure was rapidly established by simple inspection of the proton spectrum, and confirmed by a proton-proton shift correlation experiment (Chapter 8). The problem was then to establish the relative stereochemistry, and particularly that of the hydroxyl group on the seven-membered ring, which had some bearing on the enzyme mechanism. In more familiar ring systems questions of this kind can often be approached through the determination of proton-proton coupling constants, but for a system like this no precedent was available, and clearly it would be hard to guess the conformation of the large ring. Still, we could at least determine from their coupling that the two protons on the four-membered ring were *cis,* as indicated in structure **6**, and these protons provided the reference point from which to start the nOe experiments.

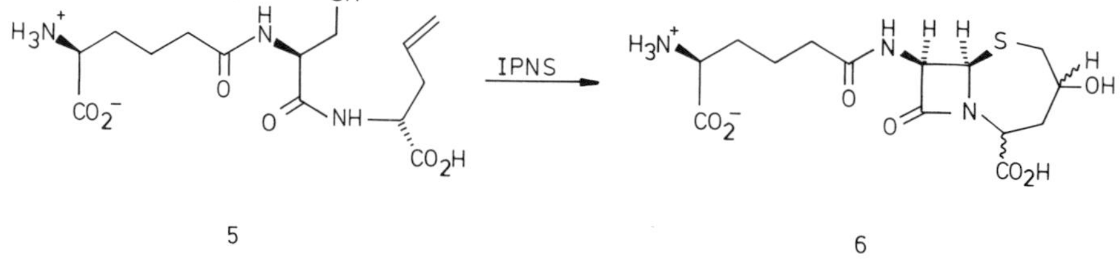

5   6

Scheme 1

Before presenting the relevant nOe data, from which the derivation of the structure may seem rather obvious, I would like to point out some of the practical problems encountered in making the measurements. The enzymatic origin of this compound, and its fairly low stability, meant that only small amounts could be made available at any one time (never more than 1 mg). So even on a 500 MHz spectrometer the experiments were at the limit of sensitivity, as can easily be seen from the selected difference spectra presented in Figure 5.22. This meant in turn that the acquisition of all the nOe's required several long spectrometer sessions of perhaps 15 hours each, so that the evidence for the structure built up quite slowly over a period of days. This is an aspect of spectroscopic work which it is hard to convey in a textbook, where the presentation of a complete set of data and conclusions drawn from it can make the subject seem dry and analytical; in reality there is a genuine sense of struggle and achievement associated with experiments of this kind. Another consequence of the low availability of the material is that the enhancements quoted below must be treated as very approximate; it is reasonable to consider a difference of a factor of two or more as being significant, but the low signal-to-noise ratio makes lesser differences insignificant.

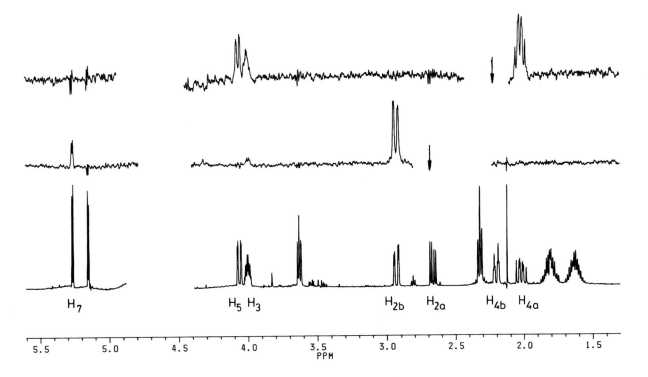

Figure 5.22 NOe experiments on compound 6.

Scheme 2 presents the results of a series of nOe measurements on compound **6**. Look initially at the experiments on protons H_{2a} and H_{2b} (at the left of the scheme). Our principal reference point is H_7, which we assume to have the stereochemistry shown, and which is known to be *cis* to the other β-lactam proton. The 5% nOe at H_7 upon irradiation of H_{2a}, combined with the absence of any enhancement here from H_{2b}, indicates that H_{2a} is also on the lower face of the molecule. (In addition, we had extensive experience of similar β-lactam containing systems, which made us quite confident of this interpretation). H_{2b} is also clearly on the same face of the ring as H_5 (3% enhancement), which fixes the relative stereochemistry at C_7 and C_5. Unfortunately the results with regard to the C_3 stereochemistry are not so clear; although H_{2b} does seem to enhance H_3 more than does H_{2a}, the accuracy of the measurements (and the complexity of the spin system) are such as to make reliance on this single observation hazardous.

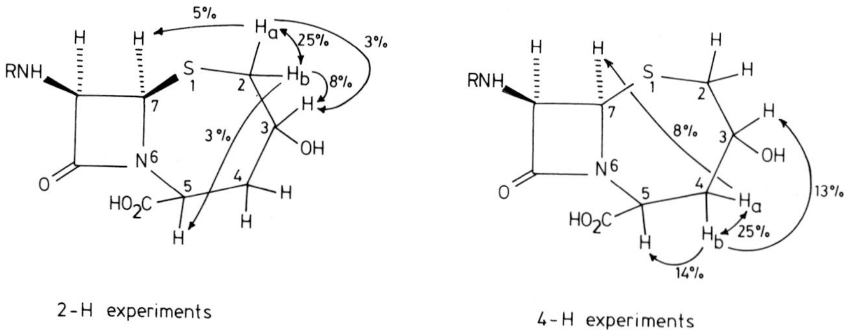

2-H experiments   4-H experiments

Scheme 2

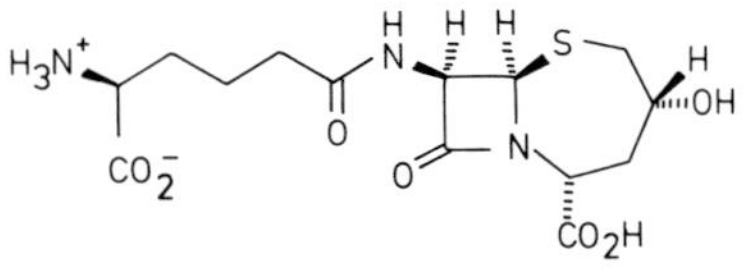

7

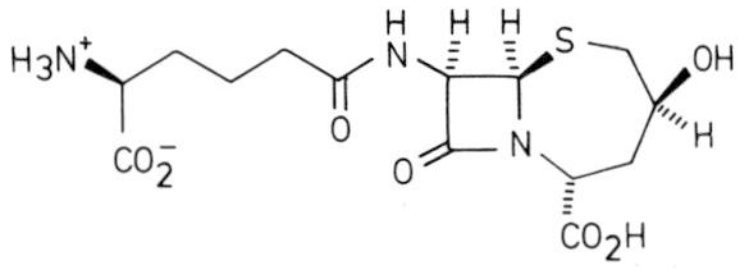

8

It happened that the experiments I have just described were performed first, leaving us initially with the problem only partly solved. Some time later a second set of measurements was made, irradiating protons H_{4a} amd H_{4b}, and these clarified the structure completely. First, large enhancements at H_5 and H_3 upon irradiation of H_{4b} associate these three protons, and since H_5 has already been related to H_{2b}, this indicates that the hydroxyl group is on the lower face of the molecule. The assignments at C_4 were further confirmed by the interaction between H_{4a} and H_7, which also indicated the conformation of the ring. The stereochemistry has thus been determined to be 7. This was later confirmed by the isolation of the isomer 8, which showed the expected reversal of the interactions between the C_2 protons and H_3.

One important conclusion to be drawn from this example is that it is nearly always worth making as many nOe measurements as possible. Even if it seems that some irradiations will be a waste of time, there is usually no extra effort involved in making them, and it is always rash to assume that the outcome is predictable. For instance, the interaction between H_{4a} and H_7 in 6 was a surprise, and contained useful information. The more confirmatory evidence that can be obtained for a hypothetical structure the better, because the scope for misinterpretation of nOe data is greater than in many other NMR experiments.

5.4.4 Heteronuclear NOe's

So far I have concentrated entirely on the homonuclear nOe, mentioning the heteronuclear form only in passing as a source of sensitivity enhancement for low γ spin-$\frac{1}{2}$ nuclei. In a *selective* experiment, the nOe between a proton and a heteronucleus can also be informative, provided certain technical difficulties can be overcome. Take the example of ^{13}C. The relaxation of protonated carbons is dominated entirely by their directly bound protons, so in this case the nOe is not very useful, containing information more readily obtained by decoupling or heteronuclear shift correlation. Interestingly, it is often possible to avoid generating this direct nOe, because it arises from the saturation of the ^{13}C *satellites* of a resonance. As proton-carbon one-bond coupling constants are quite large, selective irradiation of the central ^{12}C line without perturbation of the satellites is usually practical. In this case we expect to see only nOe's to quaternary carbons adjacent to the protonated carbon whose proton has been saturated, so the experiment may yield assignments or stereochemical relationships of an unusual kind.

The major problem with experiments of this type is sensitivity. Certainly it will be necessary to acquire the spectra with broadband proton decoupling, and so sufficient time must be left between scans to allow any generalised nOe this generates to die down (while maintaining the interesting nOe by selective irradiation). This means a delay of several times the T_1's of quaternary carbons, which are frequently greater than 10 s. However, with sufficient sample, and long accumulation times, very useful results can be obtained. Figure 5.23 illustrates the principle for a *t*-butyl group. The lower spectrum is a fully relaxed carbon spectrum obtained without nOe. The methyl resonance, to high field, is three times the size of that of the quaternary carbon, as expected. The difference spectrum (upper trace) obtained with selective saturation of the methyl groups's proton resonance (^{12}C line only) shows an 80% enhancement of the quaternary carbon, but no effect at the methyl. Because the maximum proton-carbon nOe is 200%, large enhancements of this kind are quite common.

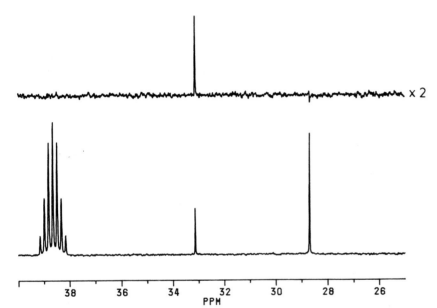

Figure 5.23 Selective heteronuclear nOe on a compound containing a *t*-butyl group. The multiplet at low field is due to the solvent.

Experiments like this have not been widely used, no doubt because of the limitations of sensitivity. However, sporadic examples do appear in the literature (see references 12-15), and as the information available in this way is difficult to obtain by other means the experiment is likely to be increasingly common in the future.

REFERENCES

1. A. Abragam, *Principles of Nuclear Magnetism*, Oxford University Press, 1961.
2. J. H. Noggle and R. E. Schirmer, *The Nuclear Overhauser Effect - Chemical Applications*, Academic Press, 1972.
3. D. Shaw, *Fourier Transform NMR Spectroscopy*, (2nd edn.), Elsevier, 1984.
4. A. A. Bothner-By, R. L. Stephens, J. Lee, C. D. Warren and R. W. Jeanloz, *J. Amer. Chem. Soc.*, **106**, 811-813, (1984); A. Bax and D. G. Davis, *J. Mag. Res.*, **63**, 207-213 (1985).
5. R. A. Bell and J. K. Saunders, *Can. J. Chem.*, **48**, 1114-1122, (1970).
6. J. D. Mersh and J. K. M. Sanders, *Prog. NMR Spec.*, **15**, 353-400, (1982).
7. For an example of the application of transient nOe's to a structural problem, see: F. Heatley, L. Akhter and R. T. Brown, *J. Chem. Soc. Perkin II*, 919-924, (1980).
8. R. L. Vold and R. R. Vold, *J. Mag. Res.*, **55**, 78-87, (1983).
9. A. J. Shaka, C. J. Bauer and R. Freeman, *J. Mag. Res.*, **60**, 479-485, (1984).
10. M. Kinns and J. K. M. Saunders, *J. Mag. Res.*, **56**, 518-520, (1984).
11. J. E. Baldwin, R. M. Adlington, A. E. Derome, H-H. Ting and N. J. Turner, *J. Chem. Soc. Chem. Commun.*, 1211-1214, (1984).
12. K. E. Kövér and G. Batta, *J. Chem. Soc. Chem. Commun.*, 647-648, (1986).
13. K. E. Kövér and G. Batta, *J. Amer. Chem. Soc.*, **107**, 5829-5830, (1985); and references therein.
14. M. F. Aldersley, F. M. Dean and B. E. Mann, *J. Chem. Soc. Chem. Commun.*, 107-108, (1983).
15. F. J. Leeper and J. Staunton, *J. Chem. Soc. Chem. Commun.*, 911-912, (1982).

Indranil & Akiko Roy
137 Friendship Court
White Plains
NY 10603
Tel: 914- 328-9822 (H)
(Cell) 914-473-02224
(Iswan- January 2004)

Indranil & Akiko Roy
137 Friendship Court
White Plains
NY 10603
Tel: 914- 328-9822 (H)
(Cell) 914-473-02224
(Iswan- January 2004)

Indranil Roy
137 Friendship Court
White Plains
NY 10603
Tel: 914- 328-9822 (H)
(Cell) 914-473-02224
(Iswan- January 2004)

Tel: 914- 328-9822 (H)
(Cell) 914-473-02224
(Iswan- January 2004)

Sanjoy Dasgupta
5217 E. Kathleen Road
Scottsdale; Az85254
Tel: 602-788-9010
(O) 602-728-3720

6

Polarisation Transfer and Spectrum Editing

6.1 INTRODUCTION

Sensitivity is the problem with NMR. From most other points of view it is the ideal spectroscopic technique for chemical applications: it has high resolving power, and the principal NMR parameters (chemical shifts and coupling constants) relate in a fairly direct and intuitive way with chemists' ideas about the electronic structure and topology of compounds. However, NMR spectroscopists too often find themselves beginning sentences with phrases like 'Well, if only you had another 50 mg ⋯', because the nuclear transitions are of such low energy in practical magnetic fields that they are not far from being saturated at room temperature. There are only tiny population differences to get hold of when performing NMR experiments, and it is only because it is a *resonant* phenomenon that we can deal with it at all. This lack of sensitivity is a great pity, because it is not too far from the truth to state that nowadays, given an *unlimited* quantity of any material with molecular weight less than, say, 3000, a combination of NMR techniques can be found which will eventually define its structure. Unfortunately interesting materials are seldom available in unlimited quantity.

Any method which has as its aim the improvement of sensitivity, then, is likely to be of prime importance for NMR. This, of course, is why chemists are happy to pay ten times as much for magnets today as they did twenty years ago; turning up the static field is a very direct means of increasing the size of NMR signals. This process must have some limit though; the technology of *persistent* (as opposed to powered or rapidly drifting) super-conducting magnets is currently bogged down in the transition from 500 to 600 MHz proton frequency, and although this problem will no doubt be solved and further progress made, one can hardly foresee improvements in field strength by orders of magnitude. Even if this were achieved, other factors, such as the rapidly approaching stage where normal molecular motions become slow on the NMR timescale, will eventually limit the usefulness of further increasing the static field. Other technical innovations, such as the use of cryogenic probes[1] will also contribute to increased sensitivity in the future, but again there must be some limit to how much improvement can be achieved in this direction.

A practical measure of the effectiveness of NMR from a sensitivity point of view might be whether it can detect the amounts of material which chemists routinely manipulate. At present this means quantities around the 1 mg level for preparations performed using conventional chromatography, and around a few μg for preparations using HPLC. With quantities as small as this *chemists* generally start to feel that, since the compound is invisible and difficult to handle, it probably counts as not being there. *Biologists,* on the other hand, can often follow substances by their biological activity and may be happier with μg amounts. Suppose we take a compromise level of 100 μg as being the smallest quantity of substance one would wish to work with. What kind of NMR can we do with that, using present day instruments, and not necessarily at the highest field imaginable?.

The answer, rather depressingly, is not very much. We can run proton spectra (and spectra of fluorine and the exotic nucleus tritium). We might be able to do some nOe experiments, and we could certainly run a proton COSY spectrum (Chapter 8), though it would take a long time to acquire. If we were interested in *any* other nucleus, however, we would be in trouble; on a very high field instrument we could probably observe ^{31}P, and with microcells and a weekend-long acquisition perhaps we could get a broadband proton-decoupled ^{13}C spectrum. If we were inorganic chemists, and wanted to look at some very low frequency metal, there would be no hope at all without *hundreds* of times more compound. For all practical purposes, the proton is the only nucleus that can be used for NMR experiments without the requirements of NMR forcing us to work with larger quantities of material than we might otherwise need.

A method which partly transferred the favourable properties of the proton to another nucleus would therefore be of enormous value. In this chapter I start by establishing the principle involved in achieving this, by looking at a very old experiment (SPI). The other two parts of the chapter then discuss two more general polarisation transfer experiments: INEPT and DEPT. The first of these is examined both because it was the original method which stimulated rapid growth in this field, and because it is more amenable to explanation in our pictorial way. DEPT is more likely to be the experiment of choice for actual problems, though, and this is covered mainly from that perspective; understanding its mechanism in detail requires an appreciation of multiple quantum coherences, which I will not try to develop for you until Chapter 8.

6.2 SELECTIVE POPULATION TRANSFER

Perhaps rather surprisingly, a method with the desirable property of imparting something of the character of the proton to other nuclei has been available for many years. This is known as *selective population transfer*, or more exactly in the form which I am going to describe, selective population *inversion* (SPI). To understand how this works we need to think in some detail about the energies and populations of a nuclear system; I will take a two-spin AX system, which we can think of as arising between a proton and a ^{13}C nucleus to which it is attached, although it could be any pair of spin-$\frac{1}{2}$ nuclei. The energy level diagram is shown in Figure 6.1, together with a representation of the spectrum we would obtain from each nucleus.

There are a number of important details in this figure, which I would like to take you through rather carefully. I have made some attempt at representing the relative energy differences of the proton and carbon transitions to scale; protons have about four times the Larmor frequency of ^{13}C, so the transitions are four times as energetic. Although the nuclei are coupled, the energies involved in this are so tiny compared with the total transition energy that the pairs of transitions (H_1,H_2) and (C_1,C_2) appear equal on this scale. We are going to be most interested in the population difference across each transition. At thermal equilibrium, which is where we start the experiment, the Boltzmann law indicates that the proton transitions will show four times as much population difference as those of carbon. For convenience the *differences* are represented $2\Delta H$ and $2\Delta C$, so $\Delta H = 4\Delta C$. The *actual* population of each level, and hence the values of these numbers depends, of course, on the total number N of nuclei present. It is convenient for our purposes to concentrate on just the *deviation* of each level from having a population $N/4$ (this is the same approach we took for a two-level system in Chapter 4 section 4.2.6, and for the homonuclear two-spin system of Chapter 5, section 5.2.2); this is also represented against each energy level in the diagram.

Now we can say something about the relative intensities of ^{13}C and ^{1}H signals obtainable from this system. There are four times as many protons

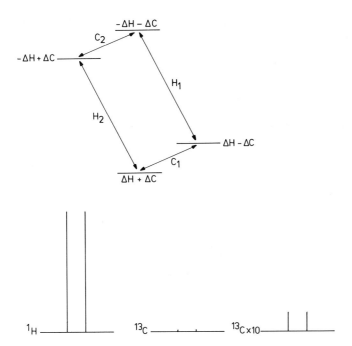

Figure 6.1 Energy levels and populations of a heteronuclear AX system.

in the excess population of the lower energy state as carbon-13 nuclei, and the magnetic moment produced by these protons is four times as big as that due to carbon (because it is proportional to γ), so when we apply a $\pi/2$ pulse to protons we get a transverse magnetisation *sixteen* times that we would obtain from ^{13}C. The signal induced in our receiver coil by this magnetisation when it precesses is further proportional to the *rate* of precession (four times as fast for protons), so in total the signal obtained from protons is 64 times that from carbon (in general, the signal available from a nucleus with gyromagnetic ratio γ is proportional to γ^3). In real life we also have the low natural abundance of ^{13}C contributing another factor of about 100, but this is irrelevant to the present discussion; we can imagine we are working with ^{13}C labelled material.

This γ^3 dependence of the available NMR signal is the reason that only proton observation is relatively favourable. The next nucleus down which is also of widespread interest (i.e. excluding ^{19}F, ^3H, ^3He and two isotopes of thallium) is phosphorus, with γ about 2.5 times less than that of ^1H, so we immediately drop a factor of almost 16 in sensitivity (which is a factor of 256 in experiment time for equal signal-to-noise ratio, assuming equal relaxation rates). By the time you get down to, say, rhodium (^{103}Rh) γ is only 3% that of protons, and sensitivity is *32 thousand* times less, even though this is a 100% abundant spin-$\frac{1}{2}$ nucleus.

The selective population inversion experiment provides a means of recovering one of these γ factors, in a fashion which is quite easy to understand. Since this is a first order coupled system, the lines we observe in the spectrum correspond directly with transitions of a single nucleus. Imagine that *one* only of the proton lines is subject to a selective π pulse; that is, the corresponding transition has its populations *inverted*. Experimentally this could be achieved using the proton decoupler with a suitably low field strength, for instance a 20 ms pulse with a 25 Hz field.

The situation that prevails immmediately after this selective inversion is illustrated in Figure 6.2; all that has happened is that the populations at the top and bottom of transition H_1 have been exchanged. Look what has happened to the population *differences* though. Across the two proton transitions they are the same as before, except that now one is negative. Across the carbon transition C_1, however, which previously had population difference $2\Delta C$, we now find $(\Delta H + \Delta C) - (-\Delta H - \Delta C)$, or $2\Delta H + 2\Delta C$. Across transition C_2 likewise we find $(-\Delta H + \Delta C) - (\Delta H$

$-\Delta C$), which is $-2\Delta H + 2\Delta C$. We have *transferred* the proton population differences onto the carbon transitions, and added them to the existing differences!. Since $\Delta H = 4\Delta C$, this means that if we sample the carbon magnetisation with a $\pi/2$ pulse immediately after this situation has been created, we get doublet components with relative intensities $+5$, -3 (as compared with direct observation without the selective inversion). Since one transition has an inverted contribution from the protons while the other does not, we say that the *net* magnetisation transfer is zero, but there is a *differential* transfer of polarisation. A final point to note is that there is no essential requirement for transitions to be inverted; any *unequal* perturbation of the populations will do; this is the origin of the SPT effects discussed in section 5.3.3 of Chapter 5. An experimental demonstration of SPI appears in Figure 6.3.

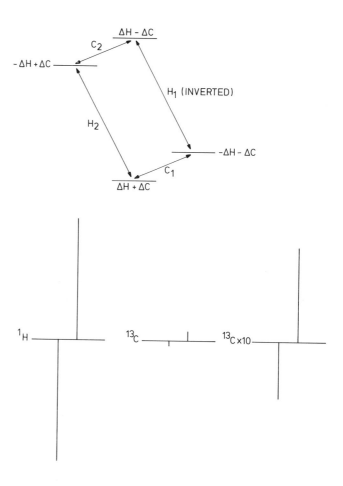

Figure 6.2 The new populations after inverting one of the H transitions.

There are two problems with this experiment. First, it is *selective,* so we would have to work through a spectrum inverting the ^{13}C satellites of proton lines as we came to them. Also, in this form it cannot be used to generate a proton-decoupled spectrum with sensitivity enhancement, because the antiphase lines will partly cancel when decoupling is turned on, restoring the normal signal intensity. These reasons mean that it is not very important in its own right, but it points the way forward to a whole class of more general polarisation transfer methods that have been developed since about 1980. The experiments discussed in the rest of this chapter share in common the transfer of polarisation from one nucleus to another, with a concomitant *change* in intensity of the detected signal by a factor γ_1/γ_2.

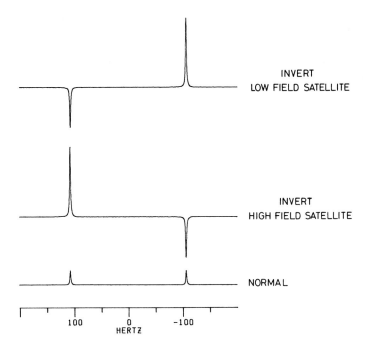

INVERT
LOW FIELD SATELLITE

INVERT
HIGH FIELD SATELLITE

NORMAL

100   0   -100
HERTZ

Figure 6.3 The SPI experiment performed on chloroform.

6.3 INEPT

6.3.1 Introduction

The basic idea

The major limitation of SPI is its lack of generality. If we could find some way to set up the appropriate polarisation of *all* proton transitions regardless of frequency, we would evidently have a very promising experiment. We need a sequence which puts pairs of proton transitions into antiphase, as in SPI, but using non-selective pulses and in a manner independent of chemical shifts. This can be done using a spin echo to prepare multiplet components along selected axes:

S: $\left(\dfrac{\pi}{2}\right)_x - \tau - \pi_x - \tau - \left(\dfrac{\pi}{2}\right)_y$

I: $\qquad\qquad \pi_x \qquad \left(\dfrac{\pi}{2}\right)_x \; Acquire \cdots$

In this notation, which was used in the first paper on INEPT[2] (Insensitive Nuclei Enhanced by Polarisation Transfer), S stands for the sensitive nucleus and I for the insensitive one (^{1}H and ^{13}C respectively in our example). Irritatingly this convention is not always followed, and it is common to see the source nucleus labelled I and the destination S in many of the huge number of publications that have been stimulated by the original idea. It is only an arbitrary convention, of course, but it would be nice if a little consistency could be achieved. Throughout this chapter, S will be used to label the source nucleus, and I that to which magnetisation is transferred (irrespective of whether S is in fact more 'sensitive' than I).

We can follow the effect of this sequence easily for an AX system with coupling constant J (Figure 6.4). The delay τ is set to $1/4J$, so doublet components (which precess in the rotating frame at $\pm J/2$ Hz) move through 1/8'th of a cycle up to the π pulses. The S spin π pulse rotates these components into the second half of the *x-y* plane, while that on the I spins ensures that the coupling is not refocused by reversing the sense of their precession. During the second τ interval chemical shifts and inhomogeneity *are* refocused of course, so they have been omitted from consideration. By the end of the second interval, the two components of the S spin doublet are aligned along the $\pm x$ axes. The final two $\pi/2$ pulses are applied in practice to both the S and I spins simultaneously, but it is easier to understand what happens by imagining that the S spin pulse occurs first. This pulse is $(\pi/2)_y$, so it rotates the components presently aligned along the $\pm x$ axis into the z direction, one up and one down. This corresponds with the desired antiphase polarisation of the doublet, so the $\pi/2$ pulse on I elicits an enhanced response as in SPI (Figure 6.5).

Figure 6.4 INEPT sequence for an AX system.

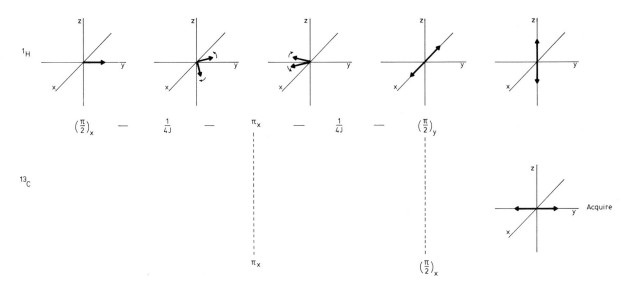

Figure 6.5 INEPT does for many doublets what SPI did for one.

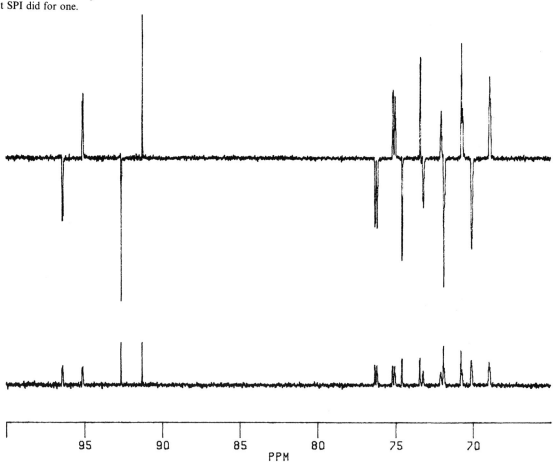

The natural I spin magnetisation

The result of the SPI sequence, and of INEPT as described so far, is to combine the transferred population differences with those which already exist across the I transitions. This causes the detected positive and negative signals to have unequal intensities, such as $+5$ and -3 for a ^{13}C doublet. In many applications of polarisation transfer this contribution from the natural magnetisation is unwanted, and steps must be taken to eliminate it.

There are various ways in which this might be achieved. Since the natural I spin polarisation is not required at any stage of the experiment, it could be eliminated before the start of the sequence by presaturation. This would require broadband irradiation (not commonly available for nuclei other than ^1H), application of one or several $\pi/2$ pulses to I, or application of a $\pi/2$ pulse followed by a B_0 field gradient (a *homospoil pulse*). In section 6.4.6 we will see a polarisation transfer experiment in which some kind of presaturation is definitely required, but in the common case of transfer from protons to a low frequency nucleus it is usually unnecessary. Instead, since the natural I magnetisation is small in comparison with the polarisation transfer component, it suffices to cancel it out using a phase cycle.

The phase which must be changed is that of the final S (i.e. usually proton) $\pi/2$ pulse. If this is $(\pi/2)_{-y}$ instead of $(\pi/2)_y$, the sense of the proton polarisation is reversed; that is, the multiplet components which had previously arrived on the $+z$ axis now find themselves along $-z$, and vice-versa (Figure 6.6). This in turn causes the phase of the polarisation transfer part of the I spin signal to be inverted, so the receiver phase must also be inverted to follow it. The natural I spin polarisation, meanwhile, gives rise to a signal whose phase is constant (and equal to that of the I spin $\pi/2$ pulse), so the phase alternation of the receiver eliminates it. Typically subtraction in this way attenuates signals several hundredfold, which is more than adequate when transfer is from protons to an X nucleus (Figure 6.7).

$\ulcorner$ ^1H STATE BEFORE LAST PULSE $\urcorner$ $\qquad\qquad$ $\ulcorner$ ^{13}C RESULT $\urcorner$

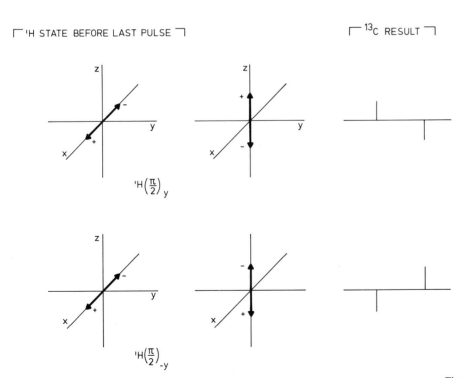

Figure 6.6 Eliminating the natural I spin magnetisation by phase alternation of the last S pulse.

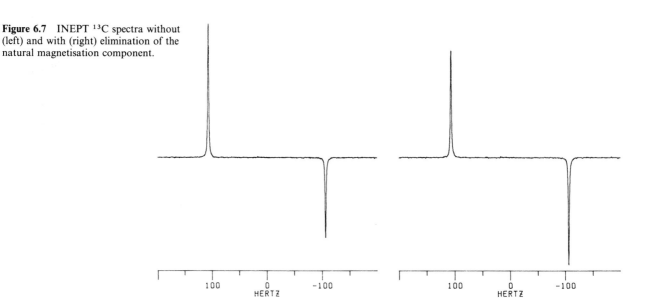

Refocused INEPT

A basic INEPT spectrum cannot be acquired with proton decoupling, because of the antiphase disposition of the components of multiplets. However this problem is readily circumvented by waiting some time Δ before starting acquisition. Suitable choice of Δ allows the I spin multiplet components to get back into phase, for instance doublets require $\Delta = 1/2J$ (Figure 6.8). To stop chemical shifts interfering with this process we also have to add another spin echo, by including π pulses in the middle of the waiting period:

$$S: \qquad \left(\frac{\pi}{2}\right)_x - \frac{1}{4J} - \pi_x - \frac{1}{4J} - \left(\frac{\pi}{2}\right)_{\pm y} - \frac{\Delta}{2} - \pi - \frac{\Delta}{2} - Decouple \cdots$$

$$I: \qquad\qquad\qquad \pi_x \qquad\quad \left(\frac{\pi}{2}\right)_x \qquad\quad \pi \qquad\quad Acquire\ (\pm x) \cdots$$

This is known as *refocused* INEPT. Varying the value of Δ gives this sequence some very interesting properties (section 6.3.3), but for the time being we will take it to be fixed to allow us to decouple INEPT spectra (Figure 6.9). While $1/2J$ is optimum for doublets, other multiplets require different Δ values for maximum enhancement (see Table 6.2 in section 6.3.3), so a compromise must be made if different multiplicities are present together. For carbon spectra, which may only contain doublets, triplets and quartets, $0·3/J_{CH}$ is a reasonable value for Δ. J_{CH} itself also varies, of course, so there is considerable scope for this delay being non-ideal.

6.3.2 Characteristics of INEPT Spectra

Effect on sensitivity

The fundamental property of INEPT is to enhance the sensitivity of the I nucleus by a factor proportional to γ_S/γ_I, i.e. to reduce the dependence of signal strength on γ to a function of γ^2. This is not necessarily quite as dramatic as it first seems because, in the common case of applying INEPT between protons and a low frequency nucleus, the proper comparison is not with the I nucleus signal obtained at thermal equilibrium, but with that obtained in the presence of proton decoupling. Broadband proton decoupling will generate a nuclear Overhauser enhancement, which also depends

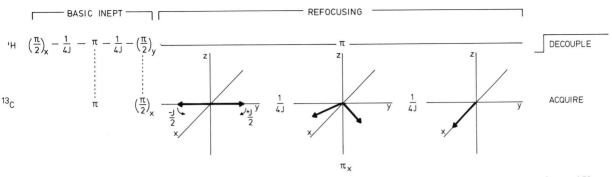

Figure 6.8 Refocusing for an AX system; this allows the use of broadband decoupling during acquisition.

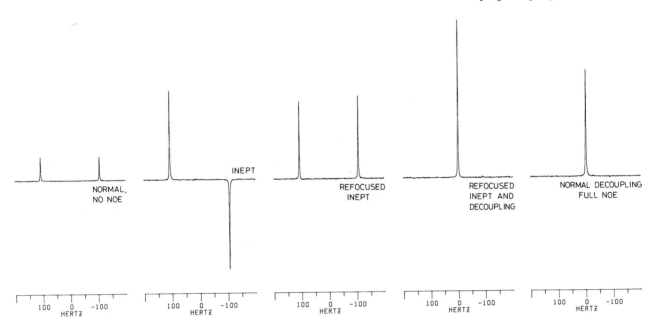

Figure 6.9 Permutations on the spectrum of chloroform. It can be seen that the refocused INEPT spectrum shows some loss of intensity as compared with simple INEPT; this is partly due to transverse relaxation during the refocusing period, and partly to imperfections in pulses.

on the ratio of gyromagnetic ratios, so the true comparison we should make is:

$$\text{INEPT}^{\dagger}: \qquad I = I_0 \left| \frac{\gamma_S}{\gamma_I} \right|$$

$$\text{NOE}: \qquad I = I_0 \left(1 + \frac{\gamma_S}{2\gamma_I} \right)$$

† This is the enhancement obtained when transfer is from one S nucleus only. With more nuclei the enhancement factor is slightly different - see Table 6.2

Here I represents the enhanced signal intensity and I_0 that without enhancement (the amount of nOe is commonly quoted as the *enhancement η* rather than the total *intensity,* as we saw in the previous chapter, but throughout this section I refer to the latter figure unless I specifically say 'enhancement'). An important feature here is that the INEPT intensity depends on the absolute value of the ratio, while the nOe also involves its *sign*. Therefore when the nOe arises between protons and a nucleus with negative γ it may cause a *reduction* rather than an increase in intensity; if the value of η happens to be close to -1 then it is possible to get complete loss of signal (Chapter 5, section 5.2.4). I will return to this point shortly.

In comparing a spectrum acquired via INEPT with one observed directly but with full nOe, we also have to take into account the relaxation properties of the nuclei involved. For the experiment with nOe the T_1 of interest is that of the observed nucleus. For the INEPT experiment, however, the only population difference that contributes to the final signal is that of nucleus S, typically protons, so it is the S T_1 that determines the repetition rate. This makes the repetition rate of INEPT experiments independent of the T_1 of the I nucleus, which in practice is often more important than the actual enhancement of signal by polarisation transfer.

Now we can apply these ideas to some actual cases. Table 6.1 compares theoretical INEPT intensities (relative to that obtained at equilibrium and without nOe) for a few nuclei with the maximum signal with full nOe. The overall conclusion that can be drawn is that, allowing that some of the theoretical INEPT enhancement may in practice be frittered away by experimental inadequacies, the contest is a close one *provided* the maximum nOe is actually obtained. This depends, of course, on the extent to which the S nucleus contributes by dipolar interaction to the relaxation of nucleus I. According to circumstances there may be little or no nOe, or for nuclei with negative γ the nOe (enhancement) may happen to be -1, leading to zero signal intensity. In these cases INEPT has a clear advantage, even without any assistance due to repetition rate acceleration.

For ^{13}C, however, it is normal for protonated carbons to show full nOe, and since their T_1's are also not usually much longer than those of their attached protons there is likely to be little advantage in using refocused INEPT to acquire decoupled spectra. For fully coupled experiments, where the lack of decoupling naturally means there is no nOe, it can be very helpful to use polarisation transfer; this is dicussed later.

Table 6.1   A comparison of signal strength available by direct observation in the presence of the full nOe from protons, against that resulting from polarisation transfer from protons to the heteronucleus. The figures are *intensities* relative to direct observation of the nucleus without nOe.

Nucleus	Maximum nOe	Polarisation Transfer
^{31}P	2·24	2·47
^{13}C	2·99	3·98
^{29}Si	$-1·52$	5·03
^{15}N	$-3·94$	9·87
^{57}Fe	16·48	30·95
^{103}Rh	$-14·89$	31·78

The really major advantage arises for those nuclei with very long T_1's, and especially those with negative γ. Common nuclei which suffer from both these deficiencies are ^{29}Si and ^{15}N. Let us consider a hypothetical ^{15}N containing compound with a proton coupled to the nitrogen. Even if we assume full nOe (relative signal intensity about -4), the INEPT experiment will be more than twice as good (relative intensity about 10). On top of this, the nitrogen T_1 could easily be 32 s, while that of the proton might be 2 s. This means we can acquire 16 times as many scans with INEPT in a given time as we could with direct observation, giving a further fourfold increase in signal-to-noise ratio. Overall we could easily gain a factor of ten in sensitivity using INEPT for this experiment, for quite realistic ^{15}N NMR parameters (Figure 6.10; see also Figure 6.14 for an example from ^{29}Si NMR). The combination of these two factors opens up a whole new prospect for the detection of low-frequency, slow relaxing nuclei.

Another area in which this sensitivity enhancement opens up new possibilities is the spectroscopy of transition metals. For many complexes it might be expected that useful information could be extracted from the metal spectrum, but even abundant spin-$\frac{1}{2}$ nuclei such as ^{103}Rh and ^{109}Ag

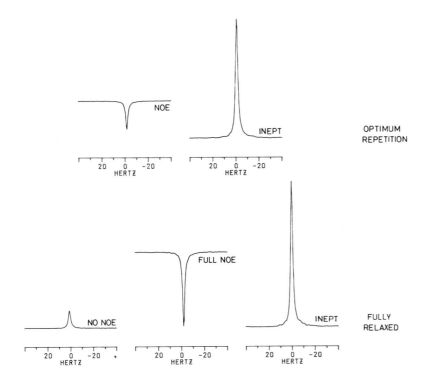

Figure 6.10 Various ways of acquiring ^{15}N spectra. The lower examples are fully relaxed (equal numbers of scans), while for the upper traces the repetition rate was optimised for the appropriate (^{15}N or ^{1}H) T_1 (equal total accumulation time). This compound (phthalimide) had a proton directly bonded to nitrogen, so the ^{15}N T_1 was relatively short (about double that of its attached proton). Even greater intensity differences can be expected for nitrogens not directly attached, but still coupled, to protons.

are so insensitive that they are extremely difficult to measure. For polarisation transfer from protons we need a reasonably large proton-metal coupling constant, which will not necessarily be available. However, what almost always *is* available is coupling to ^{31}P, because of the ubiquity of phosphine ligands. The ratio γ_P/γ_{metal} is usually large enough to make polarisation transfer in this context highly advantageous. Figure 6.11 compares direct observation of rhodium with polarisation transfer from phosphorus to rhodium via INEPT; clearly what was almost impossible has been made straightforward. Unfortunately it is still rare for commercial spectrometers to be equipped to perform this experiment (a probe tuned to two X frequencies and two broadband transmitter channels are required), but no doubt this situation will change in due course.

Effect on coupled spectra

When INEPT is used without refocusing and decoupling, so the spectra acquired retain multiplet structure, the multiplets that arise have characteristic patterns quite different from normal. We have already seen that *doublets* occur as antiphase pairs of lines. It is further possible to demonstrate that triplets lack their centre line and show as a $-1:0:1$ pattern, while quartets give rise to a $-1:-1:1:1$ structure (when you come to read Chapter 8, you might notice that these intensities follow the same 'antiphase triangle' rule described there for cross peaks between groups of equivalent spins in the COSY experiment). These patterns could be regarded as a helpful characteristic feature of the experiment (Figure 6.12), or as an annoying distraction according to taste. The DEPT experiment described in section 6.4 restores 'normal' multiplet patterns in coupled polarisation transfer spectra.

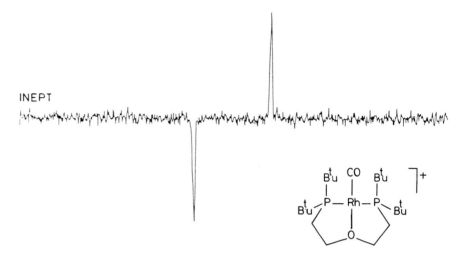

INEPT

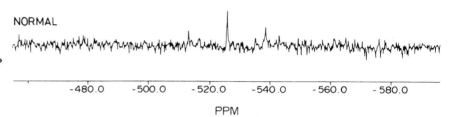

NORMAL

Figure 6.11 INEPT enhancement of ^{103}Rh by polarisation transfer from ^{31}P (courtesy of Dr. C. Brevard, Bruker Spectrospin, and Dr. D. H. M. W. Thewissen *et al.*, ITC-TNO, Utrecht, The Netherlands).

Figure 6.12 Characteristic coupled INEPT patterns.

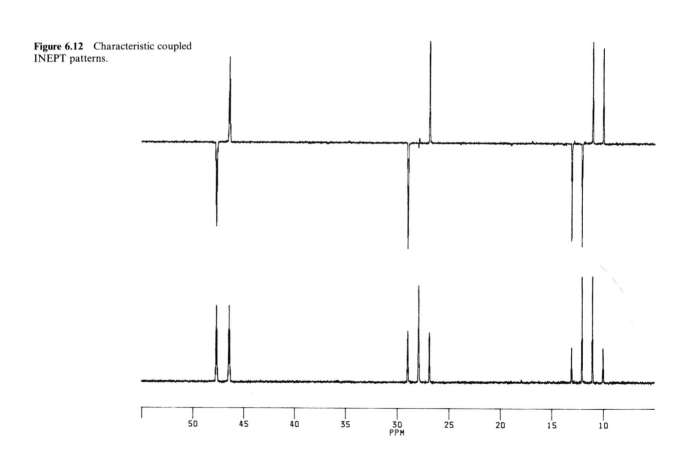

6.3.3 The Δ Delay

Choice for optimum sensitivity

A more careful consideration of the effect of the delay Δ on refocused INEPT proves to be rather illuminating. Consider first the requirements for optimum signal intensities in the decoupled spectrum. It is fairly obvious that selection of $\Delta = 1/2J$ is appropriate for a doublet, as we already saw in section 6.3.1. However, it is equally obvious that for a triplet this Δ value is completely wrong (Figure 6.13); at the end of the delay, triplet components are in antiphase and will cancel when the decoupler is switched on. In fact, this proves to be the case for all other multiplicities, so that the spectrum acquired using $\Delta = 1/2J$ *only* contains signals due to IS systems. With a little algebra[3], it is possible to demonstrate that the best Δ value for acquiring a refocused, decoupled INEPT spectrum of an IS_n system is given by:

$$\Delta = \frac{1}{\pi J} \sin^{-1}\left(\frac{1}{\sqrt{n}}\right) \qquad (6.1)$$

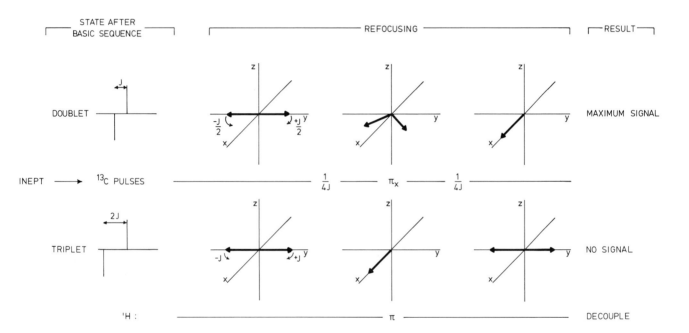

Some values of Δ are collected in Table 6.2, together with the enhancements expected, which also depend on *n*. In many applications outside ^{13}C spectroscopy only one multiplicity is present, so the ideal Δ value can be chosen. For instance, trimethylsilyl groups are commonly encountered, and their ^{29}Si spectra are advantageously obtained via INEPT with Δ selected for $n = 9$ (Figure 6.14).

Figure 6.13 The different behaviour of doublets and triplets during the INEPT refocusing period - the beginnings of spectrum editing.

Spectrum editing

Variation of the Δ delay also provides a means for distinguishing the number of S nuclei attached to I; *spectrum editing* according to multiplicity. In ^{13}C spectroscopy, for instance, it is possible to generate separate spectra for CH, CH_2 and CH_3 groups. This is the same kind of information traditionally obtained by off-resonance decoupling, but because the polarisation transfer spectra can be acquired with *broadband* decoupling both sensitivity and resolution are far superior. We will see how to do this

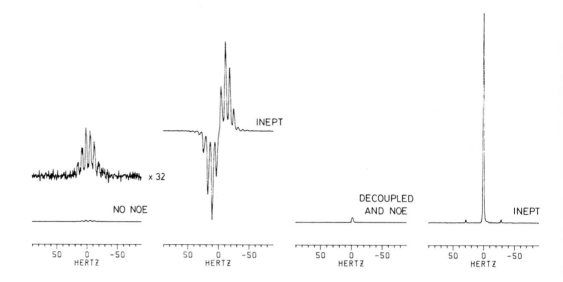

Figure 6.14 ²⁹Si via INEPT from ¹H,
showing spectacular enhancements (all
spectra acquired with optimum repetition
and equal total accumulation time).

Table 6.2 Optimum Δ values (in units of $1/J$), and the corresponding
enhancements (in units of γ_s/γ_i) for refocused polarisation transfer from n
spins-$\frac{1}{2}$.

n	1	2	3	4	5	6
$\Delta(\times 1/J)$	0·5	0·25	0·196	0·167	0·148	0·134
E ($\times \gamma_s/\gamma_i$)	1·0	1·0	1·15	1·30	1·43	1·55

in practice using the DEPT sequence (section 6.4.3), but the principle is
easier to understand in the context of INEPT.

Suppose we define an angle θ, such that $\theta = \pi J\Delta$. Then, by considering
the precession of the components of CH, CH$_2$ and CH$_3$ multiplets during
Δ, it is quite straightforward to derive the INEPT signal intensities (with
decoupling) for each group as functions of θ:

CH: $I \propto \sin\theta$

CH$_2$: $I \propto \sin 2\theta$

CH$_3$: $I \propto \frac{3}{4}(\sin\theta + \sin 3\theta)$ (6.2)

You can easily see that the case we already considered ($\Delta = 1/2J$, which
means $\theta = \pi/2$) is consistent with these formulae. In order to distinguish
the three kinds of carbon resonance, it is necessary to run three spectra
with Δ adjusted to make θ equal to $\pi/4$, $\pi/2$ and $3\pi/4$. Table 6.3 gives the
expected relative signal intensities for each group in these experiments.

The significance of these numbers is perhaps slightly easier to appreciate
by looking at some spectra (Figure 6.15). The $\theta = \pi/4$ spectrum contains
all resonances, $\theta = \pi/2$ gives CH's only, and $\theta = 3\pi/4$ again shows all
resonances, but with the CH$_2$'s inverted. Suitable combining of these
spectra then allows generation of CH, CH$_2$ and CH$_3$ subspectra, as we
shall see later.

Table 6.3 Relative signal intensities during spectrum editing.

θ $(=\pi J\Delta)$	CH	CH_2	CH_3
$\pi/4$	$1/\sqrt{2}$	1	$3/2\sqrt{2}$
$\pi/2$	1	0	0
$3\pi/4$	$1/\sqrt{2}$	-1	$3/2\sqrt{2}$

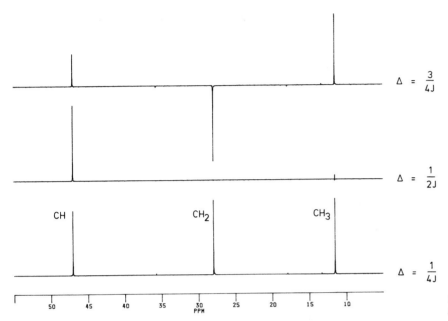

$$\Delta = \frac{3}{4J}$$

$$\Delta = \frac{1}{2J}$$

$$\Delta = \frac{1}{4J}$$

Figure 6.15 The behaviour of CH, CH_2 and CH_3 groups during editing.

Manipulation of the phases and amplitudes of signals in this way is all there is to spectrum editing, so it is not at all mysterious. In DEPT, admittedly, the mechanism by which the phases change is rather more obscure (because it involves *heteronuclear multiple quantum coherence*), but the principle involved is the same. To be honest, we could have got into much deeper water analysing INEPT too; it is only because we examined a doublet, and picked the order in which to consider the final two $\pi/2$ pulses carefully, that we avoided situations we could not describe properly using the rotating frame vector model.

For another perspective on editing, try comparing what happens during Δ in INEPT with the evolution of magnetisation during a heteronuclear *J*-modulated spin echo (Chapter 10, section 10.2). The initial states differ between the two experiments, but otherwise they are very similar.

6.4 DEPT

6.4.1 Introduction

The 'D' of DEPT stands for 'Distortionless' (you can guess the other letters) . This experiment brings about polarisation transfer in a similar fashion to INEPT, but with the important difference that all the signals of the insensitive nucleus are in-phase at the start of acquisition. Thus, decoupled spectra can be acquired without the need for an extra refocusing period Δ, and coupled spectra retain their familiar multiplet structures; this is the sense in which they are not 'distorted'. For this reason, and because of certain other advantages when it is used as an editing sequence, DEPT will usually be the method of choice when polarisation transfer is required.

6.4.2 The DEPT Sequence (as a Multiple Quantum Filter)

To reflect DEPT's most common application, I will refer throughout most of section 6.4 to polarisation transfer from protons to carbon, but of course with the understanding that we could substitute any other pair of spin-$\frac{1}{2}$ nuclei. The DEPT sequence is:

$$^1\text{H:} \qquad \left(\frac{\pi}{2}\right)_x - \frac{1}{2J} - \quad \pi \quad - \frac{1}{2J} - \theta_{\pm y} - \frac{1}{2J} - (Decouple \cdots)$$

$$^{13}\text{C:} \qquad\qquad\qquad \left(\frac{\pi}{2}\right)_x \qquad \pi \qquad\qquad Acquire\ (\pm x) \cdots$$

where $\theta_{\pm y}$ represents a proton pulse of variable length which, it will transpire, plays a similar role to the Δ delay of INEPT. The phase alternation of this pulse (combined with receiver add/subtract) serves to eliminate signals from the natural ^{13}C magnetisation, and the phases of both π pulses can also be alternated to reduce artefacts. Decoupling of protons may or may not be applied during acquisition, as required.

If you try to follow the effect of this sequence using the rotating frame vector model, you will find that the situation after the $\pi/2$ pulse on carbon takes some describing. Here we have transverse magnetisation of both the proton and carbon parts of the coupled system evolving together, and nothing we have seen to date tells us how to predict the outcome of this. In fact, it is very difficult to represent this state of affairs, known as a *heteronuclear multiple quantum coherence*, in a helpful way using classical magnetisation vectors[4]. We would also have problems predicting the effect of the θ pulse on this system without indulging in quantum mechanics, so overall the mechanism of DEPT will have to remain slightly mysterious. However, once you have read Chapter 8 (where there is some discussion, albeit of a rather metaphysical nature, of multiple quantum coherences - section 8.3.5) the following comments might make sense. Please do not read this before you have read Chapter 8; if you are reading the book in order, skip to section 6.4.3.

Consider another sequence[5]:

$$^1\text{H:} \qquad \left(\frac{\pi}{2}\right)_\varphi - \frac{1}{2J} - \quad \pi \quad - \frac{1}{2J} - \left(\frac{\pi}{2}\right)_\varphi\left(\frac{\pi}{2}\right)_{-x}$$

$$^{13}\text{C:} \qquad\qquad\qquad \left(\frac{\pi}{2}\right)_x \qquad \pi \qquad\quad - \frac{1}{2J} - Acquire\ (\pm x) \cdots$$

Essentially the way in which this differs from DEPT is that the θ pulse has been replaced by the pair:

$$\left(\frac{\pi}{2}\right)_\varphi\left(\frac{\pi}{2}\right)_{-x}$$

where φ is an *arbitrary* phase (not necessarily x or y), defined as an angle relative to the y axis. You should be able to convince yourself (at least starting from z magnetisation) that this pair of pulses is equivalent to a single pulse θ_y, provided $\varphi = \theta$, so DEPT and this experiment are actually equivalent. The group of pulses up to and including $(\pi/2)_\varphi$ can be seen to be those used for the creation of multiple quantum coherence (Chapter 8, section 8.4.3), but here extended to a heteronucleus. Thus, immediately prior to the $(\pi/2)_{-x}$ pulse, heteronuclear multiple quantum coherences exist, two-quantum for CH groups, three-quantum for CH_2's and four-quantum for CH_3's. The final proton pulse is a 'read' pulse, regenerating observable single-quantum coherence (for carbon). Varying the phase of the excitation sequence (relative to the read pulse) according to the recipe given in Chapter 8 should then allow the separation of signals by the order of multiple quantum coherence through which they have passed - *multiple quantum filtration*. Hence the potential for spectrum editing.

Spectrum editing in precisely this fashion has been implemented in the technique known as POMMIE[6] (this is just the sequence given above, with some minor changes in the details of the *rf* phases). However, since the majority of spectrometers available at present will not be able to generate the phase shifts required to separate the different orders of multiple quantum coherence, I will not discuss this approach further. If your spectrometer can do it, I recommend trying it out in comparison with DEPT to determine which works better in practice. The equivalence between the required phase shifts of POMMIE and the θ pulse of DEPT means that the differences between the experiments will depend on practical considerations rather than matters of principle.

6.4.3 Editing with DEPT

Introduction

If you have skipped to this point from the beginning of section 6.4.2, all you have missed is a rather thin case for the possibility of using the θ pulse of DEPT as a method of spectrum editing. The key to achieving this in practice is the realisation that the result of a DEPT experiment with final pulse θ_y is the *same* as performing INEPT with Δ set to $\theta/\pi J$. Thus all the discussions in section 6.3.3 can be translated directly for use with DEPT, simply by reading θ for $\pi J \Delta$. A most interesting aspect of this becomes apparent if you recall that values for Δ were themselves derived from J, e.g. for optimum sensitivity in an IS_n system, $\Delta = (1/\pi J)\sin^{-1}(1/\sqrt{n})$ (Equation 6.1). Thus, in the equivalent DEPT experiment, the required value of θ is *independent* of J. This is the principal advantage of DEPT over INEPT. There are, of course, still delays in the sequence related to J, but for editing the accuracy of the results proves to be not too sensitive to the setting of these.

Editing

To separate CH, CH_2 and CH_3 subspectra, we record experiments with θ set to $\pi/4$, $\pi/2$ and $3\pi/4$, in the same way that Δ was varied for INEPT[7]. The $\theta = \pi/2$ experiment requires twice as many scans for equal signal-to-noise ratio as the others. Just as in nOe difference spectroscopy, it is likely to be advantageous to interleave the acquisition of these three experiments over a period of time, in order to minimise errors due to gradual drift of the experimental conditions (Chapter 5, section 5.3.3). The total number of scans required will then be determined by the usual sensitivity considerations. Referring to Table 6.3, we can deduce the theoretical proportions in which these experiments must be combined to generate the desired subspectra.

The *CH subspectrum* should simply be the experiment acquired with $\theta = \pi/2$. However, this is likely to contain error peaks (particularly due to methyls), and these can be reduced once the other subspectra have been generated (see below). The $\theta = \pi/4$ and $\theta = 3\pi/4$ experiments differ only by the phase alternation of the CH_2 peaks. Thus, *subtracting* one from the other generates the CH_2 *subspectrum*. *Adding* them generates a spectrum containing only CH's and CH_3's, from which the $\theta = \pi/2$ experiment (scaled by $1/\sqrt{2}$) can be subtracted to generate the CH_3 *subspectrum*. A final refinement is to 'cheat' a little by subtracting a small proportion of the $(\theta = \pi/4) + (\theta = 3\pi/4)$ combination (generated in the course of making the CH_3 subspectrum) from the $\theta = \pi/2$ experiment, so as to reduce erroneous signals due to methyls. In my experience, however, this is seldom worth the effort involved. Figure 6.16 demonstrates the complete process.

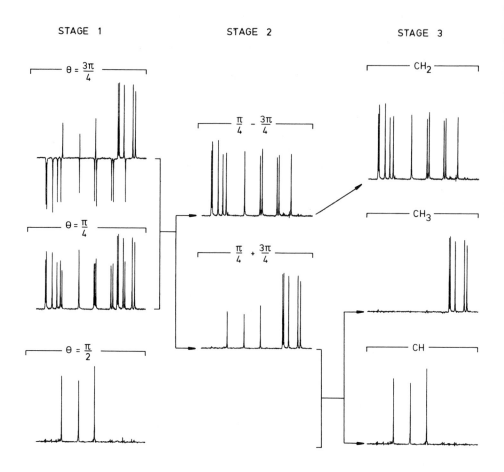

Figure 6.16 The complete process of spectrum editing using DEPT (see text).

How can all this adding and subtracting best be done? Well, conceivably we could proceed as in nOe difference spectroscopy, adding and subtracting FID's, with appropriate scaling factors where necesary. However, in practice it is much more convenient to work in the frequency domain and examine the combination spectra as you go. For this procedure to be feasible, a software package capable of interactive display and manipulation of two spectra is virtually essential; it is particularly convenient if this allows addition or subtraction of spectra with real time adjustment of the scaling factor. In this case, there is no need to remember that $1/\sqrt{2}$ is 0·707; you just combine spectra and adjust their proportions until the undesired peaks cancel. It is invariably obvious that at least some peaks are CH_2's, some CH_3's etc., so there is little problem picking out peaks to null when making the combinations. In this manner you automatically take into account any deviations from the theoretical amplitudes of the various peaks; provided all peaks of a given class are in error by the *same* amount, editing will still be effective.

Is editing accurate? (does it matter?)

It is not hard to think of reasons why editing in this fashion might not work perfectly. For instance, the amplitudes of signals depend on the flip angle of the θ pulse, so if this is wrongly calibrated, or if the $\mathbf{B}_2$ field is inhomogeneous, errors will arise. Furthermore, the three delays in the sequence have to be set according to the proton-carbon one-bond coupling, and this may vary over a range of about 125-210 Hz. Usually a compromise value is chosen, taking J to be, say, 150 Hz if olefins and carbons bearing electronegative substituents are present, or 130 Hz otherwise. Carbons with coupling constants substantially different from these values do not behave properly during the DEPT experiment, and will give rise to spurious responses in the subspectra[7]. This topic has been given a great deal of attention in the NMR literature, and solutions proposed to both problems: DEPT GL[8] for better tolerance of variation in J, and a modified DEPT with composite pulses[9] for less sensitivity to errors in flip angle.

These more sophisticated experiments involve quite complex pulse sequences and phase cycles, but fortunately from the mundane perspective of the chemist who wants to identify the multiplicities of resonances we need not be too concerned with them. The point is that, while the subspectra are quite prone to contain spurious signals, examination of the individual DEPT experiments will usually make assignments entirely unambiguous. Thus, the $\theta = \pi/4$ experiment will certainly contain the resonances of all protonated carbons. As θ is increased to approximately $\pi/2$, the amplitudes of signals due to CH_2's and CH_3's will certainly go down. Maybe they won't all go down exactly to 0; some will remain positive, some disappear and some invert, but they *will* go down. As θ is increased further, various things may happen, but CH_3's will not give rise to *negative* signals, unlike CH_2's (there is one subtle exception to this statement, which arises if the repetition rate is too high for isolated, freely rotating methyls attached to slow-tumbling molecules - see reference 7, page 283, and reference 10 for details). Comparison of these experiments allows the clarification of any strange responses; the actual subspectra can then be taken as a convenient way to summarise the results for the majority of signals.

6.4.4 DEPT for Coupled Spectra

As discussed in section 6.3, the use of polarisation transfer for sensitivity enhancement of ^{13}C is most productive in the recording of spectra with proton coupling. In this case, DEPT should ideally reproduce exactly the same multiplet patterns as would be obtained by direct observation without proton decoupling. However, in the presence of variable proton-carbon coupling constants, some lineshape distortion may arise. Two experiments have been proposed[11] to reduce these effects, known as DEPT$^+$ and DEPT^{++}; the former probably offers the best compromise of simplicity of implementation against quality of results. DEPT$^+$ consists of:

^{1}H: $\left(\dfrac{\pi}{2}\right)_x - \dfrac{1}{2J} - \ \pi - \dfrac{1}{2J} - \theta_{\pm y} - \dfrac{1}{2J} - \pi^{\dagger}$

^{13}C: $\left(\dfrac{\pi}{2}\right)_x \qquad \pi \qquad\qquad Acquire\ (\pm x)$

† This pulse to be applied every other scan.

i.e. it is just DEPT with an extra π pulse applied to protons every alternate scan. For properly set delays this is equivalent to DEPT, but in the presence of a range of coupling constants it is less prone to phase distortion of the signals (Figure 6.17). The effect of incorrect delay times is reduced to variation in the *intensity* of multiplet components by DEPT$^+$; if further improvement is required these intensity errors can be removed using DEPT^{++}, for which see reference 11.

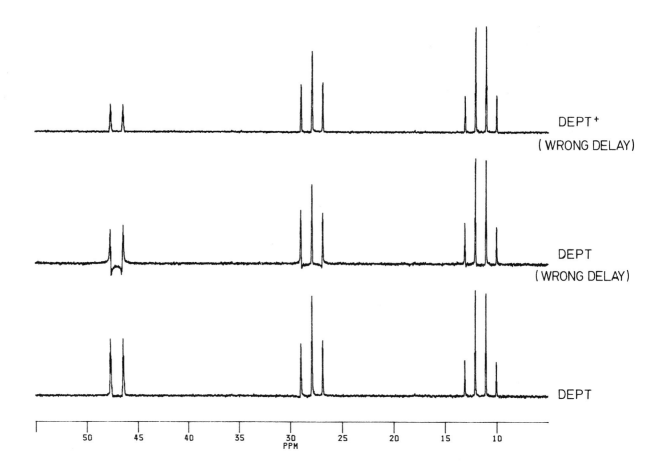

Figure 6.17 Eliminating phase errors in coupled DEPT spectra using DEPT$^+$. For the upper two traces, the inter-pulse delay was deliberately set 50% too high.

6.4.5 Quadrupolar Nuclei

Polarisation transfer is not restricted to spin-$\frac{1}{2}$ nuclei; either the source or the destination nucleus, or both, can be quadrupolar[12]. The only limitation here is that relaxation times must be long enough for the sequence to be short in comparison; clearly this will depend on the size of couplings (the bigger the better) as well as the various relaxation parameters. Assuming that this condition is met, any application discussed already (e.g. enhanced coupled spectra, editing etc.) could be attempted. However, for the most general case of editing fully coupled spectra between nuclei of arbitrary spin choice of parameters becomes quite complex[13], so I will not pursue this here. As such experiments are only useful in rather exotic circumstances anyway, it seems more productive to concentrate on sensitivity enhancement of decoupled spectra; a typical application of chemical interest might be the acquisition of deuterium with proton decoupling. To accommodate quadrupolar nuclei, another pulse in the DEPT sequence has to be made variable:

S:　　　　$\left(\dfrac{\pi}{2}\right)_x - \dfrac{1}{2J} - \quad \pi - \dfrac{1}{2J} - \theta_{\pm y}$　　　　*Decouple* $\cdots$

I:　　　　　　　　　　　$\varphi_x \qquad \pi \quad -\dfrac{1}{2J}- \quad$ *Acquire* $(\pm x)\cdots$

This sequence has been called UPT (for universal polarisation transfer), though it seems rather redundant to introduce a new name for the same sequence with one pulse width altered. Optimum sensitivity requires the choice of θ according to the spin and number of S nuclei, and φ according to the spin and number of I nuclei. In either case, if we have a set of N nuclei of spin s, the optimum flip angle is derived as follows. Each individual nucleus i can be in a state characterised by the quantum number m_i, where m_i ranges over the $2s + 1$ values $-s \cdots s$ in steps of 1. If we define a number M for the set of nuclei such that:

$$M = \sum_{i=1}^{N} m_i$$

M ranges from $-Ns$ to Ns ($2Ns + 1$ values). The optimum UPT flip angle α (which would be θ if we were considering the source nuclei and φ in the case of the destination nuclei) is found by solving the equation:

$$\sum_M D_M M^2 \cos 2M\alpha = 0 \qquad (6.3)$$

where D_M is the degeneracy of state M. It is difficult to give a general formula for D_M, but in realistic cases it is straightforward to calculate. For example, with two spin-1 nuclei the state $M = 0$ has $D_M = 3$, because you can make it three ways - (0,0), (1,-1) and (-1,1). Generally it is feasible to write out all the terms in the sum, and by manipulation get some kind of polynomial in $\cos \alpha$, which may need to be solved numerically[13]. Obviously it is not completely trivial to extract θ and φ in the most general case, but usually at least one of the nuclei involved has spin-$\frac{1}{2}$, for which the straightforward formula 6.1 for optimum flip angle can be applied.

6.4.6 Doing it Backwards

Introduction

All the applications of polarisation transfer we have seen so far have started with the magnetisation of one nucleus and transferred it to another of lower γ. There is no reason in principle why the experiment should be restricted to this 'forward' direction; transfer from, say ^{13}C to ^1H or ^2H to ^{13}C is perfectly feasible. However, such an experiment may seem quite pointless; surely it will just cause a *loss* of sensitivity as compared with direct measurement? Well, that depends on your point of view. If we compare proton and carbon NMR, a characteristic of the former is the ubiquitous nature of the proton. Proton spectra of many natural systems, such as a chemical reaction or a living cell, are completely dominated by intense signals due to water or other solvents, or abundant biochemicals like lipids. Carbon-13, on the other hand, has low natural abundance, so that it is feasible to mark species of interest (by enriching them with ^{13}C) and follow them by carbon NMR, without being hampered by the clutter of signals obtained for ^1H. That is, protons have the sensitivity, but carbon (or any other low-abundance nucleus) has the selectivity.

Now we can think again about whether 'reverse' polarisation transfer might be advantageous. A perfect experiment will eliminate the natural magnetisation of the destination nucleus completely (by phase cycling or other means). So, if we transfer magnetisation from a species labelled with a low abundance nucleus to protons, we transfer the *selectivity* of the labelling to the proton spectrum. As to sensitivity, there is certainly some loss as compared with direct observation of protons. For instance, if we start with ^{13}C magnetisation, in the worst case the population differences we transfer are four times smaller than they otherwise would have been; sensitivity is 1/4 that of direct proton observation. However, the result we obtain has the character of a carbon spectrum (in the sense that only species containing ^{13}C are visible), so the proper comparison we should be making is with direct detection of ^{13}C. Other things being equal, ^{13}C signals are 64 times weaker than those from ^1H, so with respect to direct ^{13}C detection the reverse polarisation transfer signals could be 16 times stronger. Furthermore, if we irradiate the protons between scans, the nOe at carbon will increase its population differences up to threefold, so the sensitivity could be 3/4 that of direct proton observation, or nearly 50 times that of direct ^{13}C measurements.

In practice these estimates may be over-optimistic. For instance, ^{13}C signals are always singlets, whereas protons experience homonuclear coupling. Each time a line is split in two, its signal-to-noise ratio halves; on the other hand, this coupling provides extra structural information. A further problem is that the repetition rate of the experiment is limited by the relaxation of the heteronucleus, which may be slow. This has to be assessed on a case-by-case basis; for instance the peptide whose spectra appear in the following section has very fast-relaxing carbons and is well suited to reverse polaristion transfer, but this may not be so for other compounds. Experimental design must take this factor into account. Finally, the theoretical signal intensities may not be obtained because of experimental difficulties; while most probes tuned for a heteronucleus have a coil capable of measuring protons (the decoupling coil), this is seldom optimised for sensitivity. On the plus side, it is possible to get extremely good suppression of natural proton signals using the procedure described next. In comparing reverse DEPT with other alternatives, such as the spin echo difference technique described in Chapter 10, this may be an important consideration.

Reverse DEPT

The sequence required to bring about reverse polarisation transfer is easily derived using the principles of UPT described in the previous section[14]. For most cases of interest, transfer will be from a single spin-$\frac{1}{2}$ (e.g. a ^{13}C label) to several spins-$\frac{1}{2}$ (e.g. 2 for a CH_2 group). From equation 6.3 we therefore see that the θ pulse of UPT must be set to $\pi/2$, while φ will depend on the number of destination spins. Probably we will not know this number in advance, in which case a compromise value should be used to give reasonable intensity for all expected XH_n groups; for instance $\varphi = \pi/4$ is appropriate for transfer from ^{13}C. Reverse DEPT therefore requires the sequence shown in Figure 6.18.

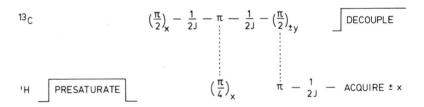

Figure 6.18 Experimental scheme for reverse DEPT.

Phase cycling is similar to ordinary DEPT, the key point being the phase alternation of the final ^{13}C pulse and receiver to attenuate 'natural' proton signals. Since in this experiment the natural signals are presumably much bigger than the polarisation transfer signal, it is unlikely that phase cycling alone will suppress them sufficiently, so this must be combined with a period of presaturation of the protons by broadband irradiation. This also generates the nOe at the heteronucleus, with its corresponding sensitivity increase, so it can be considered essential. Broadband decoupling of the heteronucleus during acquisition is an optional extra; while doubling the peak heights, it does remove some information from the spectrum, so it might not always be desirable. Of course, broadband decoupling of a nucleus like ^{13}C, with its large chemical shift range, is not a trivial problem (see Chapter 7, section 7.4).

The kind of result which may be obtained in this way is illustrated in Figure 6.19. Very good suppression of natural proton signals was achieved by combining one second of saturation prior to each scan with the phase cycling. The signal-to-noise ratio obtained in the polarisation transfer experiment was comparable to direct proton observation, but as I mentioned previously, this compound had particularly favourable relaxation properties. The potential of this experiment, or its competitors discussed in Chapter 10, seems considerable, but to date rather few examples of its use have appeared[15].

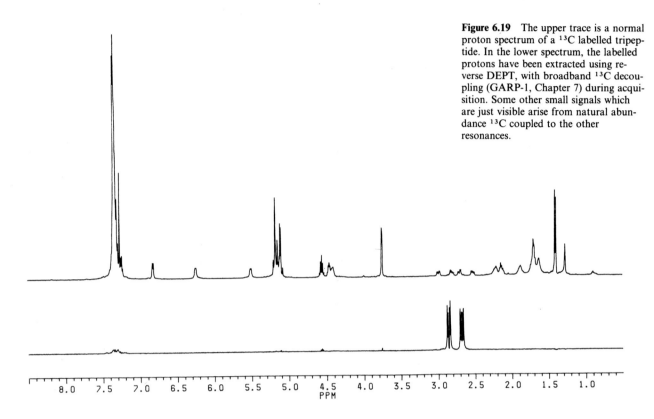

Figure 6.19 The upper trace is a normal proton spectrum of a ^{13}C labelled tripeptide. In the lower spectrum, the labelled protons have been extracted using reverse DEPT, with broadband ^{13}C decoupling (GARP-1, Chapter 7) during acquisition. Some other small signals which are just visible arise from natural abundance ^{13}C coupled to the other resonances.

REFERENCES

1. P. Styles and N. Soffe, *J. Mag. Res.*, **60**, 397-404, (1984).
2. G. A. Morris and R. Freeman, *J. Amer. Chem. Soc.*, **101**, 760-762, (1979).
3. D. T. Pegg, D. M. Doddrell, W. M. Brooks and M. R. Bendall, *J. Mag. Res.*, **44**, 32-40, (1981).
4. But it has been attempted: see R. M. Lynden-Bell, J. M. Bulsing and D. M. Doddrell, *J. Mag. Res.*, **55**, 128-144, (1983).
5. M. H. Levitt and R. R. Ernst, *Mol. Phys.*, **50**, 1109-1124, (1983); DEPT is not the main topic of this paper, but its analysis as a multiple quantum filter is mentioned in the context of identifying sources of error in pulse sequences.
6. J. M. Bulsing, W. M. Brooks, J. Field and D. M. Doddrell, *J. Mag. Res.*, **56**, 167-173, (1984).
7. M. R. Bendall and D. T. Pegg, *J. Mag. Res.*, **53**, 272-296, (1983).
8. O. W. Sørensen, S. Bildsøe, H. Bildsøe and H. J. Jakobsen, *J. Mag. Res.*, **55**, 347-354, (1983).
9. D. T. Pegg and M. R. Bendall, *J. Mag. Res.*, **60**, 347-351, (1984).
10. M. R. Bendall and D. T. Pegg, *J. Mag. Res.*, **53**, 40-48, (1983).
11. O. W. Sørensen and R. R. Ernst, *J. Mag. Res.*, **51**, 477-489, (1983).
12. D. T. Pegg and M. R. Bendall, *J. Mag. Res.*, **55**, 51-63, (1983).
13. D. T. Pegg and M. R. Bendall, *J. Mag. Res.*, **58**, 14-26, (1984).
14. M. R. Bendall, D. T. Pegg, D. M. Doddrell and J. Field, *J. Mag. Res.*, **51**, 520-526, (1983).
15. See: W. M. Brooks, M. G. Irving, S. J. Simpson and D. M. Doddrell, *J. Mag. Res.*, **56**, 521-526, (1984); D. M. Doddrell, J. Staunton and E. D. Laue, *J. Chem. Soc. Chem. Commun.*, 602-605, (1983)

7

Further Experimental Methods

7.1 INTRODUCTION

In this chapter I survey a range of experimental problems that are often encountered when practising multi-pulse NMR. As stated in Chapter 1, this is not intended to be a comprehensive manual of experimental NMR; rather I have picked those procedures which one encounters from day to day when setting up experiments, and describe how they may be performed. Also included below are some recent techniques which are not 'NMR methods' in themselves, but are intended to be incorporated in other experiments to give improved performance or reduced sensitivity to operator errors or instrumental deficiencies. Many modern NMR experiments depend very critically on correct setting of parameters such as pulse widths, delays and phase shifts, and on the spectrometer faithfully executing the operator's requests in these respects. Unfortunately it is necessary to approach most instruments with a certain basic scepticism, as very frequently what is requested and what eventually emerges from the pulse transmitter may bear scant relationship with each other; this is particularly true of methods involving sequences of events occurring on a μs time scale, such as the composite pulses mentioned in section 7.3.

Careful attention should be paid to matters such as pulse width calibration whenever multi-pulse experiments are performed, of course, but even more so the first time you try a new method. To avoid disappointment or confusion it is essential to learn how the experiment behaves, and what factors are particularly critical on your spectrometer, using samples for which you know what the results should look like, and on which the spectrum can be run rapidly. This means using strong solutions of simple compounds, just complex enough to show the expected effects. It may also be helpful to artificially reduce T_1 values by addition of chromium (III) acetylacetonate (for organic solutions) or manganese (II) chloride (for aqueous solutions), so scans can be repeated quickly. For proton work it is useful to keep at hand samples containing a single resonance (e.g. chloroform in deuterochloroform, with a little relaxation reagent to reduce T_1 to a second or two), and first order two- and three-spin systems (β-chloroacrylic acid and 1,2-dibromopropionic acid in deuterochloroform are quite good). For heteronuclear work the possibilities are endless, but ^{13}C enriched methyl iodide is handy for setting up many proton-carbon experiments; again some T_1 reduction will often be needed.

7.2 PULSE WIDTH AND FIELD STRENGTH

7.2.1 Introduction - the dB Scale

An operation frequently required when setting up a spectrometer for a pulse experiment is the measurement of the intensity of its various *rf* fields. Whether we call this a pulse width calibration or a field strength measurement depends purely on the emphasis placed on this parameter by the experiment in question. While oscillating *rf* fields can be measured in magnetic units like Gauss or Tesla (1 Tesla = 10,000 Gauss), it is often most convenient from an NMR point of view to express them in terms of the rate of precession of nuclear magnetism they induce, as mentioned in Chapter 4. In this case the relationship between a field strength B measured in Hertz (i.e. cycles-per-second) and the π pulse width t_π (i.e. rotation through 1/2 cycle) is particularly clear ($t_\pi = 1/2B$). For instance, a typical hard π pulse of 10 μs corresponds with a 50 kHz field strength, whereas a typical homonuclear decoupling field of 50 Hz would act as a soft π pulse if we applied it for 10 ms. The terms *hard* and *soft* are often used in a fairly loose way to refer to non-selective and selective pulses; obviously the exact meaning to be attributed to this will depend on the context. To measure field strength we must seek NMR phenomena sensitive to it. There is a range of possibilities, and the choice can be made according to circumstances, as surveyed below.

Transmitter pulses, and the output of the decoupler when it is used for pulsing, are usually applied with the maximum available power. In many other applications it is desirable to adjust the *rf* power to suit the circumstances, for instance to achieve a soft pulse of a particular length, or to set the selectivity of a decoupling experiment. Adjustment of *rf* power is usually made downwards by attenuation, and it is helpful to have some appreciation of how the parameter used for measuring power ratios (the deciBel or dB) relates to the NMR field strength. The dB is a logarithmic scale of power ratios, *defined* as:

$$dB = 10 \log_{10}\left(\frac{P_1}{P_2}\right) \tag{7.1}$$

where P_1 and P_2 are the two powers to be compared. Very often it is more convenient, or interesting, to measure voltage rather than power, and since the two are related by:

$$P = \frac{V^2}{R} \tag{7.2}$$

when the calculation is done using voltages the relationship:

$$dB = 20 \log_{10}\left(\frac{V_1}{V_2}\right) \tag{7.3}$$

must be used. Given that the dB represents a ratio, it is often rather puzzling to encounter its apparent use as an absolute unit, in statements such as 'the sound of the plane taking off exceeded 150 dB', or more significantly for NMR (unless your lab. is near an airport) 'the synthesiser output was +10 dBm'. The trick here is that the ratio is to some implied standard value, assumed to be universally known. For the very common units of *rf* power dBm and dBW the standards are 1 mW and 1 W respectively. Typical bench signal generators deliver a few dBm, while a typical pulse transmitter for NMR might produce +20 dBW.

If we halve the power, this corresponds with a change of $-3{\cdot}01$ dB ($10 \log_{10} 0{\cdot}5$), which is near enough to -3 dB to make no difference. The *rf* field strength, however, is proportional to the *voltage* applied across the coil, so halving the field strength requires 6 dB less power. Remembering this simple relationship allows suitable attenuations to be selected to adjust the field strength as required.

7.2.2 Pulses on the Observed Nucleus

With good sensitivity

The simplest pulse width measurement to make is that of the transmitter pulse, with a sensitive observed nucleus and a sample strong enough to give reasonable signal-to-noise ratio in a single scan. In this case we can search for the π pulse length, under the expectation that this will give rise to no signal. The starting point for this search should be obtaining a single-scan spectrum using a short pulse, which you are confident corresponds with less than a $\pi/2$ flip angle; for typical pulse transmitters on high-resolution instruments 1-2 μs should certainly fulfil this requirement. This spectrum is used to determine the phase correction required for a positive absorption signal. Repeating the experiment with increasing pulse lengths allows the null point corresponding with a π flip angle to be detected; the advantage of using a pre-determined phase correction is that if you pass the null the signal will appear inverted (Figure 7.1). Note that some spectrometer software automatically scales each new spectrum to the same size, so to get a result like Figure 7.1 it may be necessary to select an appropriate non-scaling mode.

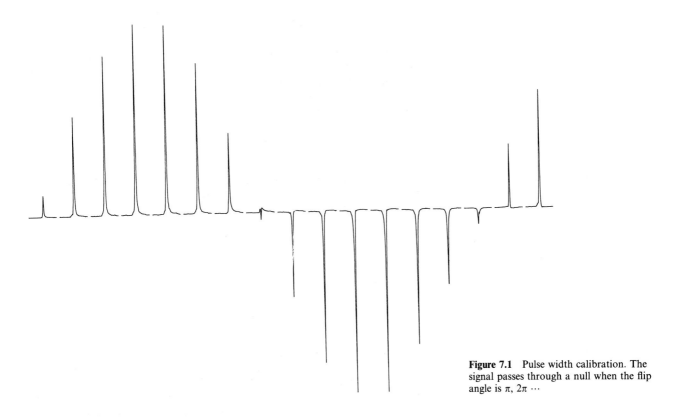

Figure 7.1 Pulse width calibration. The signal passes through a null when the flip angle is π, 2π $\cdots$

In practice there are a number of complications with this procedure. If you are unfamiliar with the spectrometer, and have no clue as to the approximate length of the π pulse, take care that the null obtained is the *first* one. 2π, 3π $\cdots$ pulses also give rise to similar nulls; for this reason keeping a log of approximate pulse lengths for different nuclei can be very helpful to casual users. Another snag is deciding what appearance the spectrum should have when the 'null' is reached. No π pulse is perfect, and residual signals will arise for all pulse lengths. One cause of this is the sensitivity of π pulses to off-resonance effects (Chapter 4), which can be simply avoided by setting the transmitter frequency exactly on a line before starting the calibration. B_1 inhomogeneity will ensure there still remains some peculiar signal even so (Figure 7.2), and you have to make a rather arbitrary decision about what corresponds with the best choice of pulse length. It is best to concentrate on the central part of the line, which (hopefully) arises from the bulk of the sample, and ignore the outer wings.

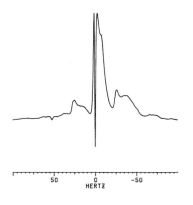

50   0   −50
HERTZ

Figure 7.2 The residual signal obtained with a 'π' flip angle. The exact appearance of this will depend strongly on probe design and construction.

With poor sensitivity

Once it becomes difficult to see the signals from the sample in a single scan, more patience and ingenuity are required to measure the pulse width. A number of strategies can be adopted according to how bad the sensitivity really is. If you can see some signals, but they disappear into the noise around the π pulse length, it may be sufficient to note the last pulse length which gives a visible positive signal and the first which gives a visible negative signal, then take a value halfway in between. Alternatively, signal averaging can be applied to a limited extent, depending on the endurance of the operator. For many nuclei it is hardly practical to wait several T_1's between scans for the z magnetisation to recover during a pulse width calibration. It is acceptable to pulse repetitively under steady-state conditions, but it is then more effective to search for the *second* null corresponding with the 2π flip angle, since the π null becomes a rather flat function of pulse duration[1].

In the limit of no signals visible at all in a reasonable time, it is necessary to resort to some form of pre-calibration. It is possible to keep a record of pulse widths found with strong samples, as suggested above; provided careful attention is paid to probe tuning (Chapter 3) these will be quite reproducible. Exceptions will occur for solutions electrically very different from those used for the calibration, such as aqueous electrolytes if organic solutions were used for the original measurements. A better procedure is to make up a sample closely similar to the sample under investigation, but having a resonance strong enough to permit determination of the π pulse duration. The value determined in this way can be used without further adjustments to the spectrometer. Some nuclei, for instance many metals, will never give enough signal to permit pulse width calibration however strong you make the solution. If you are in the unfortunate position of having to do a sophisticated NMR experiment on such a nucleus, then the remaining possibility is to exploit any heteronuclear coupling it may possess by treating it as the 'other' nucleus and using the methods described next.

7.2.3 Pulses on Other Coupled Nuclei

Many experiments require that pulses be applied to another nucleus coupled to that under observation. Foremost amongst these are the polarisation transfer methods like DEPT, and related 2D experiments. Typically the other nucleus is ^{1}H and the observed nucleus ^{13}C or another heteronucleus, with the proton pulse being applied using the ^{1}H decoupler, so finding the π pulse length is often referred to as 'measuring the decoupler pulse width'. However, I will reserve this term for occasions when we are directly interested in decoupling; there are plenty of experiments in which these are not the nuclei involved, for instance DEPT from ^{31}P to a metal, or backwards from ^{13}C to protons. In all cases the problem is the same; we want to find the π pulse length for a nucleus that is not being observed directly.

All kinds of experiments can be devised in which the intensity of the signals from the observed nucleus depends on the flip angle of the pulse applied to its coupling partner. A simple sequence is[2]:

X: $\qquad\qquad\qquad \left(\dfrac{\pi}{2}\right) - \tau - Acquire \cdots$

Y: $\qquad\qquad\qquad\qquad\qquad\qquad\quad \alpha$

where X is observed, Y is the other nucleus, both spin $\frac{1}{2}$, and an XY group (i.e. a doublet for the X signal) is required for calibration. If τ is set to $1/2J_{XY}$, then when the α pulse on Y equals $\pi/2$ (N.B. *not* π) no signal is observed. Either side of this the doublet components appear in antiphase, undergoing phase reversal as α passes through $\pi/2$ (Figure 7.3). The major disadvantage with this method is that no decoupling can be applied to Y during acquisition of X, leading to loss of sensitivity. Also the best repetition rate for signal averaging is determined by X's T_1, so for the typical case in which Y is a proton and X a low sensitivity and potentially slow-relaxing nucleus this can become tedious.

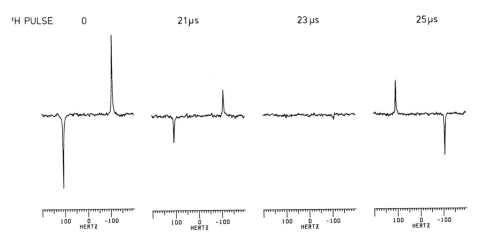

A better approach is to use the DEPT sequence (Chapter 6), when decoupling of Y is permissible and the repetition rate is determined by Y's T_1. Starting with a small value of the θ pulse to allow the phase correction to be determined, search for the value of θ which nulls signals due to XY_2 and/or XY_3 groups; this corresponds with $\theta = \pi/2$ (Figure 7.4). For transfer from protons to carbon, scanning every second or two and using 8-16 scans allows the measurement to be made quite accurately on fairly dilute samples. It is of course necessary to measure the pulse widths for carbon first, and to choose a value for the delays in the sequence according to the expected proton-carbon couplings.

These two methods are only suitable for fairly strong *rf* fields (so that the resulting pulses are non-selective). Calibrating a soft pulse on a coupled nucleus, such as may be required for the SPI experiment, can be quite tricky. The best approach is probably to use the first of the two methods described above, with the Y pulse on-resonance and of sufficient field strength to make it relatively non-selective (e.g. at least 1 kHz). Once the pulse width has been found with this field, a precision attenuator can be used to reduce the field strength to the required value.

Figure 7.3 Calibration of pulses on a coupled nucleus - first method.

7.2.4 Field Strength - Homonuclear

When we are interested in continuous irradiation of a nucleus with low power, as for homonuclear decoupling, it is more usual to refer to the field strength rather than the pulse width. It is useful to have a calibration of the field strength delivered against decoupler power setting to hand, to aid in choosing parameters for effective decoupling or optimum selectivity. While it is possible to measure the length of a soft π pulse delivered by the decoupler, calculating the effective field strength during decoupling from this may be difficult, because homonuclear decoupling on FT spectrometers involves rapid pulsing of the irradiation (*time shared* decoupling), which will cause a potentially unknown degree of reduction in its mean power.

A much better method for measuring weak, homonuclear decoupling fields is to use the so-called *Bloch-Siegert shift*. This is a displacement of a signal from its usual frequency brought about by nearby irradiation (Figure 7.5), which can be related to the field strength and offset from

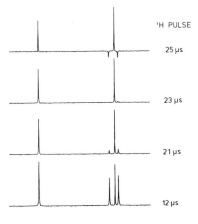

Figure 7.4 Calibrating pulses on a coupled nucleus using DEPT.

resonance of the decoupling (B_2) field by:

$$\delta v = \frac{B_2{}^2}{2(v - v_i)} \tag{7.4}$$

where B_2 is measured in Hz, v is the true frequency of the observed signal, v_i is the frequency of irradiation and $B_2 \ll v - v_i$. This can be conveniently exploited as a difference experiment, in which a normal spectrum containing a single line is subtracted from a spectrum with irradiation close to the line, allowing δv to be measured directly. The irradiation should be set a suitable distance from the test line so that the offset is considerably greater than the expected field strength. A series of experiments of this kind performed with different attenuations of the decoupler power should show a linear relationship between attenuation in dB's and field strength.

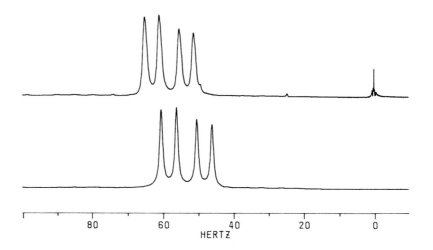

Figure 7.5 The Bloch-Siegert shift. The lower trace is a normal spectrum, while for the upper a weak *rf* field has been applied at frequency 0, other conditions remaining identical. From the observed shift, the field strength can be calculated to be about 20 Hz.

7.2.5 Field Strength - Heteronuclear

There are two common situations in which it is helpful to be able to determine the strength of a heteronuclear (i.e. not the nucleus under direct observation) irradiation field: as a preliminary to a pulse width determination as in section 7.2.3, and in order to set up broadband decoupling. The pulse width determination can just be made directly of course, but if the pulses are on some exotic nucleus or unfamiliar probe it may speed things up and reduce the chance of error if you measure the field by another method first. More significantly, modern broadband decoupling techniques require that phase shifts are applied to the decoupler in a predetermined sequence and at a rate derived from the field strength (section 7.4). As long as the heteronuclear transmitter can be switched on continuously, it is possible to avoid the need for using the procedures of section 7.2.3 by determining the field strength directly.

A simple approximate method for measuring this parameter is observation of off-resonance decoupling effects. Consider the example of a CH group; the carbon resonance without proton decoupling is a doublet with coupling constant J. If continuous irradiation with sufficient intensity is applied exactly at the proton chemical shift, then the doublet collapses to a singlet. However, if the irradiation is moved away from exact resonance, the doublet reappears, but with reduced line separation. Provided the field strength is much greater than the offset from resonance, it can be derived from the reduced splitting by:

$$B_2 \approx \frac{J \delta v}{J_r} \tag{7.5}$$

where J_r is the reduced splitting and δv the offset of the irradiation from resonance. A more exact formula is:

$$B_2 = \frac{\delta v \sqrt{J^2 - J_r^{\,2}}}{J_r} \qquad (7.6)$$

but provided care is taken that B_2 is very large compared with δv the approximation 7.5 is acceptable. The procedure for using these formulae is first to measure the coupling constant and the exact proton resonance frequency, then apply the off-resonance decoupling with δv selected so as to give a small but easily measured residual splitting J_r (Figure 7.6). This is a fairly inaccurate measurement, and should only be used as a starting point for either of the above applications, to be followed by a more refined determination by other methods. It is most suitable for medium to strong decoupling fields, in contrast to the Bloch–Siegert shift method which is more convenient at low field strengths.

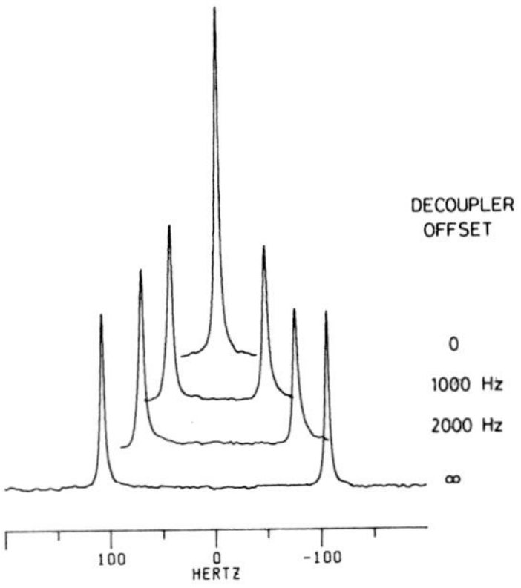

Figure 7.6 The effect of off-resonance decoupling can be used to measure a heteronuclear decoupling field.

7.3 COPING WITH IMPERFECT PULSES

7.3.1 Introduction

However much care you invest in measuring pulse lengths by the above procedures, it will prove impossible to achieve the effects of ideal $\pi/2$ or π pulses. The two principle causes of this are variation in B_1 field strength throughout the sample volume, due to imperfect construction of the transmitter coil, and tilting of the effective angle of rotation due to B_1 having finite size (Chapter 4, section 4.3.2). In routine 1D spectroscopy, involving only $\pi/2$ (or shorter) pulses, neither of these effects is particularly noticeable. The tilted effective field angle causes phase errors across the spectrum, but being an approximately linear function of offset these get lumped in with errors from other sources, and are corrected during data processing. Slight variations in effective pulse length over the sample volume due to B_1 inhomogeneity have scarcely discernible results.

Any kind of multipulse experiment, however, may be seriously impaired by these problems. A simple example of this is the inversion-recovery T_1 measurement mentioned briefly in Chapter 4. Since the object is to measure the return to equilibrium of magnetisation placed along the $-z$ axis by a π pulse, it is easy to appreciate that imperfect functioning of this pulse will lead to incorrect results. More significantly for chemical applications, many modern experiments rely on aligning magnetisation vectors along specified axes in order to bring about certain kinds of magnetisation transfer. The INEPT and DEPT sequences we have already seen fall into this category, as do their two-dimensional equivalents (Chapter 9), and other two- and one-dimensional experiments such as INADEQUATE (Chapter 8). Deviations from the ideal alignment, such as arise when pulses rotate magnetisations about a tilted axis or through the wrong angle, degrade the sensitivity of these experiments, and may introduce confusing artefacts in some cases. The loss in sensitivity can be particularly galling as many of these techniques are aimed at increasing this very parameter. All experiments involving spin echoes are also very sensitive to pulse defects, which cause a diverse set of artefacts in the 2D methods known as *J*-spectra (Chapter 10).

7.3.2 Composite Pulses

One line of attack on pulse defects is purely instrumental; probe design does gradually improve. This can help with both the problems mentioned above, as both homogeneity and strength of the B_1 field are increased. The latter seems at first sight to depend more on the pulse transmitter than the probe, since increasing the applied *rf* voltage increases the B_1 field strength, but in fact there is no real problem (aside from increased cost) in boosting the transmitter output by orders of magnitude; it is actually the ability of the probe components to take the applied voltages without arcing or insulation breakdown that limits practical field strength. With small diameter probes B_1 field strengths of 20-50 kHz (12·5 - 5 μs $\pi/2$ pulse) or more are currently quite practical, but the development of probe hardware has run parallel with a tendency to higher static fields, with a corresponding increase in spectral width which balances the improvement in B_1 field. For a 20 kHz field an offset of only 3·5 kHz tilts the effective field angle significantly (about 10°); for, say, ^{13}C observation on a 500 MHz spectrometer this is only 28 p.p.m. out of a possible shift range of around ±120 p.p.m.

There remains, then, a need for other means of dealing with pulse imperfections. A very exciting development over the last few years is the attempt to devise clusters of pulses (*composite* pulses) which have a net effect equivalent to a $\pi/2$ or π pulse, but with less sensitivity to offset and/or B_1 inhomogeneity. This is still an area of active research, and I cannot present 'the solution' to the problem, but by discussing what progress has been made so far I can at least draw your attention to the possibilities on offer. For most of the composite pulse sequences it is quite obvious that they add up to the desired effect if their component pulses are ideal. The effects of offset or inhomogeneity errors can sometimes be demonstrated graphically using the vector model, as in Figure 7.7 below, but in general you should consult the references for a proper explanation of how these pulse combinations work.

The earliest attempts at constructing composite pulses treated the two problems of offset dependence and B_1 inhomogeneity *separately*[3]. Subsequently sequences have been devised that compensate both effects together; developments have naturally concentrated more on π pulses as these are more critically sensitive to offset errors[4,5]. Two early composite pulse equivalents to $\pi/2$ and π pulses are:

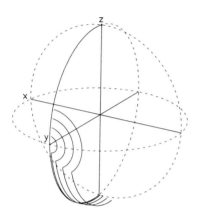

Figure 7.7 Magnetisation trajectories for a composite π pulse. Starting from the $+z$ axis, the path of the magnetisation is traced on the surface of a sphere for a series of assumed B_1 fields, decreasing the flip angle of the nominal $\pi/2$ pulse in 5° steps; each path ends much closer to the $-z$ axis than would that of a simple π pulse with the same field strength error.

$$\frac{\pi}{2}: \qquad \left(\frac{\pi}{2}\right)_x \left(\frac{\pi}{2}\right)_y$$

$$\pi_y: \qquad \left(\frac{\pi}{2}\right)_x \pi_y \left(\frac{\pi}{2}\right)_x$$

where the clusters of pulses are intended to be applied without intervening delays; in practice this may not be possible, and delays of a few μs may need to be inserted within them. The π pulse is of most interest to us, and this has found application both to inversion of z-magnetisation and to echo generation; this is mentioned again in Chapter 10. Its operation can be understood by considering the trajectories of magnetisation components starting on the z axis, with the introduction of progressively greater errors in the components of the pulse (e.g. see Figure 7.7). The composite pulse is inserted in place of its equivalent in the sequence of interest, and when phase shifts are required the phases of all its components are shifted equally, retaining their relative phase.

The 3-element composite π pulse provides good compensation for B_1 inhomogeneity (up to about ±20% deviation in pulse duration from ideality causes less than 1% change in efficiency of inversion), but only in the absence of offset effects. Compensation for both effects together requires a longer sequence; the best available to date has been named, in rather dubious taste, GROPE-16[5]. To describe ever longer composite pulses, it is convenient to introduce an abbreviated notation. Suppose a $\pi/2$ pulse around the x axis is represented X. Then an equivalent y rotation will be written Y, and 180° phase shifts represented with a bar, thus: $\bar{X}$, $\bar{Y}$. Other pulse lengths will be written as multiples of the $\pi/2$ pulse, e.g. 2$\bar{X}$ for a π_{-x} pulse. As an example of this notation, the composite π pulse mentioned above can be written X2YX. GROPE-16 can then be represented:

$3\bar{X}4XY3\bar{Y}4YX$

This will tolerate up to $\pm20\%$ error in pulse length and offsets up to $\pm0.5B_1$ with less than 1% reduction in inversion efficiency. For the example of a 50 kHz field, this would correspond with setting the $\pi/2$ pulse length anywhere between 3 and 6 μs, with inversion at least 99% efficient over ±25 kHz. This represents a tremendous advantage when setting up experiments, not least because of the considerable leeway available in calibrating the pulse width.

Composite $\pi/2$ pulses with dual compensation have particular application in the suppression of SPT artefacts in nOe experiments[6] (Chapter 5, section 5.3.3). The sequence $Y3\bar{X}4X$ behaves as a composite $\pi/2$ pulse with 99% effective creation of transverse magnetisation within similar bounds of B_1 inhomogeneity and offset as GROPE-16; this pulse was used to generate nearly all the nOe difference spectra in Chapter 5.

All is not, however, completely rosy. The sequences described here have been designed around the behaviour of the z component of magnetisation, either aiming to eliminate it ($\pi/2$ pulses) or invert it (π pulses). The error compensation only applies to this component; if transverse magnetisation is present the analysis is considerably different. To take a simple example, the sequence $(\pi/2)_x(\pi/2)_y$ takes z magnetisation into the transverse plane, with compensation for B_1 inhomogeneity. However if we applied this pair of pulses to a vector initially aligned along the $+y$ axis, the result would be to transfer it onto the $+x$ axis, not what we want at all. The π pulses mentioned here, on the other hand, still add up to 180° rotations for transverse magnetisation, but error compensation is limited[7]. It is found that, while a composite pulse such as $X2YX$ makes the *magnitude* of a spin echo less sensitive to B_1 inhomogeneity, it introduces errors in its *phase* (a simple π pulse, in contrast, creates echoes whose phase is independent of the actual flip angle). This means that composite pulses cannot simply be substituted into sequences involving transverse magnetisation without careful analysis of the problem; for a spectacular example of this, see reference 8.

Experimental factors may also intervene to reduce the effectiveness of these sequences. Although the use of a composite pulse reduces the demands on probe performance, it places greater strain on other parts of the spectrometer. The timing and magnitude of phase shifts must be accurately controlled; although many spectrometers can apparently be programmed to generate a composite pulse, what actually happens in the hardware when such a program is run may be shrouded in mystery. As well as problems with timing of phase shifts, the longer sequences may suffer from 'pulse droop', as a transmitter designed to produce 20 μs pulses begins to give up the ghost (GROPE-16, for instance, lasts 160 μs if the nominal $\pi/2$ pulse is 10 μs).

For these reasons it is important to determine whether use of a composite pulse is really improving things on a particular spectrometer. In many experiments in which the use of a composite pulse is desirable the improvement should amount to something rather ill-defined, such as an increase in the 'accuracy' of a measurement, so it becomes necessary to try some alternative where the outcome will be clear-cut. For composite π pulses a suitable test sequence is:

$$\Pi - \frac{\pi}{2} - Acquire \cdots$$

where Π is the composite pulse. Choosing a sample with a single line allows the efficiency of inversion to be investigated as a function of offset or pulse duration. If the offset is varied for the Π pulse, it should be set back close to the observed line for the $\pi/2$ pulse to avoid additional off-resonance effects. Figure 7.8 shows some experiments comparing a straight π pulse with GROPE-16 on a 500 MHz spectrometer with a 30 kHz proton B_1 field.

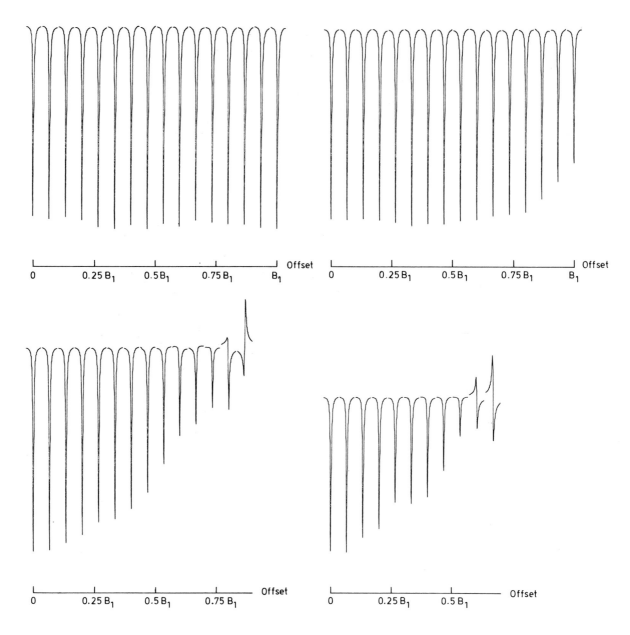

Figure 7.8   A test of the composite π pulse GROPE-16 (upper spectra), comparing it with a simple π pulse (lower spectra). The offset dependence is plotted both with optimum flip angles (left), and with all flip angles 20% high (right) to simulate severe B_1 inhomogeneity.

7.4   BROADBAND DECOUPLING

In many experiments it is necessary to eliminate the coupling between two nuclei completely, by *broadband* irradiation of the nucleus which is not being observed. Most often this is the proton, of course, but it would sometimes be convenient to do the same thing for other abundant NMR-active nuclei like ^{31}P and ^{19}F. Broadband decoupling has traditionally been brought about by the combined use of high-power irradiation and phase modulation of the *rf* to create 'side-bands' spread throughout the region of interest. Various modulation schemes have been arrived at in an empirical way, including the original random or *noise* modulation (leading to the widespread use of the term 'proton-noise-decoupling' or PND to describe broadband decoupling; this is in fact a misnomer since very few spectrometers use noise modulation nowadays).

None of the empirical modulation methods have proved entirely satisfactory. The two major problems are a requirement for quite high B_2 field levels, and hence for high *rf* power which heats the sample, and ineffective decoupling, which leads to loss of resolution. Both problems worsen on higher field spectrometers, where the bandwidth to be irradiated is greater. The line-broadening caused by inefficient broadband decoupling leads to loss of sensitivity, and is not refocused during a spin echo sequence, so it is particularly detrimental to those experiments which contain echoes (see Chapter 10).

Over the last few years a very important advance has been made in the area of broadband decoupling, which reduces many of these problems. To begin with, consider why high-power CW decoupling does not work very well. We get an interesting insight into this by imagining the effect of applying a train of π pulses to the nucleus to be decoupled (taken here to be protons, with ^{13}C observation):

^{1}H: $\qquad \pi - \tau - \pi - \tau - \pi - \tau - \pi - \tau - \pi - \tau - \pi - \tau \cdots$

^{13}C: $\qquad \dfrac{\pi}{2} \; Acquire \cdots$

Every time we pulse the protons, we reverse the direction of precession of ^{13}C multiplet components (Figure 7.9). Thus in the middle of each τ interval they realign, and if we sampled the ^{13}C magnetisation at those times it would contain no information about the coupling; it would be decoupled from the protons. Now, we already know that such a train of π pulses will be very susceptible to errors due to either pulse length inaccuracy or tilting of the effective axis of rotation, since they will accumulate (see the Carr-Purcell experiment in Chapter 4). We would therefore not be surprised to discover that this was a poor way of bringing about broadband decoupling. If we imagine the τ interval shrinking to zero this degenerates into continuous irradiation, and we can use the same argument to explain why that does not work.

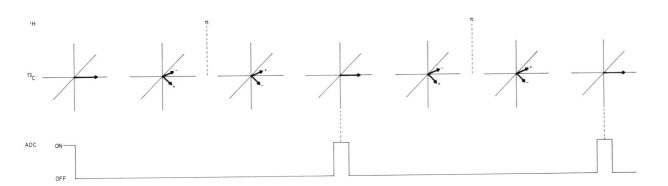

In the light of the previous section, thinking about decoupling as application of a train of π pulses should give us a clue as to how to improve it. If we replaced the non-ideal π pulses with *composite* pulses, perhaps the effects of B_2 inhomogeneity and offset could be reduced sufficiently that effective decoupling would occur. A good deal of research along these lines has taken place, the general idea being to take a basic composite π pulse and apply systematic permutation of its components and phases to generate improved decoupling sequences[9]. Although the composite π pulses mentioned earlier could be tried, a particularly good starting point has proved to be:

Figure 7.9 Broadband decoupling with a train of π pulses.

$$\left(\frac{\pi}{2}\right)_x \pi_{-x} \left(\frac{3\pi}{2}\right)_x$$

which has the advantage that it does not involve 90° phase shifts and so is compatible with existing phase-modulation hardware in decouplers. A modification of the previously described shorthand notation is convenient for representing decoupling sequences involving only 180° phase shifts. Once again we write the pulse durations as multiples of the $\pi/2$ pulse, and indicate phase shifted pulses with a bar over the duration, but since there are no 90° shifts the use of X and Y is unnecessary and simple numbers suffice. Thus the above sequence becomes $1\bar{2}3$ (this justifies the acronym WALTZ applied to the derived sequences - see Chapter 4). Although other starting points are known that are predicted to give better performance, the expansion of $1\bar{2}3$ called WALTZ-16[10] is remarkably tolerant of practical problems such as mis-calibration of the B_2 field, and appears to be the method of choice. WALTZ-16 can be represented:

$$\cdots Q \bar{Q} \bar{Q} Q \cdots$$

where the sequence of phase alternations is to be repeated without intervening delays, and:

$$Q = \bar{3}4\bar{2}3\bar{1}2\bar{4}2\bar{3}$$

This can usually be implemented in a straightforward way using existing decoupler hardware, and will probably become standard on spectrometers built after 1984. Figure 7.10 compares the efficiency of decoupling of WALTZ-16 with an older modulation scheme (frequency-swept square wave modulation).

Figure 7.10 Comparison of broadband decoupling by WALTZ-16 with a commercial modulation scheme (frequency swept square wave). The carbon line of benzene is observed while the decoupler offset is varied; the same field strength was used in each case. The linewidth in the WALTZ-16 spectra is < 0·3 Hz.

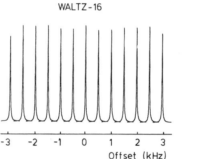
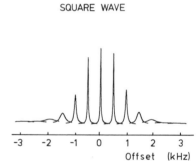

WALTZ-16

SQUARE WAVE

Offset (kHz)

Offset (kHz)

The use of composite pulse decoupling essentially eliminates problems of sample heating during broadband proton decoupling, even at 500 MHz. It also makes possible high-resolution measurements in the presence of broadband decoupling, as the linewidths achieved can be 0·25 Hz or less; in the time domain the corresponding increase in T_2 can be very beneficial to spin echo experiments. Finally broadband decoupling of other nuclei, such as ^{31}P decoupling with ^1H observation, becomes much more feasible. The effective decoupling bandwidth for WALTZ-16 using field strength B_2 is about $2B_2$; field strengths of up to 5 kHz can usually be obtained without the use of extravagant amounts of *rf* power, depending of course on the probe and nucleus. On a medium field spectrometer (250 MHz) this would allow, for instance, ^{13}C irradiation over 160 p.p.m. or ^{19}F irradiation over 40 p.p.m. with fully effective decoupling and narrow lines. If slightly greater residual broadening of the lines is acceptable, other sequences are known which provide an even wider bandwidth (e.g. GARP-1[11], with an effective bandwidth of nearly $5B_2$).

7.5 RELAXATION AND REPETITION

7.5.1 Introduction

The relationship between the two relaxation parameters T_1 and T_2 and the ideal pulse width α and repetition rate of scans T_r in a multiscan experiment is complex. However, it is often possible to make simplifying assumptions which allow an experiment to be set up in an optimum fashion, provided T_1 and T_2 can be measured or estimated. T_2 measurement is hardly practical, but for strong samples a rough measurement of T_1 can be made, as described in section 7.5.2, and of course this sets an upper bound on T_2. In this section I want to indicate some of the factors that must be taken into consideration when selecting experimental parameters; I am concerned throughout with *maximising* sensitivity and *minimising* experimental artefacts, but *not* with any regard for quantitative measurement of signal intensities, which requires establishment of *equilibrium* rather than steady-state conditions (section 7.6).

7.5.2 Pulse Width and Repetition Rate

In many experiments the pulse widths employed are determined by other considerations than relaxation, such as the need to bring about particular types of coherence transfer (see Chapter 8); in such cases the repetition rate is the only free parameter. In simple 1D spectra, however, we are free to choose the pulse width as well, and the two factors together will have considerable influence on the experimental outcome. To understand why, we have to consider two related phenomena: what is happening to the z magnetisation, and what is happening to that in the x-y plane. The relative rate of recovery of z magnetisation compared with the repetition rate will determine how much signal we can get out of the sample, while if we pulse fast enough that the x-y magnetisation does not completely disappear between pulses various complications ensue. To begin with it is easier to ignore the transverse magnetisation, i.e. to assume that $T_2 \ll T_r$, and concentrate on the z-magnetisation; unfortunately this is not often a realistic assumption, and we will have to return and examine what happens when it breaks down later.

If a sample is at thermal equilibrium, then application of a $\pi/2$ pulse elicits the maximum NMR signal, as all the magnetisation is rotated into the transverse plane. For repetitive signal averaging, however, this is not the best pulse angle to use, because the z magnetisation will not be substantially regenerated each time unless the delay between pulses T_r happens to be many times T_1. This is not usually the case, since, unless we are doing some experiment such as a T_1 measurement in which sampling of the equilibrium magnetisation is required, such slow repetition is inefficient. Instead it is better to repeat at a rate equal to the acquisition time, which is in turn determined by the required resolution, and select the pulse angle α so as to maximise the steady-state of the z magnetisation thus created. The best value for α actually depends on the offset of signals within the spectrum[12], but it is possible to define an average optimum value α_E, known as the *Ernst angle,* given by:

$$\cos \alpha_E = e^{-T_r/T_1} \tag{7.7}$$

Under conditions where $T_2^* \approx T_2 \approx T_1$, and $T_r \, (= A_t) > T_2$, this is the best pulse angle to choose. Such conditions are often approximately fulfilled in proton NMR (although in practice we might be inclined to choose a shorter pulse to allow quantitative measurements). They are rarely fulfilled for observations of heteronuclei, where resolution requirements may be low and hence A_t short, and T_2 can be reduced by poor broadband decoupling of protons, or in 2D NMR where the acquisition times are often set in the region of T_2^*.

If we are compelled to use $\pi/2$ pulses (or generally any sequence which entirely eliminates the z magnetisation) by the experimental protocol, then obviously the above formula does not allow determination of a sensible repetition rate. In this case the ideal is determined by the balance between how many scans we can accumulate in a given time and how much we reduce the steady-state magnetisation by accelerated pulsing. *Still assuming complete decay of transverse magnetisation between scans,* the optimum repetition rate can be shown to be[13] $T_r = 1\cdot27T_1$. The signal-to-noise ratio in this case is about 80% of that we would have obtained in an experiment using faster repetition and the Ernst angle, had that been possible or meaningful. For many experiments involving elimination of z magnetisation by the pulse sequence, in which the tranverse magnetisation *does* decay between scans, or in which the effects described below are unimportant, repetition at about $1.3T_1$ is the best choice.

None of the above circumstances are actually all that common. It is much more likely that our T_2^* will be significantly less than T_2, T_2 might be less than T_1, and, being pressed for time and sensitivity, we will want to acquire only long enough to give minimal resolution without undue loss of signal (e.g. $A_t = T_2^*$). Such conditions apply nearly always in 2D NMR, and often in 1D experiments with heteronuclear observation. Repetitive scanning with $T_r < T_2$ leads to the creation of *steady state echoes*[14], because transverse magnetisation which still exists by the time the next pulse arrives is refocused during the subsequent scan. The analysis of this situation is too complex to pursue here; see references 14 and 15. The practical results are as follows.

In 1D spectra acquired in this regime, signal phases and intensities become a function of resonance offset, *unless* α is set to the Ernst angle. This is, however, no longer the best angle from a sensitivity point of view; higher values are better. In the extreme case of $T_r = T_2^*$ and $T_2^* \ll T_2, T_1$ (e.g. with poor homogeneity of the static field) the optimum pulse angle becomes *independent* of T_1[15], and is about 0.4π. In practice, if a phase cycle such as the CYCLOPS procedure for reducing artefacts of quad detection is in force, steady state echo effects in 1D spectra are greatly reduced. They can be further attenuated if necessary by introducing periodic random variations of T_r[14]. In 2D NMR experiments involving coherence transfer (see Chapters 8,9), the presence of residual transverse magnetisation or other coherences at the beginning of the next scan can cause diverse artefacts, as undesirable extra transfers are brought about by the subsequent pulses. In such cases it is necessary to wait longer between scans than sensitivity considerations dictate; setting T_r around $3T_1$ should ensure sufficient decay of all coherences.

7.5.3 Quick T_1 Determination

The above discussion is not much use unless you have some idea of the appropriate T_1 values for the sample in question. The inversion-recovery method for determining T_1, which you may recall is:

$$\pi - \tau - \frac{\pi}{2} \, Acquire \cdots$$

can be used for a rough-and-ready measurement if the sample is strong enough to give visible signals in one scan. Since the z magnetisation at the end of the τ interval is given by:

$$M_z = M_0(1 - 2e^{-\tau/T_1}) \qquad (7.8)$$

the detected signal passes through a null when

$$e^{-\tau/T_1} = \tfrac{1}{2} \qquad (7.9)$$

So, if you perform the experiment repetitively, starting with very small τ to determine the phase correction (remembering that the peak will be inverted at this stage), and adjusting it for a null (Figure 7.11), T_1 will be given by:

$$T_1 = \frac{\tau_{null}}{\ln 2} \qquad (7.10)$$

This will be a rather inaccurate measurement, but is good enough *for* selection of repetition rates etc. The most probable cause of error is impatience, leading to repetition of the experiment before all nuclei have fully relaxed, and hence underestimation of longer T_1's. In a realistic sample the T_1's are likely to vary anyway, so some compromise value will have to be chosen and errors in the measurement are unlikely to be significant in comparison.

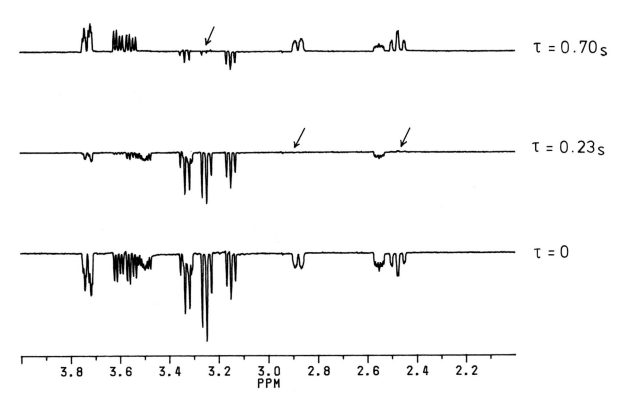

The catch with this is that the very samples for which optimisation of acquisition parameters is most important are those dilute ones on which the T_1 determination cannot easily be made. If signal averaging is attempted for the above experiment it is necessary to leave about $5T_1$ between scans, and this rapidly becomes intolerably boring. In order to cope with such situations it is advisable to develop a feel for the T_1's appropriate to the compounds and nuclei you work with, perhaps by performing quick measurements when you have the chance and keeping a log of them. Although there are many tables of T_1 values available in the literature, these are not very helpful in practice because they have usually been determined under unrealistic conditions (from the point of view of everyday spectroscopy; for instance in degassed solutions). Proton T_1's for medium molecular weight molecules in the presence of oxygen are often only a second or two, but plenty of exceptions occur both to longer and shorter values. Other spin $\frac{1}{2}$ nuclei can show extremely slow relaxation, particular offenders in this respect being quaternary carbons, and ^{29}Si and ^{15}N in a variety of environments. For unprotonated ^{15}N T_1 values of many *minutes* are common, even in non-degassed solution. Quadrupolar nuclei (spin > $\frac{1}{2}$) are efficiently relaxed by their interaction with local *electric* field gradients, and nearly always show short T_1's ranging from μs to hundreds of ms.

Figure 7.11 Finding approximate T_1's by the inversion-recovery null method. τ has been adjusted to null the arrowed peaks in each case; the corresponding T_1's are 0·3 s (middle spectrum) and 1·0 s (top spectrum).

7.6 QUANTITATIVE INTENSITY MEASUREMENTS

7.6.1 Introduction

Integration of spectra to give information about peak areas is very familiar from routine proton NMR. It is quite straightforward to obtain accurate enough integrals to allow determination of the relative number of protons contributing to a particular peak. It is very tempting to extrapolate from this application to all kinds of other areas where determining the relative amounts of species in solution would be informative, such as in kinetic experiments or assays of product mixtures. Unfortunately, while the 10-15% accuracy which is normally adequate for comparing proton counts is easily achieved, the kind of accuracy that would be considered acceptable for other quantitative applications (say better than 1 or 2%) can prove elusive. In this section I will examine briefly some of the reasons why NMR, and particularly FT NMR, is not easy to apply in a rigorously quantitative manner. This topic is covered in other practical NMR texts in considerable detail, but it is so important, and chemists' expectations of what is possible are often so exagerrated, that I feel it is worth gathering together some of the salient points here.

7.6.2 Making the Signals Proportional to the Number of Nuclei

In optimising an NMR experiment for sensitivity as in the previous section, we work with a regime in which the z magnetisation reaches a steady state removed from thermal equilibrium, in order that scans can be accumulated at a reasonable pace. For quantitative measurements, however, this is no good at all, for the signal intensity is a function of T_1, and so unless all the T_1's of the nuclei to be measured are *identical* errors will be introduced. In principle if one knew the T_1's involved the measured intensities could be corrected; however a T_1 determination of sufficient accuracy will be more difficult to make than a simple measurement of a spectrum without saturation, so this is not a practical option. However an approximate check on T_1, as described in section 7.5.3, will be useful in order to find out how long the interval between pulses T_r need be to avoid saturation.

To get some idea of how difficult it may be to keep the z magnetisation of all nuclei in thermal equilibrium, consider some specific figures. Proton T_1's are usually a second or two, as mentioned earlier, but we have to remember that we must take the *longest* T_1 in the molecule to determine the acceptable repetition rate. Suppose this has been measured at 3 s. If we use pulses having a flip angle α, and apply them every T_r seconds, then the steady state z magnetisation M_z set up is given by:

$$M_z = M_0 \left(\frac{1 - e^{-T_r/T_1}}{1 - e^{-T_r/T_1} \cos \alpha} \right) \qquad (7.11)$$

where M_0 is the equilibrium magnetisation. This assumes the transverse magnetisation is allowed to decay completely between scans, as will certainly be the case in a quantitative experiment. For the particular case of $\alpha = \pi/2$ this obviously simplifies to:

$$M_z = M_0 (1 - e^{-T_r/T_1}) \qquad (7.12)$$

Now suppose we specify that M_z must not deviate from M_0 by more than 1% (i.e. it must be at least $0.99 M_0$). For our 3 s T_1 nucleus this means that T_r must be about 14 s if we apply $\pi/2$ pulses! For shorter pulses the situation is a little better, for instance $\pi/4$ pulses require about a 10 s delay to satisfy the same condition; however we do not recover as much sensitivity from the extra scans that can be fitted in as we lose from shortening the pulse, so this is not a good choice if sensitivity is critical. A reasonable guideline is to use $\pi/2$ pulses and a delay of at least 5 times the longest T_1 to keep errors due to saturation below 1%. This can easily become a very long time; for instance ^{13}C T_1's of 10-30 s are common, necessitating several minutes between scans. In such a situation it can be extremely profitable to use relaxation agents such as chromium (III) acetylacetonate to artificially reduce T_1.

There is an additional problem in heteronuclear experiments if broad-band proton decoupling is applied. Since this saturates the proton transitions, a nuclear Overhauser effect (Chapter 5) may arise, and this can cause very large intensity variations independent of the effects described above. For ^{13}C, for instance, the maximum nOe from protons is about 200% (expressed as the *enhancement*, i.e. the signals with nOe can be 3 times as big as those without). While protonated carbons will often show such an nOe in the presence of decoupling, quaternaries may show little or none, and obviously this makes a mockery of any intensity measurement however much care is taken to ensure proper relaxation.

Fortunately the nOe does not arise instantaneously on commencement of decoupling, but builds up on a timescale comparable with the ^{13}C T_1's, so provided decoupling is switched off long enough between scans it can be suppressed (the *inverse gated decoupling* experiment, Figure 7.12). 'Long enough' means a time long compared with the ^{13}C T_1, and since we have to leave 5 T_1 between pulses anyway, but are scarcely likely to be using an acquisition time anywhere near this long, this presents no problem. Relaxation reagents help suppress the nOe as well as shortening T_1's, so there is even more incentive for their use if broadband decoupling is required. It is important to remember that because the nOe must be suppressed, and because the scan repetition rate must be low, one's expectations of the signal-to-noise ratio achievable with a given amount of sample in a quantitative heteronuclear experiment must be substantially downgraded. This is very troublesome, because as we shall see below high signal-to-noise ratio is a prerequisite for accurate quantitative measurement.

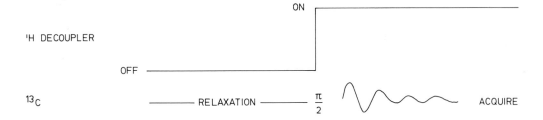

7.6.3 Sampling the Data

Figure 7.12 The experimental scheme for inverse gated decoupling.

Having taken the trouble to ensure that the signals coming from the sample reflect the relative amounts of the nuclei from which they originate, we are still only part of the way to doing a properly quantitative experiment. The digital sampling of the data which occurs in a Fourier transform spectrometer, and the numerical extraction of areas from the digitised spectrum, can introduce further errors if incorrectly performed. The signal-to-noise ratio of the spectrum also has an influence on the accuracy of the integral, which should be obvious, but sometimes it is hard to keep in mind just how significant this is. *Integrals,* after all, seldom look noisy. These questions have been analysed in detail in the context of measuring nOe's[16], and here I summarise the conclusions of that paper.

We must remember that a typical peak in an NMR spectrum is likely to be quite sharp, and may only be characterised by a few data points. The integral is normally evaluated numerically using the trapezium rule, and inadequate digitisation will make this calculation inaccurate (there are better ways to evaluate the area, such as iterative fitting of a calculated line to the observed one, but since simple numerical integration is most common I will concentrate on this). It transpires that the requirements for digitisation are not too stringent, and provided there are more than 3 points across the half-height width of the line errors from this source will be less than 1%. A practical linewidth in proton NMR might be 0·5 Hz, which matched filtration would increase to 1 Hz, so around 0·2 Hz/point would be safe digitisation. This is a little better than that commonly employed in acquiring full width spectra (typically 0·3-0·4 Hz/point), but is easily achieved. In *heteronuclear* experiments, of course, it is more common to use poor digitisation, and so care must be taken to improve this substantially for quantitative work, either by increasing the number of data points or by broadening the lines with an appropriate window function.

The next factor which must be considered is how far each side of the line the integral should extend to properly represent the area of the peak. From the point of view of a true Lorentzian line, the answer is that the integral must be extended around 20 times the linewidth each side to cover more than 99% of its area; this means at least 20 Hz each way for the above example. In real life, however, a number of purely practical problems will intrude. The line will have spinning sidebands, and possibly ^{13}C satellites; these must either *always* be included in the integral or *always* excluded to make comparisons meaningful. The first order spinning sidebands are quite likely to be around 20 Hz away anyway, so it is probably advisable to aim to include these. Higher order sidebands and satellite lines can be the subject of discretion; it is better to include them, but if there are other lines nearby the more important requirement of excluding their wings or satellites may preclude this. The important thing is to be aware of the problem, and to take a specific decision as to what to do about it.

The digitisation of the spectrum, and the manner in which we integrate it, are at least completely under our control, so the systematic errors due to these sources can be reduced as much as necessary. There remains the random error in the measurement due to noise in the spectrum, about which we can do nothing other than making the sample stronger, buying a higher field spectrometer, or waiting longer for the spectrum to accumulate. Usually the room for manoeuvre in the first two of these possibilities is limited, and our only option is to increase the accumulation time until the signal-to-noise ratio is adequate, but because this only improves as the square root of the number of scans there is usually little benefit in spending more than a few hours signal averaging. The error in the integral due to noise depends on both the signal-to-noise ratio *and* the manner in which the spectrum is digitised; taking the parameters discussed above, which are chosen to reduce *systematic* errors due to digitisation below 1%, it is possible to calculate that in order to be 99% certain that the measured integral falls within $\pm 1\%$ of the true value, we require a signal-to-noise ratio for the integrated peak of about 250:1 (defined as in Chapter 3). This kind of sensitivity can usually be achieved for proton observation without much difficulty, but for other nuclei it may be a problem. As a reference point, Figure 7.13 shows a spectrum in which the signal-to-noise ratio of the marked peak is approximately 250:1.

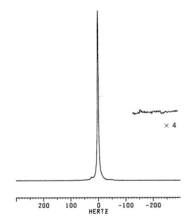

Figure 7.13 A line with a signal-to-noise ratio of 250 : 1.

7.6.4 Other Instrumental Problems

A variety of other instrumental limitations can introduce systematic errors into intensity measurements. Foremost amongst these is the difficulty of adjusting FT spectra so that all the peaks have pure absorption phase. Introduction of a proportion of the dispersion component into a peak will evidently reduce its intensity, as the integral of the dispersion part is zero. Provided this does not reduce the size of the integral so much that noise becomes a problem, it will only be *differential* phase changes between peaks that cause errors, so it is the frequency-dependent part of the phase correction that is most important. Using the normal two-parameter phase correction (Chapter 4) in a spectrum with good signal-to-noise ratio, it should be possible to adjust the phase by eye within a few degrees of ideal, *provided* the assumption of linear variation with frequency holds. If this assumption breaks down, then the phase of each individual peak should be adjusted before it is integrated. A 5° error in phase alters the integral by under 1%. For really accurate work a procedure exists which allows the phase correction to be checked very carefully[17]; this would only be useful at a level of accuracy where it would also become necessary to use iterative curve-fitting rather than numerical integration.

Another mundane problem that can cause large errors is the condition of the spectral baseline. Even in the complete absence of noise, for a variety of instrumental reasons the baseline of the spectrum (i.e. the value it adopts where there is no signal) is unlikely to be either exactly at zero or in exactly the same place throughout. A major source of baseline roll is the bandpass filter used between the receiver and the ADC, which is stimulated into oscillation by the onset of the NMR signals.

Effects due to this filter can be greatly reduced by suitable adjustment of the delays between the pulse, the activation of the receiver and the start of digitisation. Typically the delay between the pulse and the acquisition of the first data point is required to be equal to the reciprocal of the sampling rate ($1/2\mathcal{N}$, in the notation of Chapter 2), for reasons unrelated to the baseline (minimisation of the first-order phase correction). The point during this delay at which the receiver is activated then determines the type of baseline distortion generated; it has been shown[18] that, assuming the filter bandwidth equals the spectral width, turning on the receiver 0·6 of the way through the delay reduces the distortion to a simple DC offset (Figure 7.14). Spectrometers do not always do this automatically, so it is worth investigating whether or not such a scheme is in operation.

Figure 7.14 Spectra often exhibit baseline roll, particularly if they include relatively broad, intense peaks. In the left hand spectrum the receiver has been turned on half way through the post-pulse delay, while for the right hand spectrum the timing scheme described in the text was employed. The vertical scale is the same in each case.

Inevitably there will still remain a contribution from the baseline whatever steps are taken to minimise it, and integral routines always incorporate some degree of correction for this, which may be automatic or under operator control (with adjustments such as 'offset and drift' or 'bias and slope'). Many spectrometer software packages also incorporate more sophisticated baseline flattening routines, allowing the operator to subtract various functions from the spectrum in such a way as to give a flat baseline. I find that subtraction of a simple 4-term polynomial, with manual rather than automatic adjustment of the coefficients, gives the most reliable results, and that this is better than using the built-in corrections of the integral routine alone.

Finally, in comparing the integrals of peaks in different parts of the spectrum, we are making an assumption about the *linearity* of the whole chain of events which leads from the pulse to the digitised data. We assume the pulse excites signals equally over the spectral range (a fair assumption for $\pi/2$ pulses with offsets a good deal less than B_1), and that each link in the receiver chain amplifies every frequency thereby generated to an equal extent. Given the complexity of the average receiver, which might involve processing the signal through a pre-amplifier, a mixer, an IF amplifier, another mixer, an audio amplifier, an adjustable bandwidth audio filter and a digitiser, the second assumption seems questionable. The audio bandpass filters which must be used to optimise sensitivity are a prime candidate for introducing non-linear response over the spectral range. In lieu of a theoretical analysis of this problem on a given spectrometer, a fair experimental test of it can be made by using a sample with a single line and varying its position within the spectral window. This is not *exactly* equivalent to running a spectrum with fixed transmitter offset and lines in different places, but nevertheless gives some indication of the degree of error to be expected from this source.

The moral of this section is that there are numerous contributions to the error in a quantitative measurement made by FT NMR, and while each of them may be reduced to 1% or so in a practical fashion, the combined error is still likely to be significant. I am always sceptical of measurements purporting to be accurate to better than a few percent overall, *unless* they come with evidence that careful attention has been paid to the above details.

7.7 SELECTIVE EXCITATION AND SUPPRESSION

7.7.1 Introduction

There are several circumstances in which it is useful to select portions of the spectral range of a nucleus for excitation while leaving others undisturbed. According to the relative proportions of the affected and unaffected parts, we might think of this as being selective or *tailored* excitation of peaks, or conversely as peak suppression. At the 'suppression' end of the scale the most common problem is eliminating a single, intense peak, to alleviate the dynamic range difficulties discussed in Chapter 3. Examples of selective excitation include experiments which require selective pulses such as the transient nOe method or SPI, and procedures for extracting a small part of a complex spectrum. This is a very large topic, and to avoid excessive length I will simply present a few selected methods appropriate to the problems encountered in relatively routine spectroscopy.

7.7.2 Peak Suppression

By pre-saturation

A review of the literature of peak suppression would reveal a bewildering array of methods, for this has been a very popular field for experimentation. Fortunately this is a rare case where it is possible to state unequivocally that one method is best: pre-saturation. I hasten to add that this statement only applies subject to the condition that the peak to be eliminated is not involved in chemical exchange with any other interesting resonances. In the absence of exchange, pre-saturation of a resonance in order to decrease its intensity is 'best' in the sense that it combines the highest degree of suppression with minimum distortion of the rest of the spectrum. These are the two criteria we are interested in: how well can the offending peak be attenuated, and what else happens to other more interesting peaks.

Pre-saturation is brought about by irradiating the resonance to be suppressed with a weak *rf* field, for long enough to eliminate the population difference across its transition. The irradiation is turned off immediately before the application of pulses and acquisition of the spectrum, so that nearby peaks are not subject to Bloch-Siegert shifts (Figure 7.15). Typically dynamic range problems requiring peak suppression are only encountered in proton spectra, so the irradiating field is conveniently available using the decoupler. In optimising the experiment it is necessary to select the field strength to be used and the length of the irradiation. The field strength B_2 needs to be sufficient to make the 'saturation factor' $B_2^2 T_1 T_2$ (B_2 in Hz) large (e.g. > 100), while the irradiation must be applied for long enough to allow saturation to build up (e.g. a few times T_2).

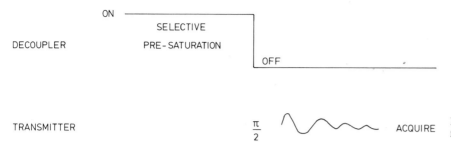

Figure 7.15 The experimental scheme for peak suppression by pre-saturation.

It happens that the need for peak suppression most often occurs in aqueous solution (because water has a strong residual solvent peak, and because many biological materials must be dissolved in water, have high molecular weight and are in short supply), so the peak to be suppressed is that due to HDO. The relevant T_1 and T_2 values (although highly dependent on what is dissolved in the solution) might then be 5 s and 1 s, so a B_2 field of 20 Hz applied for 2-3 s should be sufficient. In practice it is easy to determine suitable parameters empirically without knowledge of the relaxation times, since the sample is obviously strong and preliminary experiments can be performed quickly. The aim should be to use the minimum field strength (for maximum selectivity) for the shortest time (for fast accumulation of scans) that gives acceptable suppression. Although the fundamental reason for suppressing peaks is reduction in the dynamic range of the spectrum, which only becomes necessary for intensity ratios greater than a few thousand to one (for a 12 bit digitiser), *aesthetic* improvements arise much sooner than this (Figure 7.16). Attenuating a strong peak can make it easier to observe its weak neighbours, and may make the spectral baseline less distorted.

In 2D NMR experiments such as COSY (Chapter 8) it is desirable to turn off the saturating field before the start of the whole of the ($\pi/2$ - t_1 - $\pi/2$ - *Acquire*) sequence. However, this may lead to an unacceptable degree of recovery of the solvent magnetisation during the t_1 interval, in which case irradiation must be left on until the start of t_2 (i.e. the acquisition of data). Since the B_2 field is present during t_1 it may cause Bloch-Siegert shifts, which will appear *only* in v_1 in the transformed spectrum, disturbing the symmetry of the experiment. For the example given above of a 20 Hz field, resonances within a few hundred Hz will show appreciable shifts, so this effect can be quite annoying.

The pre-saturation scheme becomes unsatisfactory if the irradiated nucleus is involved in chemical exchange with other interesting nuclei, since the saturation propagates to them via the exchange. This is quite a common problem, because of the frequent use of peak suppression in aqueous solution; the protons of water will of course exchange with functional groups such as OH's and NH's. If we are trying to observe such groups we must also not be using *deuterated* water as solvent, so the pre-saturation experiment becomes unsuitable at exactly the time when we most need to reduce the dynamic range. This situation, which biologists often encounter, requires other methods which avoid exciting the water resonance in the first place.

By tailored excitation

Many schemes have been proposed which attempt to tailor the frequency distribution of the *rf* excitation so as to avoid stimulating a region of the spectrum. This generally involves replacing the normal non-selective hard pulse with either a group of weak pulses, or a sequence of hard pulses designed to have an equivalent effect (a kind of composite pulse). The most common 'soft pulse' method is Redfield's 2-1-4 tailored pulse[19], but although this is quite effective it requires that the spectrometer be able to

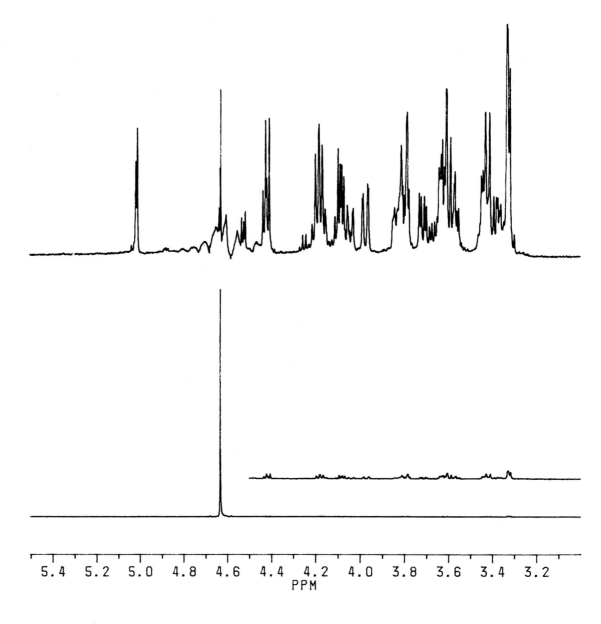

Figure 7.16 Suppression of the peak due to HDO by pre-saturation. The inset in the lower spectrum has a ×8 vertical scale. The HDO peak has been suppressed in the upper spectrum.

deliver long, weak pulses, which is often not possible. It is also a fairly demanding experiment to set up if very high degrees of peak suppression are required. Recently a number of methods have been suggested which use composite hard pulses to achieve a similar effect, and as these are likely to be of more widespread use I will restrict my attention to them.

The sequence:

$$\left(\frac{\pi}{2}\right)_x - \tau - \left(\frac{\pi}{2}\right)_{-x} Acquire \cdots$$

known as 'jump and return'[20] (JR), is particularly easy to understand, and is the parent of a whole family of experiments. If the transmitter frequency is set exactly on the peak to be suppressed, this peak will remain static in the rotating frame during the τ interval, so the second pulse will return it to the z axis. Other peaks, however, will precess during τ, so that some or all of their magnetisation will be unaffected by the second pulse (Figure 7.17). Clearly any line which precesses through exactly π, 2π $\cdots$ during τ (i.e. a line with offset $1/2\tau$, $1/\tau$ $\cdots$ Hz) is also suppressed, while for offsets in between 0 and $1/2\tau$ there is a variable band of excitation with a maximum at $1/4\tau$. τ is selected to put the region of maximum excitation over the spectral area of interest.

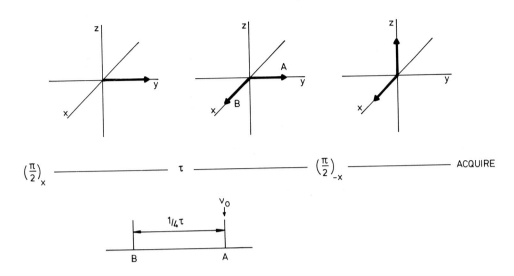

$\left(\frac{\pi}{2}\right)_x$ ──────────── τ ──────────── $\left(\frac{\pi}{2}\right)_{-x}$ ──────── ACQUIRE

The disadvantage of this experiment, which it shares with others of its type, is immediately apparent from this description. The excitation is not uniform, so that quantitative comparisons between different parts of the spectrum become impossible; this appears to be an unavoidable cost of not exciting the solvent resonance. The width of the null at the transmitter frequency is also small, which means that if the solvent resonance is broad, or has spinning sidebands, or the transmitter frequency is incorrectly set, suppression will be poor.

It proves possible to construct sequences with better properties, particularly with respect to the width of the null, by adding more pulses with relative durations given by binomial coefficients[21]. If we use the short-hand notation already introduced for decoupling sequences, so that JR is written $1\bar{1}$, the next few members of the series are $1\bar{2}1$, $1\bar{3}\bar{3}1$, $1\bar{4}6\bar{4}1$ and so on, it being understood that all pulses are separated by the delay τ. In these sequences the '1' duration need not correspond to a $\pi/2$ pulse, as long as the correct relationship between the elements is maintained. The sum of the pulse angles of the components is the effective flip angle experienced by nuclei with offset $1/2\tau$, which is commonly chosen to be $\pi/2$ or less; thus JR in its original form is atypical in having an effective flip angle of π (i.e. a null) at this offset.

The best compromise between complexity of pulse sequence and performance as a peak suppression method appears to be offered by $1\bar{3}\bar{3}1$. In a practical application one might choose to aim for a net flip angle of $\pi/4$, and select τ according to the spectral range to be excited. For example, to observe most of a normal proton spectrum the region of maximum excitation needs to be set around ± 4 p.p.m. from the water resonance, so for a 500 MHz spectrometer τ would need to be about 250 μs. Since the length of the '1' component should equal one-eighth of the $\pi/4$ pulse width, this can become an inconveniently short pulse. The solution is to reduce the B_1 field by attenuating the transmitter signal so as to make the pulse length manageable. For a $\pi/4$ pulse length without attenuation of 4 μs the '1' duration would be 0·5 μs. On most spectrometers this would be better increased to at least 2-3 μs (to avoid peculiar effects due to the imperfect edges of the pulse), which requires about 12-15 dB of attenuation. Some spectrometers allow the transmitter pulse to be attenuated under software control; otherwise fixed value attenuators which plug in series with the transmitter output are readily available. When using the latter method be careful to place the attenuator between the transmitter and the pre-amp,

Figure 7.17 The effect of jump-and-return on two lines. one on-resonance and the other offset by $1/4\tau$.

not between the pre-amp and the probe, since otherwise the NMR signals will also be attenuated.

Spectra acquired with a $1\bar{3}3\bar{1}$ excitation pulse typically show a several thousand-fold reduction in the intensity of the solvent resonance, comparable with the results obtained by pre-saturation. The phase of the spectrum is a strong function of offset, but this is nearly linear and so can be corrected in the usual way. The phase also inverts at the transmitter frequency, so unless you have a special additional phase correction to account for this half the spectrum will be upside down (Figure 7.18).

Figure 7.18 Solvent suppression using a $1\bar{3}3\bar{1}$ pulse (for an unsuppressed spectrum to compare, see Figure 7.16). Note that the single peak to low-field of the solvent appears inverted.

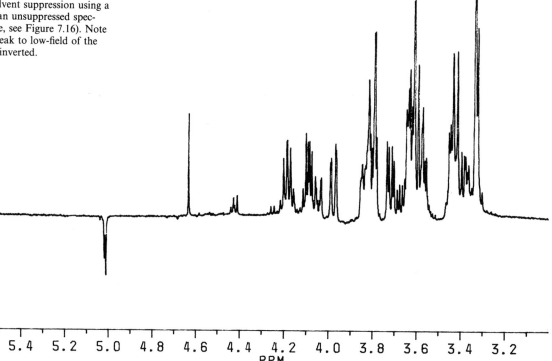

7.7.3 Selective Excitation

With soft pulses

The most straightforward way to excite a limited spectral region is to reduce the strength of the B_1 field. How can this be related to the effective bandwidth of the pulse? Notice that we wish to convert the variation of the *rf* amplitude as a function of *time* to its variation as a function of *frequency*, that is to convert from the time to the frequency domain. Provided the nuclear system is *linear* (i.e. provided its response to a combination of stimuli is the sum of its responses to the individual stimuli), this can be achieved by Fourier transformation of the time variation of the *rf* amplitude. The Fourier transform of a rectangular pulse (a good approximation to the signal envelope produced by turning the transmitter on then off again) is the ubiquitous function $(\sin x)/x$, or sinc x (see Chapter 2, Figure 2.16).

For a pulse of duration τ the corresponding sinc curve has nulls every $2/\tau$ Hz; the useful area for uniform excitation is well inside the first pair, say $\pm 0{\cdot}2/\tau$ from the centre frequency. However, because there is appreciable intensity outside the first pair of nulls, from the point of view of selectivity (i.e. not perturbing other resonances), the excitation must be considered to extend some distance, perhaps $\pm\ 10/\tau$ from the centre frequency. Thus to avoid affecting signals 200 Hz away we should select a pulse at least 50 ms long; the field strength can then be adjusted to set the flip angle at resonance to the desired value (e.g. 5 Hz for a $\pi/2$ pulse). Note that the selectivity is much worse than might be supposed from examination of the field strength alone; this is a consequence of the *shape* of the pulse. This 50 ms pulse, despite its poor selectivity, would only be useful for *exciting* resonances uniformly in a narrow band around ± 4Hz from resonance.

This kind of soft rectangular pulse is widely used, despite its evident disadvantages. Typically the pulse is required at proton frequencies, perhaps for the SPI experiment described in Chapter 6, or to invert a resonance for a transient nOe measurement (Chapter 5). In this case the decoupler can be used as the source of the irradiation, so there is usually little difficulty adjusting the field strength and length of the pulse. If the additional feature of continuous control of the *rf* amplitude *during* the pulse is available, then a much more selective result can be obtained by suitable choice of the pulse shape. Giving the pulse a Gaussian envelope eliminates the sidelobes in the frequency domain characteristic of the rectangular pulse, and ensures that the excitation falls smoothly and rapidly to zero as offset increases[22] (this is extremely similar to the manner in which apodisation removes the wiggles from the transform of truncated data - Chapter 2). Figure 7.19 compares the experimental offset dependence of the excitation produced by equivalent Gaussian and rectangular pulses.

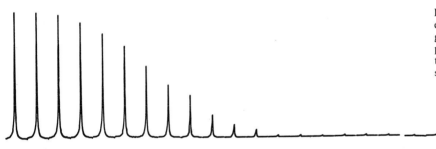

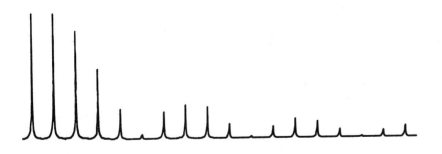

Figure 7.19 Offset dependence of the excitation produced by equivalent rectangular (bottom) and Gaussian (top) pulses. The pulse length was 80 ms, and the offset was incremented in 2·5 Hz steps.

With DANTE

While nearly all spectrometers have a source of proton irradiation which can be switched on for a long time and adjusted in power (the decoupler), it is much less common for this facility to be available at other frequencies. The broadband tranmitter output can always be attenuated, of course, but often pulse amplifiers are designed to operate only for a few hundred μs and will be damaged if turned on for longer. If a selective pulse is required on a heteronucleus (for instance to do a ^{31}P magnetisation transfer experiment), then a different strategy must be adopted. The DANTE sequence[23] (Delays Alternating with Nutations for Tailored Excitation, a name with a particularly fiendish derivation) consists of a train of m hard pulses of very small flip angle, with constant phase and constant separation:

$$\alpha - \tau - \alpha - \tau - \alpha - \tau - \alpha - \tau - \alpha - \tau - \cdots$$

The effect of this can be understood by considering the Fourier transform of the whole pulse train as for the rectangular pulse, or alternatively using the vector model. In the latter representation, a resonance exactly at the transmitter frequency remains static in the rotating frame during the τ intervals, so the small pulses have a cumulative effect, leading to a net flip angle on resonance of $m\alpha$. Other resonances move in between pulses, and so are generally less excited, unless they happen to rotate through an integral number of cycles. Thus the excitation has sidebands every $1/\tau$ Hz, and a detailed analysis shows that each sideband has a sinc-like shape.

The selectivity of irradiation at each sideband is determined by the total duration of the sequence $m\tau$, the distance between nulls in the sinc curve being approximately $2/m\tau$. Thus it is desirable to make α a very small fraction of the desired pulse width, subject to $m\tau$ remaining small in comparison with relaxation times. Once again, attenuation of the transmitter output will be needed to avoid inconveniently short pulses. τ itself can be selected to keep the sidebands away from peaks which should not be excited, with the transmitter frequency set on the region of interest.

DANTE can be used for any experiment requiring selective excitation; Figure 7.20 illustrates a method for unscrambling a complex coupled ^{13}C spectrum. During the DANTE train broadband proton decoupling is in effect, and m and τ are chosen so as to apply a $\pi/2$ pulse to a single line. The decoupler is then turned off for the acquisition, so that the coupled multiplet structure is recorded.

7.8 SOME SPECTROMETER TESTS TO TRY

The sensitivity, lineshape and resolution tests described in Chapter 3 are taken as the basic criteria of spectrometer performance, and when buying an instrument much time is spent investigating these parameters. However, in reality there is not much to choose between the various makers in these matters; even when one or other instrument does seem to have an edge on the test samples, it is often found that under more realistic conditions the difference disappears. Since investing in a high-field NMR spectrometer is usually quite a big decision, even for a major industrial concern, I thought I would take this opportunity to mention a few less obvious experiments which can shed some light on instrument quality. None of these is in standard use, so to make comparisons between different spectrometers it will be necessary to specify very precisely what you want doing, and ensure that identical experiments are run on each spectrometer.

Not all the suggestions made here relate to *quantitative* measurements of instrument performance. At the limit, it can prove quite hard to devise quantitative tests of important features such as short-term stability of the field/frequency ratio or B_1 homogeneity; however qualitative differences can be detected. Other areas, such as software style, are entirely subjective, but no less important for that. One of the hardest things in setting out to assess an instrument is deciding what you would like it to do; in a University the answer is usually 'everything', but this is not an adequate starting point. In industry the application may be much better defined.

The relative weight given to different aspects of an instrument should vary according to its intended use. For instance, if the object is to run routine proton spectra, all of the conventional tests, and what follows here, can be ignored. Instruments from all four major manufacturers (JEOL, Bruker, GE NMR Instruments and Varian) have adequate performance for this application. Instead the area of competition should be the software package; how simple, quick and convenient is it? Can non-specialists learn

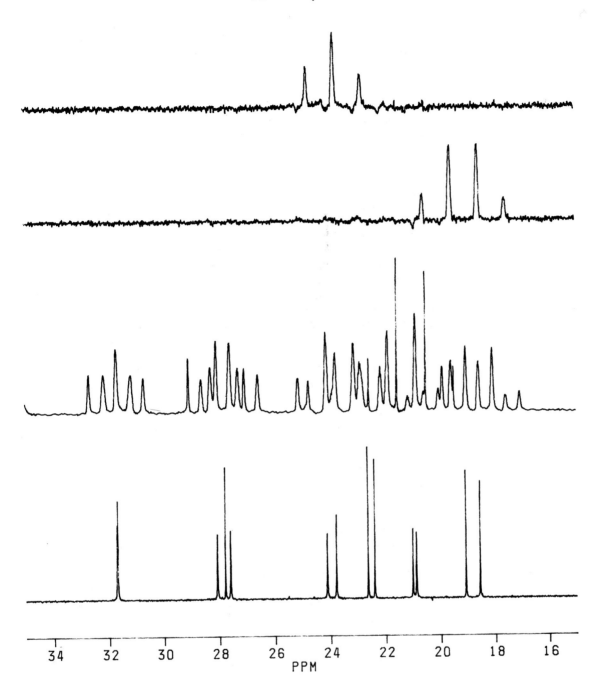

to use it in an hour or two? What provision is there for automated operation? *Is the documentation adequate* (this is often a major deficiency). These are some of the important questions to be answered; however it is difficult or impossible to address them during short demonstrations. A better approach is to talk to a cross-section of existing users of the instruments; they will all have complaints and criticisms, but maybe some will be distinctly more critical than others.

A very different approach is needed if the instrument is to be used in a non-routine way, and pushed to its limits, whether of sensitivity, resolution or sophistication. Here the software is still important, but perhaps a little inconvenience can be tolerated; *flexibility* becomes the key criterion. Many of the experiments described in the rest of this book require considerable control over sequences of events occurring on a μs timescale. Two-dimensional experiments need powerful data handling facilities, and the ability to define the processing mode as freely as possible. These things are

Figure 7.20 The proton coupled ^{13}C spectrum of cholesteryl acetate is a mess even at 125 MHz (middle), but use of DANTE as described in the text allows single multiplets to be extracted (top two traces).

not provided in an entirely adequate way on most current spectrometers; progress is constantly being made however, so careful comparisons are necessary.

However excellent the *control* of the hardware, it remains to see whether it produces the desired results. Matters such as the stability of the field/frequency ratio, stability, accuracy and range of phase shifts, duration and reproducibility of pulse widths, pulse shape and B_1 homogeneity are not easy to test, but may have a crucial impact on multi-pulse techniques. The following paragraphs suggest a series of experiments which can provide at least qualitative information about some of these parameters.

Measure the π pulse width, and then re-measure it with other transmitter phases (there will be at least 4, and maybe more); there should be no variation. Examine the residual signal following a π pulse; this is indicative of the homogeneity of the B_1 field. Check that nulls occur at exactly double, triple $\cdots$ the π pulse length (2π, 3π $\cdots$ pulses). If this is not the case it is symptomatic of an inferior pulse transmitter, which is not able to deliver the longer pulses. Assuming this test was acceptable, try comparing the normal π pulse width with that obtained with 12 dB of attenuation in the transmitter output. This should be increased exactly four times; if not it may mean the edges of the pulse are deficient (but make sure the attenuation is accurate).

Difference spectroscopy is a sensitive test of overall performance. Try acquiring two scans of a spectrum containing a single line, using constant transmitter phase and subtracting them (be careful to allow complete relaxation between scans). The amplitude of the residual peak will depend on the short term stability (see Chapter 5). Do this several times, to get an indication of the range of variation. Try the same experiment, but using alternating transmitter phase and addition. The result should be the same; if not it is symptomatic of poor phase shifting.

Phase shifts other than 180° are harder to test; for recent spectrometers offering small phase increments try:

$$\left(\frac{\pi}{2}\right)_x \left(\frac{\pi}{2}\right)_\varphi \quad Acquire \;\cdots$$

with variable φ; the signal amplitude should depend on sin φ. Phase shifts on the decoupling channel can be tested similarly using a modification of the 'other nucleus' pulse width calibration (section 7.2.3):

$$^{13}C: \qquad\qquad \frac{\pi}{2} - \frac{1}{2J} - \qquad\qquad Acquire \;\cdots$$

$$^{1}H: \qquad\qquad\qquad \left(\frac{\pi}{2}\right)_x \left(\frac{\pi}{2}\right)_\varphi$$

Providing the ^{1}H $\pi/2$ pulse length is accurate, the intensity of the components of the ^{13}C doublet should vary as cos φ. Further indication of quality of phase shifts and overall stability can be obtained by performing a multiple-quantum filtration (not necessarily as a 2D experiment, but see Chapter 8, section 8.4.3 and the last part of section 8.5.1) of the highest possible order; examine the quality of suppression of unwanted peaks and the degree of excitation of desirable ones.

Two-dimensional spectroscopy (Chapters 8,9 and 10) is generally very sensitive to spectrometer defects. Try running a COSY experiment (Chapter 8) on a sample strong enough that thermal noise is negligible. Examine the 't_1 noise' (Chapter 8 section 8.3.6), which is a sensitive indicator of total system performance, but make sure that all conditions (spectral widths, pulse angles, delays, processing mode and window functions) are *identical* on each spectrometer tested, and that the spectrum is not 'symmetrised' (8.3.6 again). *J*-spectroscopy (Chapter 10) is particularly touchy; have some *J*-spectra run *without* EXORCYCLE and see what the 'ghost' and 'phantom' peaks are like[24,25].

Finally, run some experiments of the kind you really want to do. Try this with easy samples, which are known to work, and with difficult ones which have never come out right. Try it with very dilute samples, and with very strong ones (perhaps surprisingly, deficiencies in the receiver electronics can show up more with intense signals). Look for unwanted glitches at the transmitter frequency, other spikes, quad images etc. Try running spectra without the CYCLOPS phase cycle and see how much worse they are. Try running spectra with a lot of data processing activity going on (e.g. a 2D transform); it is quite common for there to be interference between the computer or disc system and the *rf* electronics.

REFERENCES

1.   H. Günther and J. R. Wesener, *J. Mag. Res.*, **62**, 158, (1985).
2.   D. M. Thomas, M. R. Bendall, D. T. Pegg, D. M. Doddrell and J. Field, *J. Mag. Res.*, **42**, 298, (1981).
3.   R. Freeman, S. P. Kempsell and M. H. Levitt, *J. Mag. Res.*, **38**, 453-479, (1980).
4.   M. H. Levitt and R. R. Ernst, *J. Mag. Res.*, **55**, 247-254, (1983).
5.   A. J. Shaka and R. Freeman, *J. Mag. Res.*, **55**, 487-493, (1983).
6.   A. J. Shaka, C. J. Bauer and R. Freeman, *J. Mag. Res.*, **60**, 479-485, (1984).
7.   M. H. Levitt and R. Freeman, *J. Mag. Res.*, **43**, 65, (1981).
8.   M. H. Levitt and R. R. Ernst, *Mol. Phys.*, **50**, 1109-1124, (1983).
9.   J. S. Waugh, *J. Mag. Res.*, **50**, 30-49, (1982).
10. A. J. Shaka, J. Keeler and R. Freeman, *J. Mag. Res.*, **53**, 313-340, (1983).
11. A. J. Shaka, P. B. Barker and R. Freeman, *J. Mag. Res.*, **64**, 547-552, (1985).
12. R. R. Ernst and W. A. Anderson, *Rev. Sci. Inst.*, **37**, 93-102, (1966); (this, incidentally, is the original paper proposing the application of FT methods to NMR).
13. J. S. Waugh, *J. Mol. Spec.*, **35**, 298-305, (1970).
14. R. Freeman and H. D. W. Hill, *J. Mag. Res.*, **4**, 366-383, (1971).
15. P. Waldstein and W. E. Wallace, *Rev. Sci. Inst.*, **42**, 437-440, (1971).
16. G. H. Weiss and J. A. Ferretti, *J. Mag. Res.*, **55**, 397-407, (1983).
17. F. G. Herring and P. S. Phillips, *J. Mag. Res.*, **59**, 489-496, (1984); and references therein.
18. D. I. Hoult, C-N. Chen, H. Eden and M. Eden, *J. Mag. Res.*, **51**, 110-117, (1983).
19. A. G. Redfield, in *NMR - Basic Principles and Progress*, (P. Diehl, E. Fluck and R. Kosfeld eds.), **13**, 137-152, Springer-Verlag, (1976).
20. P. Plateau and M. Guéron, *J. Amer. Chem. Soc.*, **104**, 7310-7311, (1982).
21. P. J. Hore, *J. Mag. Res.*, **55**, 283-300, (1983).
22. C. J. Bauer, R. Freeman, T. Frenkiel, J. Keeler and A. J. Shaka, *J. Mag. Res.*, **58**, 442-457, (1984).
23. G. A. Morris and R. Freeman, *J. Mag. Res.*, **29**, 433-462, (1978).
24. G. Bodenhausen, R. Freeman, R. Niedermayer and D. L. Turner, *J. Mag. Res.*, **26**, 133-164, (1977).
25. G. Bodenhausen, R. Freeman and D. L. Turner, *J. Mag. Res.*, **27**, 511, (1977).

8

Homonuclear Shift Correlation

8.1 INTRODUCTION

When planning how to introduce the subject of two-dimensional NMR, I found it hard to choose between two classes of experiment. The *J*-spectra, described in Chapter 10, can be understood in detail (for first order systems) using our pictorial vector model, and from that point of view seemed attractive. On the other hand, those experiments are relatively limited in their applications, and it is easy to get the feeling that one is going to rather a lot of trouble for not much gain. Homonuclear shift correlations of various sorts, however, are so obviously useful that there should be no risk of disillusionment setting in; for that reason they might be thought very suitable as introductory examples. Unfortunately, we will not quite be able to get to the bottom of the key step in these experiments using arguments about the net magnetisation, so there is a risk of introducing a fog of residual confusion about what exactly is going on. As you can see from the chapter title, shift correlations won through in the end, and I hope the lack of complete rigour in describing the experiments will be sufficiently compensated for by the excitement of seeing some realistic problems tackled very soon.

With this in mind, I begin the chapter with an explanation of the process of *frequency labelling*, which is the basis of all two-dimensional NMR experiments; we can understand this completely in terms of the net magnetisation. I then ask you to take as a fact without justification an interesting consequence of applying a second pulse to a system which already has some transverse magnetisation, so that we can proceed immediately to examine some applications of the experiments. In fact, we have encountered special cases of this phenomenon already in Chapter 6, so it should not be too hard to accept its generalisation here. We will return to look again at what happens during this second pulse, and to try to get some feel for the concepts of *coherence* and *coherence transfer* later. Before that, though, a number of technicalities arising from the basic concept of a two-dimensional experiment will need to be discussed; parts of this section will have analogies with the latter parts of Chapter 4 section 3. Several sections in that chapter contained the suggestion that reading them might be postponed until later; this is the time when an understanding of that material will become necessary. The discussion of technical details is much more extensive here than in the remaining chapters of the book, because most of it applies to all two-dimensional experiments. The fundamental experiment on which this chapter is based has spawned a large number of variants; the next section will survey some of these that seem to me to be the most useful. Finally two other shift correlation experiments that give somewhat different information are discussed.

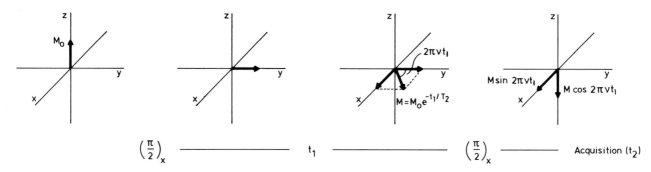

Figure 8.1 Amplitude modulation of an NMR signal can be brought about by varying the interval between two pulses.

8.2 FREQUENCY LABELLING

I want you to imagine an experiment. It will seem, at first, to be a remarkably silly experiment, but in fact it illustrates the basis of a technique of extreme importance to NMR spectroscopy. Consider a sample which only has one resonance line, for instance a solution of chloroform in a deuterated solvent with proton observation; the line has chemical shift v. We will follow the fate of the line in the rotating frame after a $(\pi/2)_x$ pulse, as we have done many times before, and for simplicity we will ignore longitudinal relaxation completely (but we will include transverse relaxation to give the signals a realistic exponential envelope). Figure 8.1 shows the line having precessed for a certain time at v cycles-per-second; the time is called t_1 for reasons that will become clear later (note that this should not be confused with T_1, the longitudinal relaxation time). At the end of the interval t_1, a second $(\pi/2)_x$ pulse is applied and the NMR signal is measured as a normal FID. What is the consequence of this?

The best way to find out is to think of the projection of the components of the magnetisation onto the x and y axes. The vector has precessed through an angle $2\pi v t_1$ during the interval, so, if its length is M, then the component along the y axis is $M \cos 2\pi v t_1$ and that along the x axis is $M \sin 2\pi v t_1$, by simple trigonometry (Figure 8.1 again). The value of M itself is related to the initial magnetisation M_0 by:

$$M = M_0 e^{-t_1/T_2} \tag{8.1}$$

by the definition of T_2 (actually this is T_2^*, but I omit the $*$ for reasons of typographical convenience).

The second pulse was specified as $(\pi/2)_x$, so this rotates the y axis component through a further 90° to place it along the z axis, while leaving the x axis component unchanged. So the amount of magnetisation left in the x-y plane, which of course determines the size of the NMR signal we see, is $M \sin 2\pi v t_1$. The spectrum obtained is perfectly normal apart from this change in its amplitude.

Now think about what would happen if we performed a series of experiments with different values of t_1, say starting from 0 and increasing to a few seconds; *this corresponds with discrete sampling of the t_1 interval.* If we took the data from each experiment and transformed it into a spectrum we would evidently see a peak each time, but the size of the peak would vary as a function of t_1. In fact, it would oscillate sinusoidally with frequency v, since the transverse magnetisation at the end of t_1 is given by $M \sin 2\pi v t_1$. Figure 8.2 illustrates exactly this experiment, or at least the first few t_1 values (for the record, these are 500 MHz proton spectra of chloroform!). The experiment was set up so that v was 80 Hz and the interval between t_1 values was 1 ms.

Remembering that the NMR data consists of discrete sets of numbers, now imagine selecting the point from each spectrum which corresponds with the top of the chloroform peak. Assuming the lock system of our spectrometer is functioning correctly, it will be the same point each time. What will we get if we plot these points as a function of t_1? Simply a graph of the amplitude of the signal, which is oscillating with frequency v and

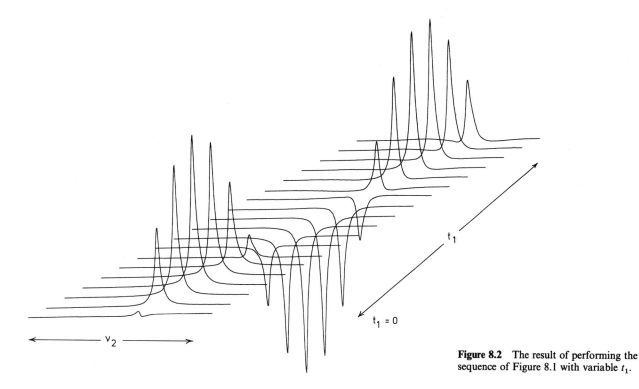

Figure 8.2 The result of performing the sequence of Figure 8.1 with variable t_1.

decaying exponentially with time constant T_2; there it is in Figure 8.3. I hope it strikes you immediately that this looks just like an FID; *of course it does* - it is a sinusoidal oscillation decaying exponentially, and we have digitised it by making increments in the value of t_1. In fact it *is* a kind of

Figure 8.3 A slice from the data of Figure 8.2, taken parallel with t_1 through the tops of the peaks.

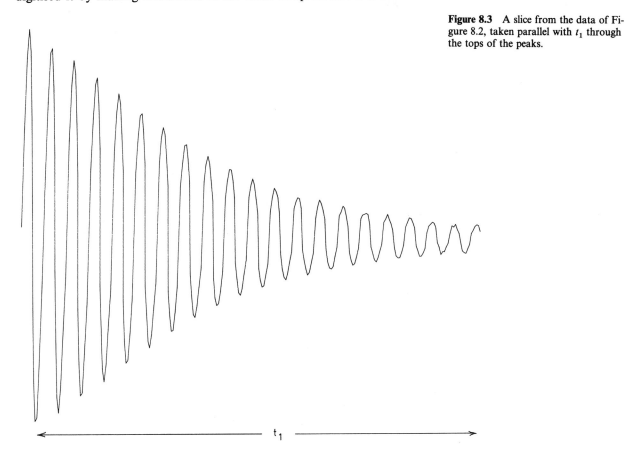

FID, but instead of existing in 'real time' like the signal we monitor in a normal NMR experiment, it has been generated point by point as a function of the variable t_1. To emphasise this, we usually refer to it as an *interferogram*.

We know very well by now that if we perform a Fourier transform on such a set of data, we obtain a frequency spectrum consisting of a Lorentzian line with width $1/\pi T_2$ and frequency v, in other words exactly what we obtained already from the 'normal' FID's that made up the experiment. To make this experiment more general, rather than just picking the set of points corresponding with the top of the chloroform peaks we could perform Fourier transforms on *every* column of points taken from the complete set of FID's. This amounts to treating the data as a *two-dimensional* data set, i.e. instead of being a function of just one time variable like a simple FID, it is a function $f(t_1,t_2)$ of two variables. The first time parameter t_1 is the interval between the two pulses (it is the first time in the experiment, hence t_1), and the second is 'real time' during the acquisition of the data. The two-dimensional Fourier transform converts the data to a two-dimensional frequency spectrum $f(v_1,v_2)$, and we need to work out the meaning of the two coordinates v_1 and v_2.

v_2 is evidently just the chemical shift v of the peak, and I think it is easy to appreciate that *so is* v_1 in this case, because the interferogram we generated was oscillating with frequency v. So what we get is a square spectrum with both axes representing chemical shifts, with a peak appearing in the frequency domain at (v,v), i.e. on the diagonal (Figure 8.4).

Figure 8.4 The result of transforming in the second dimension - a two-dimensional absorption line. The alternative contour representation, at the right, is explained later in the text.

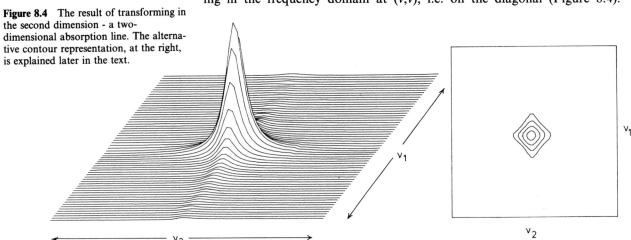

Cross-sections through the peak along either the v_1 or v_2 axes are Lorentzian with linewidth $1/\pi T_2$. This is our first 2D NMR experiment, and not, I will rapidly admit, a very impressive one as it contains no more information than a simple spectrum. However, it has the elements of a prototype 2D experiment (Figure 8.5), in which a signal, modulated as a function of the variable t_1, is later detected as a function of t_2; *all two-dimensional experiments are performed in this way.* I want to stress this point strongly so as to emphasise that the idea behind two-dimensional spectroscopy is really extremely simple. Designing the 'something' and the 'something else' of Figure 8.5 may be a complex matter, but the *concept* of modulating the magnetisation before detecting it is not.

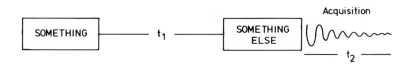

Figure 8.5 The 'prototype' 2D experiment.

The reason this particular experiment seems rather pointless is that the magnetisation is experiencing exactly the same modulation during t_1 and t_2 (precession at its Larmor frequency); of much more interest are experiments in which different things happen during the two time intervals. In general we arrange experiments such that the magnetisation which evolved with some frequency during t_1 (i.e. was 'labelled' with this frequency) evolves with a *different* frequency during t_2; the latter frequency will invariably be a normal chemical shift plus couplings. This gives rise to peaks at (v_1, v_2) with v_1 and v_2 different, that is off-diagonal or *cross* peaks. To understand how to interpret an experiment, we need to know what the two frequency axes represent (in this example they are both proton chemical shifts, but they could be shifts of different nuclei, or coupling constants and shifts and so on) and how the t_1 and t_2 magnetisations are related, in other words what gives rise to the cross peaks.

When we perform a 2D experiment such as this, we need to choose the range over which t_1 varies and the increment between individual t_1 values. We will discuss this in more detail later, but I want to point out at this stage that the digitisation of the t_1 interval is completely analogous with that of a normal FID. Thus we use the spectral width (which is the range of expected modulation frequencies during t_1) to determine the increment according to the Nyquist criterion, and we use the desired digital resolution to determine the total number of sampled points. However, we at once encounter serious practical problems. Recall the example from Chapter 2 in which we digitised a 10 p.p.m. proton spectrum at 0·2 Hz/point on a 500 MHz spectrometer, which led to a 5 s acquisition time and 50,000 data words (25,000 complex points with quad detection). All these are practical values on an instrument with a modern data system. If we try to carry out the 2D experiment described above with the same spectral range and the same digital resolution in *both* dimensions, we would need 50,000 words for each set of t_2 data, and 2 by 25,000 sets to characterise t_1, a total of 2,500,000,000 words! Also, the total acquisition time would be 7·5 s (that is the mean t_1 value plus the t_2 acquisition time) times 25,000 t_1 increments times 8 (which for technical reasons would probably be the minimum number of scans per t_1 increment as we shall see later), which is 17 days! Obviously we need to revise our ideas about digitisation when we start doing 2D NMR.

Lest you start thinking that all that 2D spectroscopy introduces is a lot of bothersome technical problems, I am now going to move rapidly on to examine situations where this experiment is highly informative. It is true that a certain number of experimental difficulties arise regarding the manipulation of 2D data sets, and we will return to examine these later, hopefully in an enthusiastic frame of mind.

8.3 JEENER'S EXPERIMENT

8.3.1 Introduction

The sequence we have been examining was actually the first 2D NMR experiment ever proposed, suggested by Jeener in 1971. Its usefulness becomes apparent when we consider what happens during the second pulse to systems which possess homonuclear coupling. Before we do that, however, I want to talk about an experiment which, although highly impractical, would obviously be useful if only it could be done. It will turn out that the major defects of this imaginary experiment are absent from the Jeener experiment, while the information content of each is similar.

8.3.2 A CW-FT 2D Experiment

Homonuclear decoupling is an old and very informative experiment. As you know, it lets us demonstrate which groups of nuclei in a molecule share *J*-coupling, which in turn often eliminates or supports hypotheses about the structure. The detection of coupling patterns by this experiment is taken to be more 'solid' structural evidence than simple examination of chemical shift and coupling constant data, and considering why this should be is rather enlightening. The reason is our confidence that, in proton spectra for example, large couplings will rarely be observed between nuclei separated by more than three bonds. In comparison the information gleaned from chemical shift values is rather vague, since many factors can influence shifts and we do not understand them all properly. It is very dangerous to rely on shift data alone as evidence for a structure; a theme of this book is that we no longer need to do so in many cases.

There are certain problems with the determination of coupling patterns by homonuclear decoupling. If we are faced by a complex spectrum, it may not be evident which irradiations are likely to be informative; this may cause us to spend much time trying experiments which are ultimately fruitless. Even when we *can* see what should be irradiated, it is not always possible to perform the irradiation with sufficient selectivity in a crowded spectrum, or conversely the result of the decoupling may be lost in a complex group of lines. The method of decoupling difference has been suggested as a solution to the latter problem, but this itself suffers from many disadvantages, particularly the effects which result from Bloch-Siegert shifts.

The decoupling experiment we would like to be able to do, in an ideal world, would probably go something like this. We would set the spectrometer to increment the decoupling frequency across the region of interest automatically. The decoupling at each frequency increment would be entirely selective, and the result would manifest itself as a clear-cut change elsewhere in the spectrum. All the decoupling experiments would be presented together, so that the coupled systems could be mapped out directly. The whole experiment should be performed quickly and with high sensitivity.

Probably the closest we could come to this in practice would be to perform a series of decoupling-difference spectra with very low (and hence selective) decoupler power (these would be FT INDOR experiments, but do not worry about the details of what happens, just take it that perturbations would appear at coupled protons as their coupling partners were irradiated). It would not be straightforward to present such a series of spectra in an informative way, but set that aside and consider a more fundamental problem. The experiment would give rise to a two-dimensional NMR spectrum, but it is a spectrum *which is CW in one dimension*. The decoupler frequency is being varied across the spectrum in the same way that we might vary the transmitter frequency in an old-fashioned CW spectrometer, and this carries the same penalty of slow extraction of information. The time it would take to move the decoupler from one side of the spectrum to the other would depend on how fine we required the increments in its frequency to be; in other words on the *resolution* in that dimension. This is analogous with the requirement for a slow scan speed to give high resolution on a CW spectrometer. Even without its other practical disadvantages, this would be an intrinsically inefficient experiment because the measurement is being made in the frequency domain. *If only we could introduce the FT advantage into the second dimension!*

8.3.3 Magnetisation is Transferred

No doubt it will not surprise you to learn that the Jeener sequence introduces precisely this advantage over the imaginary 'CW-FT 2D' spectrum. The reason for this is that in a coupled system the second pulse of the experiment causes magnetisation which arose from one transition during t_1 to be redistributed amongst all the others with which it is associated. As I said in the introduction, I want to leave the mechanism of this for later, and I hope you can just take on trust for the time being that this is what happens. We have already seen magnetisation transfer experiments (SPI, DEPT and INEPT) in Chapter 6, but those cases are rather easy to understand, because the systems are heteronuclear (allowing the separate consideration of the effects of the pulses on the source and destination nuclei), and because we choose special arrangements of the multiplet components on which to perform magnetisation transfer. The more general and homonuclear experiment we have here is a little harder to visualise, but we need at least to be sure what is meant by 'all the others with which it is associated', so I remind you of the energy level diagram for an AX system, which we have seen several times in other chapters (Figure 8.6).

This system has four transitions, and correspondingly four lines appear in the NMR spectrum. Each of the transitions, for instance A_1, is the source of each of the lines in the spectrum, for instance the one at $v_A + \frac{1}{2}J_{AX}$; this is the basic characteristic of a first-order system. If the second pulse of the Jeener experiment is $\pi/2$, then portions of the magnetisation arising from transition A_1 are converted into magnetisations of all the other transitions, that is A_2, X_1 and X_2. Some also remains in transition A_1. Likewise, components from all the other transitions are redistributed. This means that a line detected in t_2 may have components of its amplitude modulated as functions of the frequencies of all the other lines in t_1; it will therefore show cross peaks to these lines in the two-dimensional spectrum. This is a little tricky to conceive, as everything is being mixed with everything else all at the same time, but the resulting spectrum is very easy to appreciate (Figure 8.7).

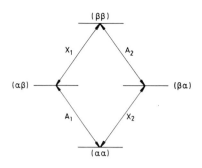

Figure 8.6 Energy levels of an AX system.

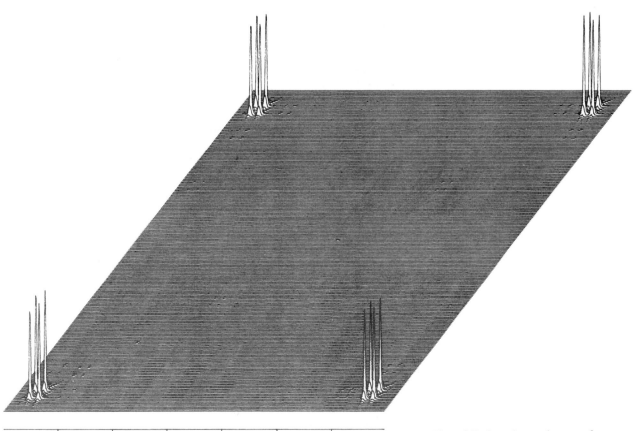

6.8   6.7   6.6   6.5   6.4   6.3
PPM

Figure 8.7 Jeener's experiment performed on an AX system (β-chloroacrylic acid).

The peaks along the diagonal in this spectrum arise from magnetisation components which have the same frequency during both t_1 and t_2, that is from the portion of magnetisation that was *not* transferred elsewhere during the second pulse. So, looking along the diagonal we see in essence a normal spectrum of the system, in this case just two doublets. Off the diagonal we see cross peaks between all lines which are either part of the same multiplet (the square patterns near the diagonal), or which are parts of different multiplets that share a coupling (the groups of four lines further from the diagonal). Note that since magnetisation is transferred in both directions between transitions, a cross peak at (v_1, v_2) has a symmetrical partner at (v_2, v_1). The correlations *within* multiplets are more or less of a nuisance, and we shall see later that they can be eliminated. The important correlations are those *between* multiplets, for these are diagnostic of coupling patterns. In this simple example there is only one spin system, so everything is correlated with everything else; in a complex system we can directly trace the couplings via the off-diagonal peaks. It is as though the coupling between protons provides pathways along which the magnetisation can travel during the second pulse.

8.3.4 Two Real Examples

There are many interesting details about how we perform this experiment and exactly what its advantages and disadvantages are that we will need to discuss. However, to whet your appetite for the more technical sections which follow I thought it would be worthwhile examining two real problems, so you can see how easy it is to make spectral assignments using Jeener's experiment. The first is relatively simple, and could well be tackled by more traditional methods, while the second would have been considered a major problem ten years ago; nowadays it can be solved in an afternoon. Before looking at the spectra, though, there are two other points that need to be cleared up. The first regards the name of this technique, which I have been referring to so far as Jeener's experiment. In the literature this description is sometimes used, but far more commonly the term COSY (from COrrelation SpectroscopY) is encountered. Unfortunately, since nearly all experiments involve correlating *something*, COSY is also applied to various related methods, often in combination with other modifying terms, and this can lead to confusion. We are stuck with this, though, and from now on COSY is the name I shall use for homonuclear shift correlation through *J*-coupling brought about by the Jeener sequence.

The second point we need to deal with is how the result of the COSY experiment is presented. The rows of spectra (known as stacked plots) we have been using so far illuminate the nature of the data, but rapidly become cluttered and confusing to interpret as spectral complexity increases. A far clearer representation of the relationship between diagonal and off-diagonal peaks is obtained if we plot contours of the data, in exactly the same way as a mountain range would be represented on a map. To illustrate this, Figure 8.8 reproduces the same data as Figure 8.7 as a contour plot. For a realistic data set it is often sufficient (and quicker) just to plot the lowest contour level, rather than several as here, since the relative heights of the peaks are represented by the size of their cross-sections anyway. When working with COSY data, it is helpful to have a plot of the normal spectrum to hand in order that the diagonal peaks can be readily identified. It takes a little practice to relate the contours on the diagonal with peaks in a normal spectrum, but this soon becomes straightforward.

Now to the examples. In both cases we had an idea of the structure already, and the object of assigning the spectrum in detail was to check its consistency with this structure. This is very commonly the position, even when performing structure elucidation from scratch, since many inferences about sub-units of the structure can be made directly from the simple

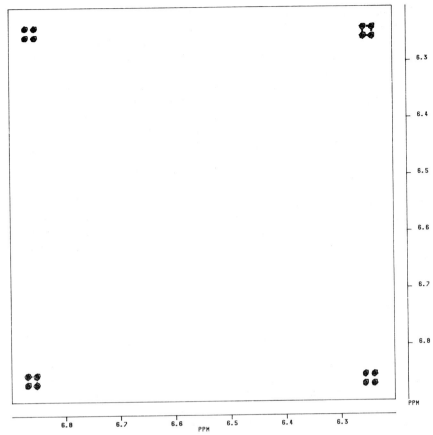

Figure 8.8 This alternative contour representation of a 2D spectrum is generally more convenient.

NMR spectra (and of course from other spectroscopic methods). Hypothetical arrangements of the inferred sub-units can then be tested, traditionally against more detailed interpretation of the spectra. The assignment possibilities of 2D NMR experiments make this testing stage much more rigorous.

The first compound (1) is a natural product extracted from a bean[1], and was of interest because its structure was similar to other compounds known to be glycosidase inhibitors. The normal proton spectrum is reproduced in Figure 8.9, and the contour representation of the COSY spectrum in Figure 8.10. It is best at first to spend a moment relating the groups of

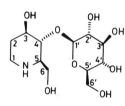

1

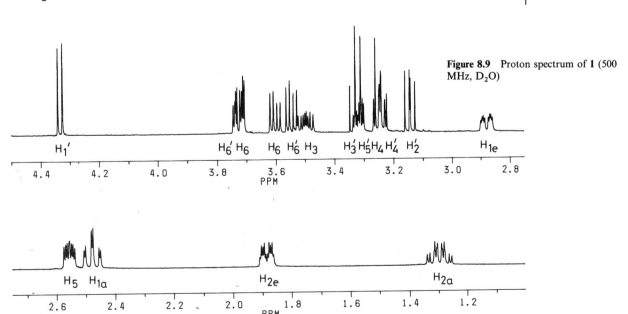

Figure 8.9 Proton spectrum of 1 (500 MHz, D_2O)

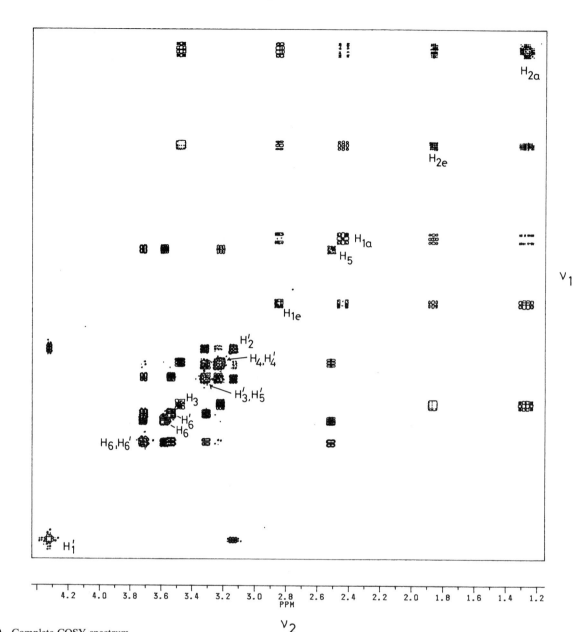

Figure 8.10 Complete COSY spectrum of **1**.

peaks in the 1D spectrum with the diagonal of the COSY experiment, to orientate yourself with respect to the latter. Note that, typically, 2D NMR experiments are performed with much lower digital resolution than 1D spectra, so that not all the fine structure visible in the normal spectrum can be identified in the COSY contour plot. This is rarely a disadvantage, as we will see when we examine this question in more detail later.

In interpreting the spectra, the most important thing is to identify resonances whose assignment is obvious. An example of this would be the anomeric proton H_1' of the glucose ring, which is the only signal we would expect to be a doublet at low field. This provides a starting point for tracing the coupling patterns, and in this case allows us to assign the glucose ring completely. Figure 8.11 illustrates tracing the path from the anomeric proton out to its only cross peak, then back to the diagonal to locate the signal due to H_2', then out to another cross peak and so on. Notice how, although the region of the spectrum where the glucose resonances fall is quite crowded even at 500 MHz, the cross peaks are well resolved. This is because *two* chemical shifts contribute to the position of a cross peak, so there is correspondingly a lower probability of accidental overlap. In this spectrum it happens that we *can* identify the multiplets

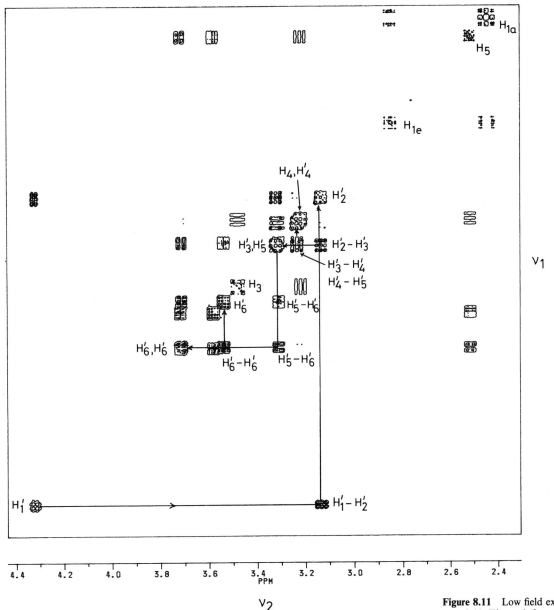

ν_1

ν_2

Figure 8.11 Low field expansion of Figure 8.10. The path for assigning the glucose ring is traced out.

corresponding to each proton by carefully examining expansions of the 1D spectrum. When the spectrum becomes more complex so that this is no longer possible, one can often still proceed *by making identification of cross peaks the object*, rather than identification of peaks in the 1D spectrum. This requires a change in the way one thinks about spectra, but is a perfectly valid approach to structure confirmation.

A good starting point for assigning the signals of the other ring is either of the two high-field multiplets, which must arise from H_{2a} or H_{2e} (a = axial, e = equatorial) as these are the only protons not adjacent to a heteroatom. Notice that though we are still making use of chemical shifts to help with the assignment, it is gross features such as this rather than fine details which are important. We do not, for instance, need to make any assumptions about whether a proton adjacent to nitrogen or one adjacent to oxygen should be at lower field. I leave tracing the assignment of this ring as an exercise; Figure 8.12 is an expansion of the region of interest.

Having convinced ourselves that both rings of the suggested structure can be accommodated by the spectrum, the remaining problem is to identify the point of attachment of one ring to the other. There is certainly no concrete evidence for this in the 1D proton spectrum, but a remarkable

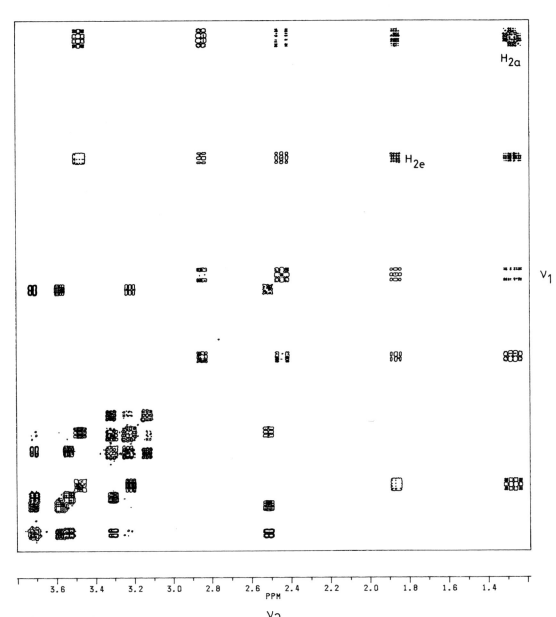

Figure 8.12 High field expansion of Figure 8.10. Try tracing the assignment of the other ring without reference to Figures 8.9-8.11. The starting point can be either of the H₂'s as marked.

feature of the COSY experiment enables us to identify at once where they are joined. This is the ability to detect cross peaks due to couplings that are less than the observed linewidth, and hence unresolvable in the 1D spectrum. In proton spectra, 4 and 5 bond couplings often fall in this region (0·1-0·5 Hz; geometrical factors such as the well known W arrangement may make them larger in some cases). In order for cross peaks due to small couplings to be detected, certain conditions must be satisfied. This is discussed further in section 8.4.2, but in essence amounts to ensuring that the acquisition times in the two dimensions are long enough (the usual factor controlling resolution). Figure 8.13 shows the same region as 8.11, taken from a COSY spectrum run so as to optimise the detection of long-range couplings. The anomeric proton H_1' now shows a number of cross peaks to other protons in the glucose ring, but most importantly it also shows a cross peak to H_4 of the other ring (arrowed in Figure 8.13). This is due to the unresolved 4-bond coupling between these protons, and indicates the position of attachment unambiguously.

The second example moves up a degree in complexity, to a compound (2) with molecular weight around 1000. By comparison with what it is possible to achieve with 2D methods this is still a fairly simple problem though.

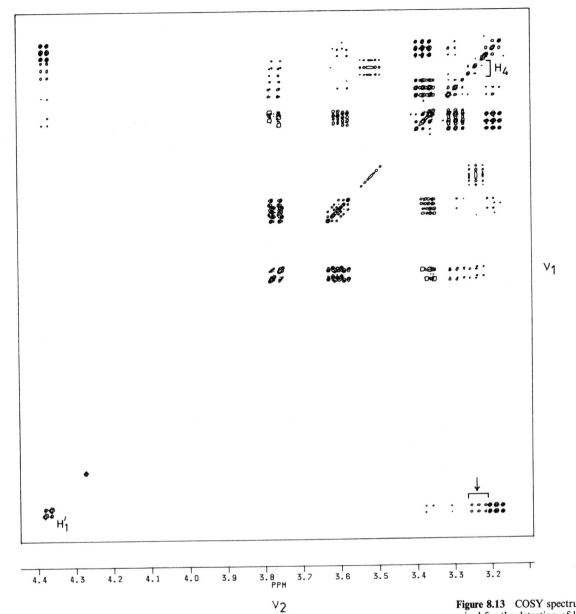

Figure 8.13 COSY spectrum of **1** optimised for the detection of long-range couplings.

The detailed use of the COSY spectrum is similar to that of the previous example, and I leave you to verify for yourself that it is consistent with the assignments given in the caption (Figure 8.15). The point I want to make in this case is the importance of not forgetting that more traditional methods may still be used to good effect. It is easy to become so mesmerised by the possibilities of 2D NMR as to neglect simple ways in which the problem may be approached.

Examining the full proton spectrum of the compound (Figure 8.15) illustrates this point well. There are signals over a wider shift range than in the previous example, and some of the outlying ones look likely to be important starting points for the assignment (for instance, the doublets A, B and C at low field which can be attributed to amide NH protons). It would be possible to perform the COSY experiment so as to encompass the full range of shifts, but this would lead to certain problems. Either the experiment would have to be performed with very low digital resolution, which would not be acceptable in this case, or it would take a long time to accumulate and lead to a large dataset. Whether or not the size of the dataset was problematical would depend on the computing power available; the long accumulation time however is a fundamental obstacle. The

Figure 8.14a Complete COSY spectrum of **2**.

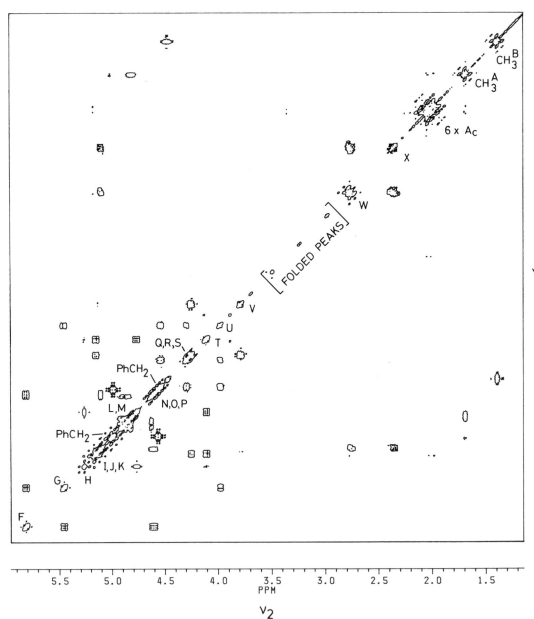

ν_1

ν_2

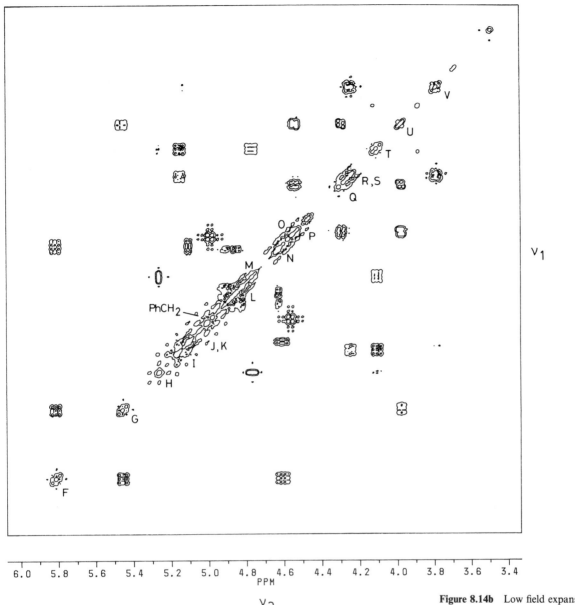

Figure 8.14b Low field expansion of
Figure 8.14a.

COSY experiment can be made quicker to run and easier to process by
first performing a few straightforward decouplings to identify the neighbours of the low-field protons (Figure 8.16). With hindsight I could have
done the same thing with the two methyl doublets at high field as well,
making the region for the 2D experiment smaller still. At the time I had
other reasons for wanting to include these peaks in the 2D spectrum, and
eliminating the NH region had already reduced the size of the experiment
to a level which I knew the data system could handle easily, so these
decouplings were not carried out.

8.3.5 Details of the COSY Experiment

Introduction

Now that we have some idea of how useful an assignment tool a COSY
spectrum can be, I am going to examine in much more detail how the
experiment is performed, how various artefacts may arise and can be
eliminated, different approaches to representing the data and how to

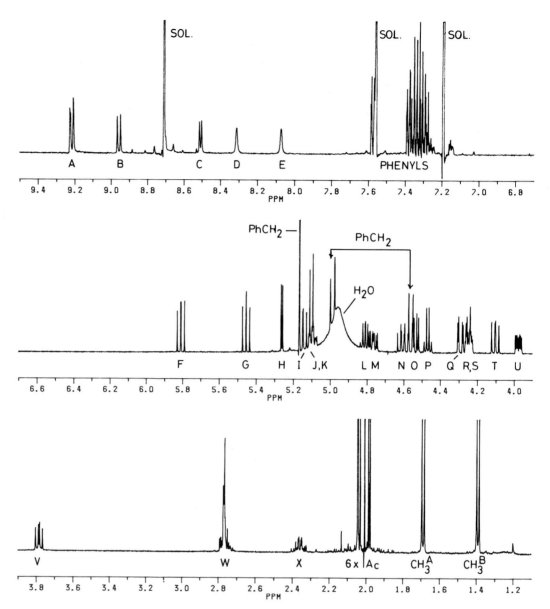

Figure 8.15 Proton spectrum of **2** (500 MHz, pyridine-d_5). Assignments are as follows (see formula **2** for numbering scheme): A, ring A NH; B, ring B NH; C, Ala (region D) NH; F, A_3; G, A_4; H, B_1; I, B_4; J, Glu (region E) α proton; K, A_1; L, Ala (region D) α proton; M, B_2; N, A_2; O, A_6; P, Ala (region C) α proton; Q, A_6; R, B_5; S, B_6; T, B_3; U, A_5; V, B_6; W and X, Glu (region E) CH_2's; $CH_3{}^A$, Ala (region C) Me; $CH_3{}^B$, Ala (region D) Me.

choose suitable experimental parameters in practice. This is a fairly technical section, and will be of most interest if you want to set up COSY or other experiments on a spectrometer yourself. If you are not technically inclined, or have no interest in operating a spectrometer, I would recommend omitting most of this section, except the parts 'Relative phases in phase-sensitive COSY' and 'Digital resolution and acquisition times'. A large proportion of this discussion applies, in principle if not in detail, to all 2D experiments, not just to COSY, and the remaining chapters assume that you have read this one. I also assume in the following that you are familiar with section 4.3.5 of Chapter 4.

Effect of longitudinal relaxation

In our preliminary discussion of the COSY sequence I specifically ignored longitudinal relaxation. The reason for doing this was that it gives rise to extra peaks in the spectrum, which it takes a little added complication to eliminate, and I did not want to distract you from the main point, the concept of frequency labelling. It is not difficult to understand how relaxation back to the *z* axis influences the resulting spectrum, however,

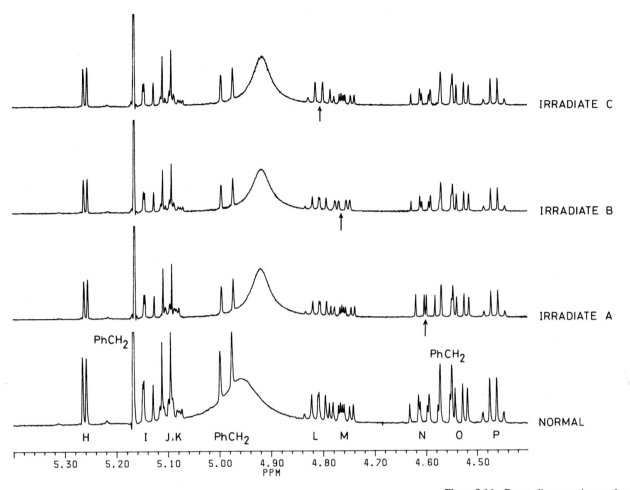

Figure 8.16 Decoupling experiments for compound **2**.

and this is illustrated in Figure 8.17. Here we have just the same diagram as Figure 8.1, except that the inevitable reappearance of z magnetisation during t_1 has been added. The second pulse, aside from its effects on the transverse magnetisation which we have already discussed, must return this z magnetisation to the x-y plane, where it will generate a signal. Since this is a magnetisation component which was not precessing during t_1 (because it was along the z axis), it will give rise to a peak after the 2D transform with a v_1 frequency of 0. So we will obtain a copy of the spectrum along the line $v_1 = 0$; these unwanted signals are called *axial* peaks. If the spectrum was acquired using quad detection in v_1 (see below) the $v_1 = 0$ line runs across the middle, so this is a rather objectionable effect (Figure 8.18).

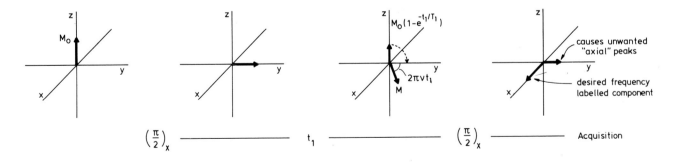

Figure 8.17 The origin of 'axial peaks': longitudinal relaxation during t_1

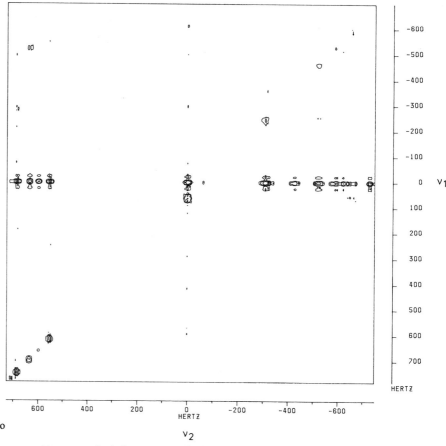

Figure 8.18 Without the phase cycle to suppress axial peaks, an objectionable band arises along the line $v_1 = 0$.

Fortunately it is easy to eliminate these axial peaks. To see how this can be done, consider the effect of changing the phase of the second pulse by 180°. For the magnetisation that gives rise to the axial peaks this inverts the phase of the signal, as the vector now arrives along the $-y$ axis (Figure 8.19). The magnetisation which gives rise to the signal we want, however, is that which is left along the x axis; evidently it is still left there whether the second pulse is $(\pi/2)_x$ or $(\pi/2)_{-x}$. So if we perform pairs of scans, phase alternating the second pulse, and add the data each time, the axial peaks will be cancelled and the desired peaks reinforced. This is the first component of the phase cycle for the COSY experiment.

Eliminating v_2 artefacts

Figure 8.19 Phase alternation of the second pulse inverts the phase of the unwanted component, while leaving the frequency labelled part unaffected.

We can add some more stages to the phase cycle at once, by realising that the problem of suppressing quad images and other artefacts in v_2 is identical to that which we encounter in 1D experiments. The solution is the same: we apply the CYCLOPS procedure to the whole experiment. This

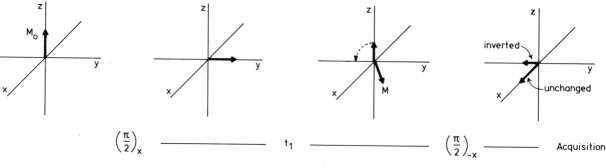

involves, remember, 90° phase shifts to eliminate receiver imbalance, and 180° shifts to cancel other non-coherent signals, both combined with appropriate data routing. In the case of the COSY experiment (and generally for multi-pulse experiments) we cycle the phases of *all* the pulses and the receiver together. Combined with the two-step axial peak suppression this gives us an eight step cycle so far (Table 8.1).

Table 8.1 The first stage of the COSY phase cycle, combining axial peak suppression with CYCLOPS. 'Phase 1' and 'Phase 2' are the phases of the first and second pulses respectively.

Scan	Phase 1	Phase 2	Receiver
1	x	x	x
2	x	$-x$	x
3	y	y	y
4	y	$-y$	y
5	$-x$	$-x$	$-x$
6	$-x$	x	$-x$
7	$-y$	$-y$	$-y$
8	$-y$	y	$-y$

Quad detection in v_1

The same reasons that, in Chapter 4, led us to wish to place the receiver reference frequency at the centre of the spectrum when performing 1D experiments apply in the 2D case. I have been tacitly assuming all along that quad detection was in effect during t_2 (i.e. during the normal acquisition period); we must also deal with the problem of distinguishing positive and negative modulation frequencies during t_1. This can be achieved, in direct analogy with the 1D experiment, by measuring *two* signals during t_1 whose reference phases differ by 90° (one particularly nice thing about 2D NMR is that nearly all the problems are analogous to ones we have already discussed, so if you have worked with and understood 1D techniques the extension to two dimensions should be easy). The only problem with this idea is that it is not completely obvious what is defining the 'reference phase' during t_1; however a little thought should convince you that it is the relative phases of the two pulses (also see Figure 8.20).

Now, as t_1 is a discrete variable, the possibility of simultaneously measuring two signals as we do for normal quad detection does not exist. Instead we have to perform two experiments with the same t_1 value, introducing the appropriate phase shift into the second and storing this as the imaginary part of the t_1 data. This is the method of Ruben, States and Habekorn[2] for v_1 quad detection (in the absence of a name for this technique offered by its inventors, I will refer to it henceforth as RuSH). The necessary phase change can be brought about *either* by shifting the phases of both the first pulse and the receiver by 90°, *or* equivalently by shifting the phase of the second pulse by $-90°$. Justifying this procedure by examining the mathematics of the 2D Fourier transform is not difficult, but does require familiarity with a few relationships between trig functions and exponentials of complex numbers. Rather than pursue this, I refer you to a paper which discusses this question in detail[3]. What we *can* do without getting embroiled in complex algebra is persuade ourselves that the two alternatives are equivalent, and are analogous in t_1 with the two-phase method of quad detection we have already met, by reference to Figure 8.20.

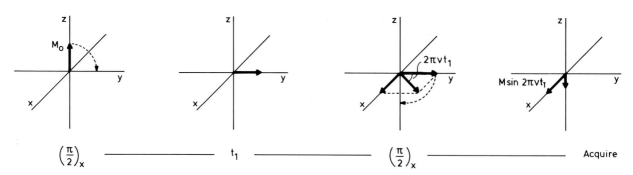

$$\left(\frac{\pi}{2}\right)_x \underline{\hspace{6em}} t_1 \underline{\hspace{6em}} \left(\frac{\pi}{2}\right)_x \underline{\hspace{8em}} \text{Acquire}$$

Figure 8.20a The basic COSY sequence gives rise to sinusoidal modulation of the signal.

EITHER

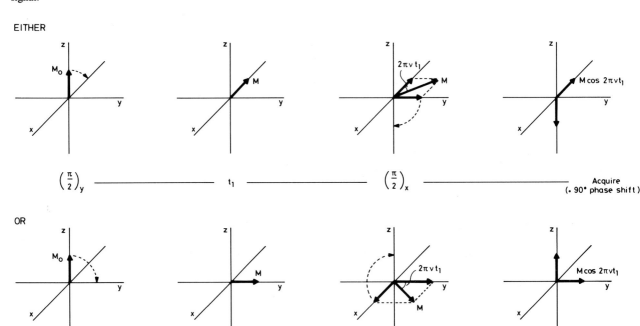

$$\left(\frac{\pi}{2}\right)_y \underline{\hspace{6em}} t_1 \underline{\hspace{6em}} \left(\frac{\pi}{2}\right)_x \underline{\hspace{6em}} \begin{array}{c}\text{Acquire}\\ \text{(+ 90° phase shift)}\end{array}$$

OR

$$\left(\frac{\pi}{2}\right)_x \underline{\hspace{6em}} t_1 \underline{\hspace{6em}} \left(\frac{\pi}{2}\right)_y \underline{\hspace{8em}} \text{Acquire}$$

Figure 8.20b To get the alternative cosine modulation required for v_1 quad detection, either pulse can be shifted in phase by 90°.

The first thing to call to mind is the requirement for detecting signals modulated as both sine and cosine functions in order to distinguish positive and negative frequencies by a complex Fourier transform. This is explained in detail in section 4.3.5 of Chapter 4. Evidently when the phases of the two pulses of the COSY sequence are the same (Figure 8.20a) the signal at the end of t_1 is given by $M \sin 2\pi v t_1$, while when they differ by 90° it is given by the corresponding cosine (Figure 8.20b). Having arranged that the modulation of the signal is of the correct type, we also need to shift the received phase 90° (this is to ensure that the appropriate quadrant of the data contains the phases of interest, see below), which occurs automatically if it is the second pulse which is varied but requires a phase change of the receiver if it is the first one. The resulting complete phase cycle for the experiment, supposing we take the option of shifting the first pulse, can be found in Table 8.2.

So for a complete phase cycle we need a multiple of *sixteen* scans to accumulate each complex point in t_1, giving a minimum of sixteen scans per increment whether or not we need so many from a sensitivity point of view. This is frequently the factor which determines the minimum time we can spend accumulating the data for a 2D experiment. In practice it is often acceptable to omit the CYCLOPS component of the phase cycle, as v_2 artefacts are usually not too bad if the spectrometer is properly

Table 8.2   The complete phase cycle for phase-sensitive COSY with v_1 quad detection by the RuSH method.

Scan	Phase 1	Phase 2	Receiver	Store As (t_1)
1	x	x	x	Real
2	x	$-x$	x	Real
3	y	y	y	Real
4	y	$-y$	y	Real
5	$-x$	$-x$	$-x$	Real
6	$-x$	x	$-x$	Real
7	$-y$	$-y$	$-y$	Real
8	$-y$	y	$-y$	Real
9	y	x	y	Imaginary
10	y	$-x$	y	Imaginary
11	$-x$	y	$-x$	Imaginary
12	$-x$	$-y$	$-x$	Imaginary
13	$-y$	$-x$	$-y$	Imaginary
14	$-y$	x	$-y$	Imaginary
15	x	$-y$	x	Imaginary
16	x	y	x	Imaginary

adjusted. This reduces the minimum number of scans required to four (two for each component of the complex points), usually plus two or more 'dummy' scans to establish a steady state before accumulation begins. Phase cycles for other multipulse experiments are built up in a similar way, and I will not analyse these in such detail.

It is also possible to bring about v_1 quad detection by a method analogous to that of Redfield. In this case we change the relative phases of the two pulses, by incrementing the phase of the first pulse, in 90° steps together with the incrementation of the t_1 value. Just as in the 1D case, we sample t_1 twice as fast as for the RuSH method, but we only store one experiment at each t_1 value, so the total number of sample points is the same. This is referred to as TPPI[4] (for Time Proportional Phase Increment). Often you will be forced to use this method if your spectrometer uses the Redfield approach for its normal v_2 quad detection, as the Fourier transform software will be written with this form of data in mind. If you do have the choice of which to use, there is no difference in principle (see reference 3), but in practice you may find one or other version of the phase cycle easier to express with your particular software.

The reason I have spent some time on what is admittedly a fairly technical matter is that use of these methods for v_1 quad detection is not yet widespread. Instead it is common to generate spectra in which the absorption and dispersion components become inextricably mixed, by a procedure which is discussed next. A large number of complications and disadvantages of 2D NMR which arise from this alternative approach to v_1 quad detection are absent from the spectra generated by either of the above two methods, and it is to be expected that they will gradually come into universal use. However, at the time of writing the vast majority of spectra in the literature, and indeed a large proportion of the 2D spectra in this book, have not been obtained in this fashion. In this period of transition, if you are a user of a spectrometer yourself, you may find you want to implement these *phase-sensitive* experiments, for COSY or for other 2D techniques. The approach described here works for *any* 2D experiment in which purely the *amplitude* of the signal is modulated as a function of t_1.

Echo or anti-echo selection

The two methods for quad detection in v_1 described above were only devised fairly recently (around 1981-2). Prior to this, an alternative and in many ways inferior approach was adopted, and as this is still in widespread use at the time of writing we must examine it. The essential component required for discriminating the signs of the frequencies, that is the sampling of two signals with a 90° phase change, remains the same, but the signals are not kept separate. Instead they are either added or subtracted to give a single component (technically, this converts the amplitude modulation of the signal into modulation of its *phase)*, which is then transformed as though generated by single-phase detection. In other words, the basic phase cycle for COSY (without CYCLOPS) becomes either Table 8.3 (subtracting, otherwise known as echo selection) or Table 8.4 (adding, otherwise known as anti-echo selection). Note that the pairs of scans 1-2 and 3-4 bring about quad detection, while 1-3 and 2-4 are to cancel axial peaks.

Table 8.3 COSY phase cycle for v_1 quad detection by selection of the coherence transfer echo.

Scan	Phase 1	Phase 2	Receiver
1	x	x	x
2	x	y	$-x$
3	x	$-x$	x
4	x	$-y$	$-x$

Table 8.4 COSY phase cycle for v_1 quad detection by antiecho selection (not preferred).

Scan	Phase 1	Phase 2	Receiver
1	x	x	x
2	x	y	x
3	x	$-x$	x
4	x	$-y$	x

I want first to dispose of the difference between adding and subtracting the signals, which although rather interesting from an NMR point of view is not too important if you only want to interpret the spectra, and then examine other effects of the use of this technique, which are of great consequence. When the signals are *subtracted,* then in effect the signs of the modulation frequencies are reversed between t_1 and t_2. This is rather similar to the effect of the spin echo sequence described in Chapter 4, where signals which moved ahead of the rotating frame during the first half of the sequence lag behind it after the second pulse, leading to refocusing. In the case of an experiment like COSY, involving magnetisation transfer, the outcome is considerably more complex, as the magnetisation components may find themselves precessing with different frequencies before and after the second pulse, but nevertheless echoes do form. These are known as *coherence transfer echoes,* and may arise in any experiment in which the detected magnetisation is modulated as a function of a time interval. Subtracting the signals is correspondingly known as echo detection, or sometimes as N-type peak selection.

The other component of the signal, obtained by adding the 0° and 90° parts of the experiment, has the same signs for v_1 and v_2 modulation frequencies. Thus no echoes form, and the component is known as the anti-echo or P-type signal (anti-echo always sounds rather ominous to me; a more descriptive term would perhaps be 'non-echo'). There are complex differences between the lineshapes of the echo and anti-echo components, and generally the echo component is preferable[3]. Therefore the phase cycle of Table 8.3 should be used, combined with CYCLOPS if necessary.

A very unfortunate consequence of this mode of sign discrimination is that the absorption and dispersion parts of the transformed spectrum become intermingled. That is, it is impossible to produce lines that are entirely absorptive or dispersive, no matter what combinations of the real and imaginary parts of the transform we take. Instead the phase of each line in one dimension varies as we pass through it in the other; it has a *phase twist* shape. This is illustrated in Figure 8.21. The line has both positive and negative going components, and because of the contribution from the broad dispersive Lorentzian it has a wide base. It is inconvenient to represent such a lineshape either as a contour or stack plot, so usually steps are taken to convert it to one having only positive components. This is achieved by computing the *magnitude* spectrum, formed by combining the real part $\mathscr{R}$ and the imaginary part $\mathscr{I}$ thus:

$$\mathscr{M} = \sqrt{\mathscr{R}^2 + \mathscr{I}^2} \tag{8.2}$$

This obviously eliminates the negative parts, but the resulting lineshape has a very wide base indeed (Figure 8.22). In two dimensions the line has a star shape, and overlap between the wings of adjacent stars can cause great confusion.

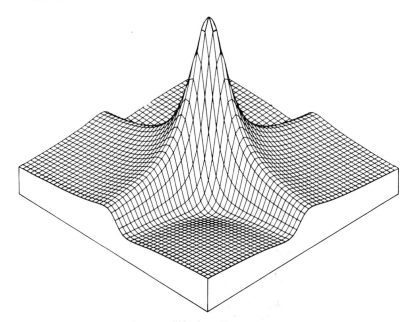

Figure 8.21 The dreaded phase-twist line shape, which arises in experiments involving detection of the coherence transfer echo.

In order to produce an acceptable resolution of neighbouring lines in the magnitude spectrum, strong resolution enhancement must be applied. Diverse window functions suited for this have been proposed; for a detailed discussion of several see Bax[5] (Chapter 1). The most popular are the Lorentz-Gauss transformation we have already encountered in Chapter 2, and the *sine-bell* window. The latter entails multiplying the FID's by a half cycle of a sine function, optionally phase shifted and/or squared according to the degree of line narrowing required and the degree of signal-to-noise ratio degradation acceptable. Window functions which start and finish close to zero and are symmetrical about the mid-point of the FID produce the desirable result of a completely absorptive lineshape in

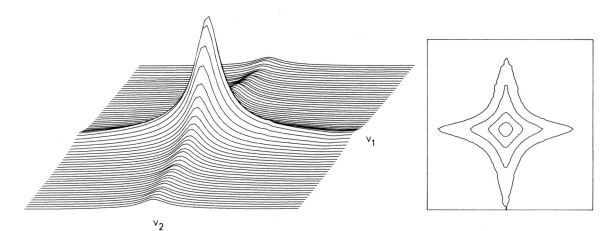

Figure 8.22 After a magnitude calculation the phase-twist has gone, but the resulting lineshape is extremely broad. Compare this with Figure 8.4, which is a pure absorption line (from the same sample) plotted with the same horizontal and vertical scales and contour levels.

the magnitude spectrum. This can be achieved using the Lorentz-Gauss transformation with suitable parameters (called in this case a *Gaussian pseudo-echo*), but the sine-bell function has a similar effect and is conveniently parameter free (Figure 8.23). However, both these manipulations are far more radical than anything one would contemplate doing to a 1D spectrum, and it must be expected that very large variations in signal intensity will result (see below).

The particular details of the effects of resolution enhancement vary in a subtle way with the function used and the parameters selected, but there is a general problem that can be identified. All the enhancement functions operate by cancelling to some extent the natural decay of the early part of the FID, while still ensuring a smooth decay to zero at the end in order for proper apodisation (Chapter 2). Cancelling part of the decay necessarily involves amplifying the later, noisier parts of the FID, and so must degrade the signal-to-noise ratio. In processing 2D data so as to give an acceptable magnitude spectrum the enhancement used is rather strong, so we must expect to lose a good deal of sensitivity. A further (and often worse) problem is the *differential* loss in sensitivity encountered if the spectrum contains lines with different widths. This is easy to appreciate if we recall that the rate of decay of signals in the time domain corresponds with their widths δv in the frequency domain according to $T_2^* = 1/\pi\delta v$.

Suppose that we have two lines, one with width 1·0 Hz and the other with width 0·1 Hz, and that we accumulate time domain data from these signals with acquisition time 2 s. If we multiplied the resulting FID by a function such as a sine-bell which strongly emphasises the central part, around $t = 1$ s, there would still be plenty of signal in this region from the 0·1 Hz wide line, as its T_2^* is about 3 s; in fact for optimum detection of this signal we should really have a longer acquisition time. The signal from the 1·0 Hz wide line, on the other hand, would have decayed to around

Figure 8.23 Strong resolution enhancement before the 2D transform (in this case by means of multiplication by a sine window in each dimension) restores a more or less absorptive shape in the magnitude spectrum, but at the cost of degraded sensitivity. The ridge running through the line parallel with v_1 is 't_1 noise' (see later in the text).

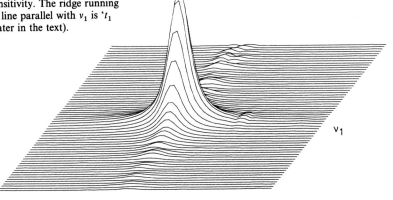
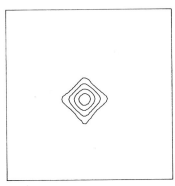

0·04 $(e^{-\pi})$ of its initial amplitude by this time; we would simply be throwing away most of this signal by using such a window function. The problem is that neither the acquisition time nor the window function can be optimised for the detection of both lines together. Variations in linewidth by such a large amount' occur quite commonly, for instance due to unresolved couplings, coupling to quadrupolar nuclei or chemical exchange, and cause large relative intensity changes in the resolution enhanced magnitude spectrum. This is a very common reason for failure to detect cross-peaks in echo-detection COSY spectra with the magnitude mode of display. It is most important to examine the range of linewidths present in a spectrum when choosing the acquisition times for the experiment, and to try out various parameters for the window functions when processing the 2D data.

The need for strong resolution enhancement and magnitude calculation is absent from the phase-sensitive experiments described previously, and for this reason alone it is always preferable to use these when possible. There is also more information available from COSY spectra which include phase, as we shall see when we compare features of the experiments later. Nonetheless, most spectra in the literature are presented in magnitude mode at the moment, and you may find that your spectrometer or data system limits you to working with this type of spectrum.

Phase in two dimensions

When spectra obtained using either the RuSH method or TPPI are transformed, the results are analogous with familiar 1D spectra. Thus the v_2 transform has real and imaginary parts, which after appropriate numerical phase correction correspond with the absorption and dispersion components of the signal. Transforming in v_1 also generates real and imaginary components, so there are in total four *phase quadrants*, corresponding with the (v_1, v_2) modes being (real,real), (real,imaginary), (imaginary,real) and (imaginary,imaginary). We generally aim to use the (real,real) part for plotting. Figure 8.24 illustrates the four types of lineshape generated by the

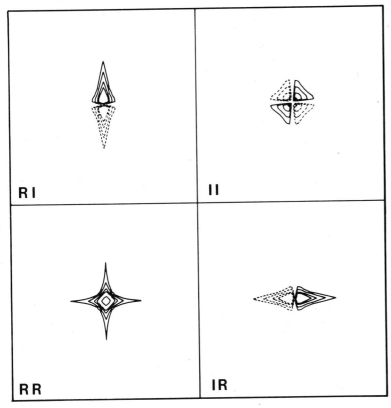

Figure 8.24 Four types of lineshape which may arise in phase-sensitive 2D spectra.

2D transform. Phase correction of the 2D data will be required to make the (real,real) part correspond with pure absorption and/or dispersion components just as for 1D experiments, and there are various ways that this can be tackled depending on the circumstances.

In direct analogy with procedures for 1D spectra, one might perform the complete 2D transform and then attempt interactive phase correction using linear combinations of the phase quadrants. Thus phase adjustment in v_2 would involve combining the (real,real) and (real,imaginary) parts (to give the v_1 real part) and the (imaginary,real) and (imaginary,imaginary) parts in the same proportion (to give the v_1 imaginary part). No current NMR data system has enough arithmetic capacity to perform this interactively on a realistic data set, although no doubt in due course this will become possible. For the time being, we need to find alternative ways of determining the phase correction factors.

A simple approach is to examine single slices out of the transformed spectrum and determine what phase correction they require, and then apply this to the complete spectrum. There are certain practical difficulties here, however, which one would not encounter in phasing a 1D spectrum. A slice taken parallel with the v_2 axis through a diagonal peak of the COSY experiment will contain the diagonal peak and its corresponding cross peaks. Typically there will be only one or two cross peaks, which will not necessarily be conveniently disposed to allow judgment of the two phase parameters. In other words, we would like peaks spread over the whole spectral range to be certain that the chosen phase correction is right, but in a cross section of a COSY experiment we will rarely get this. The problem is compounded by the fact that, as we will see shortly, diagonal and cross peaks differ in phase by 90°, and pairs of cross peaks may be in antiphase. The best procedure if you are forced to use this form of phase correction is to try out the parameters on several slices taken from different parts of the spectrum.

There would be certain practical advantages if we could determine the v_2 phase parameters before starting the 2D transform, as after making the phase correction the imaginary part of the v_2 data is no longer needed, and could be discarded to save time and space during the v_1 transform. Unfortunately in some cases there is no part of the 2D data suitable for determining the phase parameters before the complete transform, because no individual set of v_2 points gives rise to purely positive absorption signals. In the COSY experiment, for instance, only the $t_1 = 0$ spectrum would be phaseable, but if $t_1 = 0$ then the sequence $\pi/2$-t_1-$\pi/2$ degenerates to a π pulse and should give no signal. If the RuSH method is being used then the imaginary part of the $t_1 = 0$ point can be taken, as being generated by the sequence $(\pi/2)_y$-$(\pi/2)_x$ this does contain signals. If quad detection is being achieved by TPPI then there is no such possibility, as t_1 is incremented with the phase change. However, it is possible to phase later t_1 points, provided careful account is taken of the amplitude modulation of the signal (in other words, some peaks may be inverted when t_1 is not 0; this may cause confusion when adjusting the first-order phase correction). Unless the receiver phase cycle has been very carefully chosen, the zero-order phase correction derived in this way may then need altering by $\pm 90°$; trial experiments with a particular spectrometer and phase cycle will quickly determine the required procedure.

In cases where the 2D data itself provides no direct information about the phase, the only remaining possibility is to try and devise a calibration experiment that can be performed separately. For COSY a straightforward solution is to perform the $t_1 = 0$ experiment with the pulse widths set slightly shorter than $\pi/2$, so that signals are recorded. The phase parameters needed for this spectrum can then be used for the v_2 corrections. In other experiments it may be necessary to run a completely different model spectrum to extract the phase parameters, which can be a non-trivial problem as the required phase correction depends on the precise lengths of some rather short time intervals. This is surveyed in reference 3.

Phase correction in v_1 is often unnecessary, as the sources of the phase errors that are encountered in v_2 are absent. Thus, the phase relation between the two pulses, which defines the form of the modulation of each component, is exactly 0° or 90° (assuming the spectrometer is built properly!), so there is no frequency independent phase error. Also, there need be no delay in the start of acquisition in the v_1 dimension, as t_1 can start from 0; this eliminates the frequency dependent phase error. In practice, some spectrometers may not be able to start with a t_1 value of 0, in which case only a frequency dependent correction will be needed (in the notation of Chapter 4, section 4.3.5, α will be 0 and β will need to be adjusted). This can be determined by examining slices across the v_1 dimension, and will be less problematical than the v_2 adjustment because only a small correction should be needed.

Relative phases in phase-sensitive COSY

Aside from any practical difficulties of phase correction, a fundamental problem is that of knowing what phases are to be expected for different peaks. In a 1D spectrum under normal circumstances all the peaks have the same phase, and we choose to adjust this so they are all pure absorption mode for convenience. This is also sometimes possible in 2D experiments, but in a phase-sensitive COSY spectrum it transpires that the situation is a little more complex. The reason for this is that the magnetisation transfer process fundamental to the experiment causes phase shifts of some peaks but not of others. It is not possible to explain this without using a detailed quantum-mechanical description of the experiment, but if we take the results of such a calculation for granted then it is easy to understand the relative phases of different types of peaks.

Referring back to the energy level diagram for the AX system (Figure 8.6), we can distinguish three relationships between transitions. Transitions within the same multiplet, for instance A_1 and A_2, are described as *parallel* for obvious reasons. The remaining pairs are called *connected* transitions, and pairs like A_1 and X_1 are described as *progressively* connected, while pairs like A_1 and X_2 are described as *regressively* connected. The rules for phase shifts during magnetisation transfer are then as follows. Magnetisation which is *not* transferred, or is transferred between parallel transitions, retains the same phase. Magnetisation transferred between progressively connected transitions is phase shifted by $-90°$, while that transferred between regressively connected transitions is phase shifted by $+90°$.

One consequence of these rules should strike you immediately, and that is that the phases of diagonal and cross peaks must always differ by 90°. This is because the peaks on or close to the diagonal all arise from non-transferred magnetisation or from transfer to parallel transitions, while the cross peaks all arise from transfer between connected transitions. It is therefore not possible to adjust the phase of the spectrum so that all the peaks are in pure absorption; we can only have the diagonal peaks absorptive and the cross peaks dispersive or vice-versa. As the cross peaks are the informative ones, we choose to adjust so that they are absorptive, and therefore must accept a dispersive diagonal. This is sometimes quite disadvantageous as the wide dispersion lines overlap severely, but it can be partially avoided by using a variation of the experiment (section 8.4.3).

The second consequence is that pairs of cross peaks arising by transfer from one transition to two others with which it is connected progressively and regressively (sorry, that was a bit of a mouthful, but it is simple enough: for instance A_1 to X_1 and X_2) differ in phase by 180°. Thus having adjusted the cross peaks into absorption mode, we get antiphase pairs for components of the multiplets (Figure 8.25 shows this for the AX system of Figure 8.7, and by now you should realise that I cheated by presenting Figure 8.7 in magnitude mode to simplify things). The antiphase nature of pairs of cross peaks has advantages and disadvantages. The main disadvantage is that overlapping positive and negative peaks may cancel,

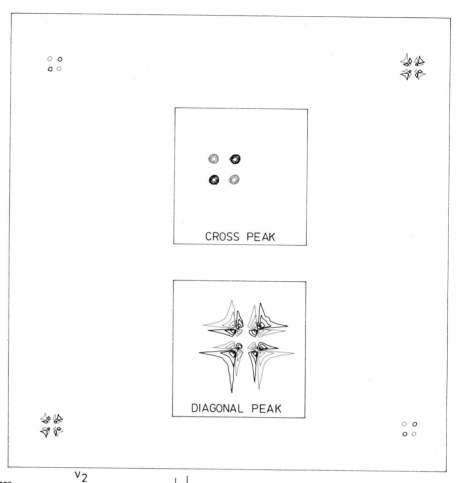

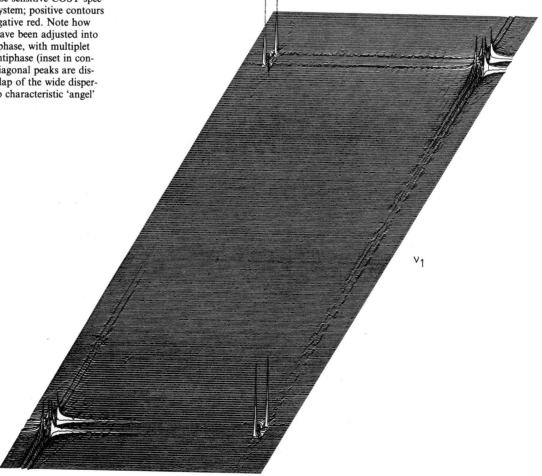

Figure 8.25 Phase-sensitive COSY spectrum of an AX system; positive contours are black and negative red. Note how the cross-peaks have been adjusted into pure absorption phase, with multiplet components in antiphase (inset in contour plot). The diagonal peaks are dispersive, and overlap of the wide dispersion lines leads to characteristic 'angel' shapes.

and it is important to realise that this potential cancellation is operative even when we subsequently throw away the phase information by doing a magnitude calculation. This is a major factor determining whether cross peaks due to small couplings can be detected. The advantage is that the phase relations between peaks are helpful in interpreting and assigning the spectrum; this is discussed further in section 8.3.6.

Digital resolution and acquisition times

One-dimensional proton spectra are generally acquired with a digital resolution similar to, or slightly less than, the observed linewidth. For instance, typical conditions might be sampling a 10 p.p.m. spectral width with between 16K and 32K data points, which leads to acquisition times of a few seconds and a digital resolution of 0·2-0·4 Hz/point, depending on field. Linewidths are typically 0·5-1·5 Hz for medium size molecules in non-degassed solution, so there are usually a few points across each line. This may be insufficient for some operations such as accurate quantitative measurement, but it is quite generous if the only object is resolution of multiplet structure. Since T_2^* is in the region of 0·2-0·6 s, the transverse magnetisation will have decayed naturally almost to zero during each scan, so there are no problems with creation of steady state echoes, and pulsing with the Ernst angle gives optimum sensitivity (Chapter 7).

Such generous digitisation is not possible for two-dimensional experiments, as we already saw from the ridiculous example of aiming for 0·2 Hz/point resolution in both dimensions of a COSY experiment with a spectral width of 5000 Hz. One of the two obstacles to achieving such resolution, the capacity of the data system to store and process the data, could be overcome if we were prepared to spend enough money, or pending the continuing rapid increase in the power of computers. For this reason, and because the level at which this becomes a problem varies by orders of magnitude depending on what kind of computer you have, I am going to ignore this limitation. Even if I did try to give guidelines about what is generally possible now, they would surely be meaningless in a few years time. Much more important is the fundamental limitation of how long it takes to accumulate the data, particularly as it is supposed to be speed of information retrieval that is the main advantage of Fourier methods. The reason the example experiment required 17 days to accumulate was that we did not approach the problem correctly; we simply transplanted ideas about what is normal for a 1D spectrum onto the 2D case.

Choice of acquisition times and digital resolution in the two dimensions, then, is a most important aspect of setting up 2D experiments, and requires a change in the way one thinks about resolution. The basic idea that must be dropped is that the aim of the experiment is to resolve individual lines in the spectrum; rather it is correlations between *groups* of lines that are interesting. This point becomes much clearer if you remember that what the COSY experiment is to be compared with is homonuclear decoupling. The 'resolution in v_1' for a set of homonuclear decouplings is determined by the selectivity of irradiation which can be achieved while still bringing about clear-cut changes elsewhere in the spectrum. This is likely to be 40-50 Hz or more, so that even a 'badly digitised' 2D experiment with a resolution of 10 Hz/point has a substantial advantage over its 1D competitor. In fact it is rarely lack of discrimination between peaks that sets the lower limit for acceptable digitisation of a COSY experiment, but more complex questions relating to sensitivity and whether cross-peaks due to particular couplings can be detected.

I am going to take as a starting point for this discussion the assumption that our aim is to detect all couplings larger than a certain size in the minimum time. To this end we must determine how much we can reduce the acquisition times A_{t_1} and A_{t_2} in t_1 and t_2 and the repetition rate of the scans without substantially reducing sensitivity or information content.

The choice of repetition rate is straightforward if you have some idea of the range of T_1 values exhibited by the sample; as $\pi/2$ pulses are being used optimum sensitivity is achieved by scanning about every $1 \cdot 3 T_1$ seconds (assuming this does not cause problems due to echo formation or spurious coherence transfers - Chapter 7, section 7.5.2). Under these conditions the magnetisation will be in a steady-state rather far from thermal equilibrium, so it will be necessary to include several dummy scans to establish this steady state at each new t_1 value. It may also be helpful to program the spectrometer so that the inter-scan delay is reduced each time t_1 is increased, so as to give a consistent repetition rate throughout the experiment.

The acquisition time in t_2 can be reduced to around T_2^* or slightly less without substantial loss of sensitivity, except when it is desired to detect couplings so small that $JT_2^* \ll 1$; this case is discussed further in section 8.4.2. Actually, because the repetition rate is determined by T_1, which is usually much larger than T_2^*, the acquisition time in t_2 often does not affect the total duration of the experiment (but it does affect the size of the dataset, and hence the time taken in processing and plotting). It is choice of A_{t_1} (and hence the number of increments to be made) that becomes the key factor determining the length of the experiment. What we need to know is how the choice of t_1 acquisition time affects the ability of COSY to detect couplings, in other words how it influences the intensity of the cross peaks. We should examine therefore the several factors which affect these intensities.

First, the cross peak intensities naturally reflect the overall signal intensity available from the sample. This is the same as the normal signal-to-noise ratio variation within a spectrum, and if one lacks basic signal strength here the position can be improved by time-averaging as usual. In this regard there is an important distinction to be made between the t_1 and t_2 dimensions, which can be illustrated by imagining two ways of using up the available time for an experiment. One possibility is to perform rather few scans for each t_1 increment, but a large number of increments, while the alternative is to have few increments each containing a large number of scans. The total number of scans is the same in each case, but is the amount of recovered signal? No, because as we increase the length of t_1, transverse relaxation eventually causes the detected signal intensity during t_2 to be reduced. So a basic aim when setting up an experiment should be to use the minimum possible number of t_1 increments with the maximum number of scans for each increment so as to occupy the available time. (Strictly speaking things are slightly more complex than this, because the envelope of the signals in the COSY experiment may not be purely a decreasing exponential in each dimension; for our purposes though it is sufficient to assume this. For a detailed analysis, see reference 6). It remains to determine the minimum number of increments that can be used without loss of information.

An extremely important factor here is the antiphase nature of multiplet components. If lines in antiphase are separated by around their linewidth or less, then some cancellation will occur. Recalling from Chapter 5 (section 5.3.3) the discussion about subtracting Lorentzian lines, we find that the line intensity is reduced to 50% if the lines are separated by about $0 \cdot 4$ times their linewidth. I am going to take this as an 'acceptable' degree of reduction; obviously whether this is so is entirely dependent on circumstances, and you may need to make a different assumption. Other line-shapes will behave differently, for instance the situation should be rather better for Gaussian lines as they are relatively narrower at the base, but the aim here is simply to establish rough guidelines rather than make a detailed analysis.

We now need to decide what value to use for the linewidth in determining how close together the lines will need to be before they are 'unacceptably' reduced. This will rarely be the actual linewidth, because we can rarely afford to digitise the spectrum well enough to manifest this. Rather, it is the linewidth we must force the signals to have so that they are properly apodised for the acquisition time we have used. A rule of thumb for determining whether Lorentzian lines will show distortion due to truncation of the time domain signal is that the acquisition time must be at least $3T_2^*$. If this is not so, then we must arrange that the window function suitably broadens the lines (i.e. increases the apparent T_2^*) until the criterion is satisfied. In other words, for an acquisition time in either dimension of A_t we can take our effective linewidth to be $3/\pi A_t$, i.e. about $1/A_t$. So if we anticipate a minimum likely separation of antiphase components of Δv Hz, according to this criterion we should have an acquisition time of about $1/2 \cdot 5 \Delta v$ s (as we require the effective linewidth to be $\leqslant 0 \cdot 4 \Delta v$). This is a less stringent condition than that of Bax[5], who suggests using an acquisition time of $1/J$ seconds to ensure detection of lines separated by J Hz, because of the inclusion of a degree of 'acceptable' cancellation.

The remaining unknown quantity is the range of separations of antiphase line pairs. In general this is completely unpredictable, because even if all the coupling constants are large *differences* between them may not be. However, it may not matter if occasional pairs of lines in a multiplet cancel as long as some remain to show the correlation, so what we can do is take the minimum separation as being equal to the minimum coupling constant we consider to be 'significant' (this assumption most often fails for complex multiplets with many closely spaced lines). You can choose any value you like for the smallest significant coupling; a practical value for proton spectra is somewhere around 3 Hz. COSY can detect *much* smaller couplings than this with suitable choice of parameters, but I am concerned here with routine application of the experiment. Putting this value into the formula for the acquisition time given above leads to a minimum value for A_{t_1} of around 130 ms. For typical proton T_1 and T_2^* values of about $1 \cdot 5$ s and $0 \cdot 3$ s (1 Hz wide lines, not uncommon in medium size molecules), we might then choose A_{t_1} around 300 ms and the repetition rate around one scan every 2 s.

These figures are only intended to illustrate the line of thought needed when setting up the experiment; it is easy to imagine reasons for either increasing or decreasing any of the parameters. For instance lineshapes will rarely be Lorentzian, as most likely some sort of Gaussian window will be applied even to phase-sensitive data; this should make it possible to reduce A_{t_1} still further without unacceptable loss of intensity. The values suggested for 'acceptable' reduction and minimum likely line separation may need revision either way for particular problems; for a strong sample more than 50% reduction may be quite acceptable. The essential thing is to plan out in advance what you are trying to detect, and to try to anticipate the necessary digital resolution. What one usually wants to do in practice is run the fastest possible spectrum as a starting point for the investigation; more refined experiments, perhaps run over a smaller spectral range, can then be devised in the light of the results from the first one.

One cautionary word is in order here. Many of the practical details of 2D NMR have been worked out by groups studying proteins. For such molecules, although the total spectrum is exceedingly complex, it is built up from well defined sub-units: the spectra of amino acids. These contain a rather limited range of multiplet structures, mostly with quite large line splittings and short T_2's (when in a protein); very short values for A_{t_1} are therefore appropriate. Everyday organic molecules show a much wider variation of line separation and width, and therefore simply applying a recipe for the COSY parameters derived from protein experiments is unlikely to be a good approach. Planning according to the appearance of the 1D spectrum is required.

How long will it take?

We are now in a position (at last!) to work out how long a typical COSY spectrum might take to accumulate. Let us look again at the 10 p.p.m./500 MHz case we examined previously (but bear in mind that it is uncommon to need to scan a full 10 p.p.m. range). According to the Nyquist criterion we need a t_1 increment of 0·2 ms (with quad detection), so to get an acquisition time of 130 ms in that dimension we must make 650 increments. With 6 scans per increment (2 dummy plus 4 real) and a repetition rate of 2 seconds per scan, the experiment will last a fraction over 2 hours (a substantial improvement on 17 days!). This can be taken as the worst case, as the spectral range is wide and the spectrometer frequency high; often results can be obtained much faster than this. For instance, take the spectral range to be 5 p.p.m. (in practice this is very often acceptable, particularly if some outlying peaks are excluded as discussed in section 8.3.4), and the spectrometer frequency to be 200 MHz. We need 130 1 ms increments, so the experiment lasts about 25 minutes. With a strong sample and Gaussian lineshapes this could no doubt be reduced by 50% or more (e.g. 64 t_1 increments, 10 minutes experiment time) without loss of information. How many homonuclear decouplings could you set up and accumulate in that time? The COSY experiment becomes superior to its 1D competition very rapidly as the complexity of the problem increases.

This discussion has assumed that it is acceptable to reduce the number of scans for each t_1 increment to the bare minimum. There are two circumstances in which this might not be possible. First, on some spectrometers failure to include the CYCLOPS component of the phase cycle may cause excessive artefacts to appear; attention from a service engineer is indicated in this case. More importantly, the actual sensitivity of the experiment may be too low, so that further signal averaging is required. At first sight it may seem that strong samples will be needed to allow one to 'get away' with only four scans per increment, but in fact this is not the case. We must not forget that signal from *all* the t_1 FID's contributes to the eventual transformed spectrum; the first experiment described above contains a total of 2,600 scans. This is not precisely equivalent to running a 1D spectrum with the same number of scans, because later t_1 increments contain less signal as mentioned already, and because the total signal is distributed over more peaks in the 2D spectrum, but nevertheless sensitivity is rather high. For proton experiments on a modern, high-field spectrometer, a few mg of a medium molecular weight compound is usually sufficient to allow use of the minimum number of scans. As soon as you start to increase the resolution, though, or try to detect very small couplings, sensitivity rapidly falls off as it becomes necessary to accumulate many spectra with large t_1 values; this is an example of the ever-present balance between sensitivity and resolution encountered in all forms of spectroscopy.

We can now also get an idea of the kind of data processing problem that may be presented by such a COSY experiment. Taking the 200 MHz/5 p.p.m. case, to get a 300 ms t_2 acquisition time we need to accumulate 300 points. As this time is not critical to the total experiment time, most likely we would round this up to the nearest multiple of 1024 (1K), i.e. 0·5K. These are complex points with quad detection (in t_2), so this represents 1K words of actual computer memory. We have 130 t_1 increments, and accumulate a real and imaginary pair for each increment, so we need to store $2 \times 130 \times 1$K sets of data somewhere, probably on disc rather than directly in memory. During processing we might like to zero-fill once or more in each dimension to improve the definition of the peaks; for instance zero-filling to 1K (complex) points in t_2 and 0·25K (complex) in t_1 means we need to process a transform containing potentially 1024K (actual) words if we keep all four phase quadrants; the (real,real) part used for plotting would contain 256K words. Had we been running an equivalent experiment using echo selection slightly less data storage would have been required: 130 increments as before, but only one spectrum stored at each increment, leading to half as much time domain data. The total acquisition time would be unchanged though, because twice as many scans are needed for each increment to get the same signal-to-noise ratio. After the transform the magnitude calculation halves the data size again by eliminating the v_2 imaginary part, so we end up with a data set the same size as the (real,real) part of the phase sensitive data.

The digital resolution of either experiment is 3·9 Hz/point in v_1 and 1 Hz/point in v_2. Note that the v_2 digital resolution is not so bad, and could easily be increased further at little cost in experiment time; this is usually the most productive way to proceed if detailed resolution of multiplet structure is required. The v_1 dimension can simply be given enough resolution to unscramble the correlated multiplets, and then slices through the v_2 dimension used to examine them. Using different degrees of digitisation in v_1 and v_2 like this is most efficient from the point of view of speed and sensitivity, but prevents the sensible use of a popular data processing trick known as *symmetrisation;* advantages and disadvantages of this are discussed in section 8.3.6.

Another look at magnetisation transfer

Finally in this 'technical' section, I want to return as promised to examine in more detail the process that occurs during the second pulse. So far we have been speaking about magnetisation and magnetisation transfer rather freely, but we have skirted around the concepts without trying to get to the bottom of them. It is certainly impossible to make a detailed description of these ideas, and particularly of the wider concept of *coherence,* without use of a proper quantum mechanical model, and for this reason non-mathematical NMR texts usually give up at this point and leave an air of mystery surrounding this area. However, while the details will elude us, a certain physical appreciation of the phenomena need not. These ideas are of such importance to modern NMR that I feel it is worth the effort of trying to explain them in simple terms; I hope you will realise though that I only aim to *stimulate* rather than *satisfy* your curiosity. Many full and detailed explanations of NMR theory exist for those that find this subject interesting.

The central concept we need to understand is *coherence.* This is a generalisation of the idea of *magnetisation,* which we can take to be a manifestation of relationships between states across a single nuclear transition, to encompass certain relationships between the states of different transitions. A good starting point for developing a feel for coherence is to consider the difference between an NMR transition which has been saturated (or has only just been put in the magnet) and one which has just experienced a $\pi/2$ pulse. In neither case does any z magnetisation result; this must mean that in each case the populations of the α and β states are equal. What, then, is the difference? For the *saturated* transition, there is no magnetisation of any kind. For the transition which has been subject to a $\pi/2$ pulse, a magnetisation component is precessing in the x-y plane. This is the resultant of the x-y components of nuclei in the sample *precessing together with the same phase*; they have been given this phase by the pulse. We say that in the saturated sample the nuclei are precessing incoherently (i.e. with random phase); in the sample which has experienced a pulse *phase coherence* has been created between the α and β states.

A coherence like this across one transition is referred to as a *single quantum coherence,* and gives rise to precessing net magnetisation which can be detected as an NMR signal. Once such a phase relationship between two levels exists, it may be transferred elsewhere by the application of further pulses. This is most easy to understand for pulses applied selectively to individual transitions, as then we only have to think about one thing at a time. Let us take as an example the AX system (Figure 8.6) again, and suppose that initially a single-quantum coherence has been created across transition A_1 by application of a selective $\pi/2$ pulse. If we follow this pulse with a selective π pulse across transition X_1, we transfer the phase information which was present in the $(\alpha\beta)$ state up to the $(\beta\beta)$ state; this state therefore gains coherence with $(\alpha\alpha)$. The two states differ by 2 in the quantum number M; the coherence is therefore referred to as a *double quantum coherence*. Coherences which might be created with ΔM values of 0, or (for more complex spin systems) 3,4 etc. are likewise referred to as zero-, triple- or higher quantum coherences. None of these gives rise to observable magnetisation; only a single quantum coherence may do that as a result of the quantum-mechanical selection rule $\Delta M = \pm 1$.

I will return to discuss the significance of multiple quantum coherences shortly, but first let us pursue the consequence of the second COSY pulse a little further. The π pulse applied to the X_1 transition in the above example is a rather special case, as it transfers *all* the coherence present across transition A_1 to the double-quantum coherence. For other pulse lengths, such as $\pi/2$, not all of the phase information present in the $(\alpha\beta)$ state is moved elsewhere, so some (single quantum) coherence remains in the original transition, some goes into the double-quantum coherence, and some new single-quantum coherence is created across transition X_1. It is this last component that is the result of what we have been calling 'magnetisation transfer', but should preferably call *coherence transfer*, and which gives rise to the cross peaks.

Of course, in the real COSY experiment we do not apply pulses selectively across single transitions, but use instead normal non-selective pulses. We can accommodate this situation by imagining that the non-selective pulse consists of a sequence (usually referred to as a *cascade*) of selective pulses applied to each transition in quick succession. Thus at the same time as the $(\alpha\beta)$ state is having its phase information partly transferred to $(\beta\beta)$, the same thing is happening between all the other pairs of states. The second pulse therefore potentially redistributes magnetisation amongst all the possible coherences of the spin system, which for the AX system includes not only the four single quantum coherences, but also zero and double quantum coherence. How much coherence is transferred to each place depends on the values of t_1 and J, and on the duration of the second pulse; a relatively simple discussion of this can be found in Bax[5] (section 2.3.3 and Appendix III). The details of this, though, do not matter too much to us, so long as we remember qualitatively the nature of the redistribution which occurs.

Coherences other than single-quantum may seem a little mysterious at first encounter, as they have no direct physical manifestation. However, they have perfectly well-defined properties, and can often be manipulated to good effect. By applying a *third* pulse to the spin system after some time we can reshuffle the coherences yet again, and as a result zero or multiple quantum coherence which has meanwhile been undetectably evolving according to its own rules may be converted back to measurable single-quantum coherence. This idea forms the basis of a range of experiments, some of which will be discussed later. In any such experiment the fundamental problem is separating from the variety of signals present at the end of the pulse sequence those which have passed through the desired level of multiple-quantum coherence. To do this we exploit the fact that the phase of signals resulting from transfer back to single-quantum from p-quantum coherence is p times as sensitive to phase shifts in the pulses which excited it than is the phase of directly excited single-quantum coherence. Suitable phase cycles of pulses and receiver can therefore select between various *coherence pathways* through which the system has been sent on its way towards the eventual precessing x-y magnetisation which we measure. Tracing coherence pathways has been developed into a particularly elegant and practical method for devising the proper phase-cycles for multi-pulse experiments, proposed independently by Ernst and Bain; regrettably it would take us too far afield to discuss this (but see Chapter 4, references 9 and 10).

8.3.6 Working With COSY

I think the basic idea of tracing the pattern of couplings via the off-diagonal peaks of the COSY spectrum is so simple that I will not bore you by discussing it further. The best way to learn to use the experiment fluently is to get some spectra run as soon as possible, preferably of compounds whose signals are readily assigned by other means at first. If you work with minimal digitisation as discussed in the previous section it need not take long to perform such experiments, so obtaining examples with which to practise should not be difficult. Therefore rather than look at more routine spectra such as those in section 8.3.4, what I want to present in this section is a survey of some specialised features of, or problems with, COSY which you may expect to encounter. In the first category, a particularly important and, until recently, rarely exploited aspect of the experiment is the information contained in the relative phases of cross peaks in the phase sensitive version. In the 'problem' category we will see a variety of artefacts peculiar to 2D spectroscopy, together with some more familiar ones.

Using the phase information

The two examples presented in section 8.3.4 were both run with echo selection and presented in the magnitude mode. If we compare this experiment with a set of homonuclear decouplings, and forget about any practical problems of selectivity with the latter, it appears that the 2D experiment is inferior in one important respect. This is that the spectrum contains no information about *which* coupling it is that gives rise to any particular cross peak. That is, if each of the peaks shown to be coupled by the presence between them of a cross peak is also split by further couplings elsewhere, there is no way to determine which of the couplings is responsible for the correlation (unless it happens that this one has a size significantly different from all the others and also is visibly resolved; in this case we could just have inspected the 1D spectrum anyway). This would not be the case if we performed an equivalent decoupling experiment, because (for first order systems) the relevant splitting would disappear from the observed multiplet when we irradiated its partner.

This is a substantial disadvantage, but fortunately it is absent from the phase-sensitive experiment; another reason why this should be preferred. To see how to extract this extra information we need to have a slightly more complex example than the AX system of Figure 8.25. A favourite test compound for two-dimensional NMR spectroscopists, 2,3-dibromopropionic acid (3), is suitable, and its phase sensitive COSY spectrum is shown in Figure 8.26. This spectrum was run with high and, unusually, equal digital resolution in both dimensions to avoid certain confusions which we will examine later. As this is a 3-spin AMX system, each pair of protons is correlated, and each cross peak is further split by coupling to the third proton. What we want to determine in each case is which coupling is responsible for the cross peak; we refer to this as the *active* coupling, while the remainder are called *passive*.

In order to do this, we take the phase relations present in the AX system as a basis, and then imagine the prototype AX cross peak (with its antiphase pairs) being split by the passive couplings *without further phase changes*. Then we can pick out which splitting belongs to the active coupling by looking for which components in the cross peak appear in antiphase. Take, for instance, the cross peak between H_1 and H_3 (Figure 8.26, bottom right hand corner). Running through the multiplet parallel

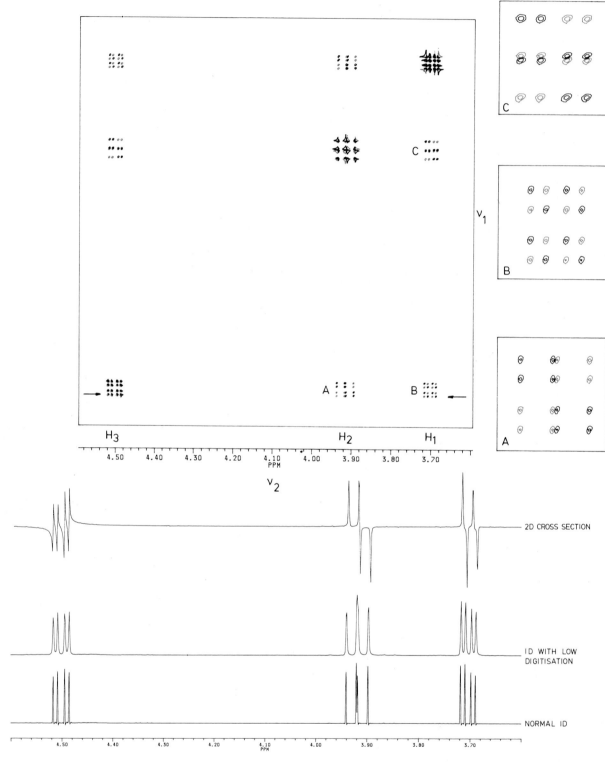

Figure 8.26 Phase-sensitive COSY for an AMX system (positive and negative contours in red and black). The part of the figure showing 1D plots compares a normal high-resolution 1D spectrum (bottom) with a 1D spectrum run with the same acquisition time and window function as the 2D experiment (middle) and a slice through the 2D spectrum along the row which is arrowed in the contour plot (top).

with the v_2 axis, the phase alternates with the smaller splitting and stays constant with the larger one; the small coupling is therefore active (i.e. it is J_{13}), and the large one must be the coupling to H_2 (in fact, as we are looking parallel with v_2, it must belong to H_1, and therefore is J_{12}). In this case the 'prototype AX pattern' is clearly evident as each of the four quarters of the multiplet (dividing parallel with v_2 and v_1). Naturally the same result is obtained by running through the peak parallel with v_1 (except that now the other splitting is J_{23}), or by examining its partner in the top left hand corner, but note that this is only so when the digital resolution is the same in both dimensions (see below for what happens when it is not). As an exercise, convince yourself by looking at the remaining cross peaks that the evidence for which coupling belongs where is self-consistent.

In a more realistic case we would not be able to afford such good digitisation, and probably we would have lower resolution in v_1 than in v_2. Figure 8.27 shows a phase-sensitive spectrum of compound **1** from section 8.3.4, run under such conditions. Although it is still possible to pick out phase relations from the contour plot here (see the expanded multiplets to the right of the figure), as the complexity of the problem increases it becomes more useful to look at cross sections through the data plotted as 1D spectra. The v_2 dimension is selected as the direction for taking cross sections, of course, because the resolution is higher. As an example the figure includes a section taken through H_{2a} parallel with v_2. All the active couplings are readily identified in this cross-section, as indicated in the figure. The only slight difficulty encountered is in the cross peak between

Figure 8.27 A more realistic phase-sensitive spectrum (partial spectrum of compound **1**), run with the following parameters: Acquisition time 487 ms in t_2 and 122 ms in t_1 (512 increments); spectral width 2100 Hz in both dimensions; zero-filled to 4K (actual) words in v_2 and 1K in v_1, leading to digital resolutions of about 1 Hz/point (v_2) and 4 Hz/point (v_1). The t_1 acquisition time is rather low for this compound, and the effect of this can be seen in the weakness of the cross peaks of H_{2e}, which is a multiplet with small line separations. Note the confusing effect of the dispersive diagonal peaks, and also see Figure 8.37. A v_2 slice taken at the arrowed position is plotted above the spectrum.

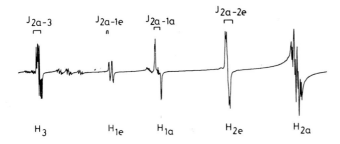

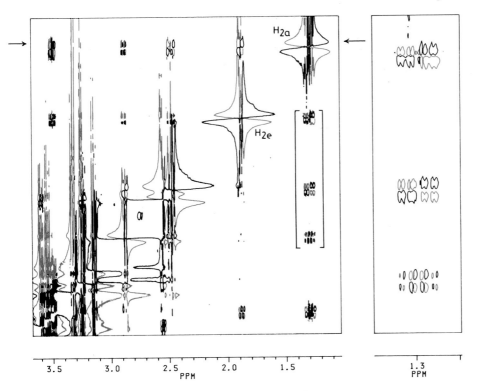

H_{2a} and H_{1a}. Here near-equality of two couplings has caused the central lines of the multiplet virtually to cancel, leading to a deceptive antiphase pair. The separation between these is *twice* the coupling constant, so the true partner of the low field line should be approximately in the middle of the multiplet, beneath the bracket drawn in the figure. The phase sensitive experiment thus provides directly a complete assignment of the spin systems; this spectrum was accumulated in 1·5 hours.

When couplings to groups of equivalent nuclei are present, as for instance in a cross peak correlating a single proton and a methyl group (an A_3X system), the pattern of phase alternation arises as if only one of the nuclei in the equivalent group were the active one, the remainder being passive. For example, taking this A_3X case, if we imagine the X part split once into an antiphase pair by the active coupling, and then each of these lines split twice more without phase changes by the other two AX couplings (with the same coupling constant, of course) we expect a pattern of four lines with intensity ratio $1 : 1 : -1 : -1$. This is illustrated in the spectrum of **4** in Figure 8.28.

Just as the intensity ratios for first order coupling with groups of equivalent nuclei in normal spectra are predicted by Pascal's triangle:

$$
\begin{array}{ccccccccccc}
 & & & & & 1 & & & & & \\
 & & & & 1 & & 1 & & & & \\
 & & & 1 & & 2 & & 1 & & & \\
 & & 1 & & 3 & & 3 & & 1 & & \\
 & 1 & & 4 & & 6 & & 4 & & 1 & \\
1 & & 5 & & 10 & & 10 & & 5 & & 1 \\
\end{array}
$$

and so on, we can write an 'antiphase triangle' for use with COSY spectra:

$$
\begin{array}{ccccccccccc}
 & & & & & 1 & & & & & \\
 & & & & 1 & & -1 & & & & \\
 & & & 1 & & 0 & & -1 & & & \\
 & & 1 & & 1 & & -1 & & -1 & & \\
 & 1 & & 2 & & 0 & & -2 & & -1 & \\
1 & & 3 & & 2 & & -2 & & -3 & & -1 \\
\end{array}
$$

etc.

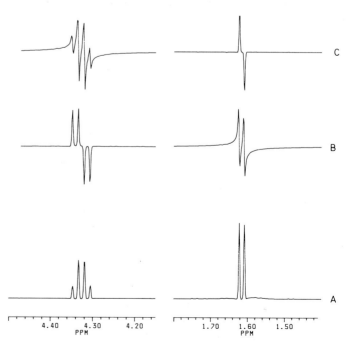

Figure 8.28 Normal proton spectrum of **4** (A), and cross sections through the phase-sensitive COSY spectrum taken parallel with v_2 at the shifts in v_1 of the methyl group (B) and the CH (C). Row B shows the characteristic pattern for a cross peak arising by coupling to three equivalent nuclei.

Measuring coupling constants

Providing the digital resolution in at least one dimension is adequate, the COSY experiment provides a convenient means of obtaining line separations from spectral regions that would otherwise be too crowded to interpret. Here we are exploiting the dispersion of signals into two dimensions to improve our chances of separating overlapping multiplets. The measurement can be made in a straightforward way from cross sections parallel with v_2 *provided that the lines are separated by more than the effective linewidth.* In spectra in which all lines have the same phase (e.g. normal 1D spectra), lines separated by much less than the linewidth are not resolved anyway; however for antiphase line pairs as we already know the lines remain resolved and only their intensity is reduced. Unfortunately it is dangerous to take the peak separation of such a close antiphase pair as an estimate of the true line separation, as the lines tend to 'repel'. The peak separation tends to a limiting value of $\delta v/\sqrt{3}$, where δv is the half-height linewidth (see Chapter 5).

As the effective linewidth can be quite large in COSY spectra because of the limitations of digitisation this may be an inconvenience. It is possible to use the graph presented as Figure 8.29 to correct the measured value provided the linewidth is known, but it must be expected that the measurement will be subject to considerable error, and that this will increase rapidly as the line separation decreases. Since the v_1 digital resolution will generally be lower, it makes no sense to try to make the measurement from sections taken parallel with that dimension. Homonuclear J-spectroscopy, discussed in Chapter 10, is an alternative method for unravelling overlapping multiplets which can also be considered, but it has the disadvantage that it only works in a simple way for entirely first-order systems.

Disappearing cross peaks

Observation of a cross peak between two nuclei is very good evidence that they are coupled (the only exception is discussed below under 'symmetrisation'). *Lack* of a cross peak, though, must be interpreted with some caution. Absence of coupling is naturally the most likely cause, but there are several other possibilities. We have seen most of them already, but since resisting the temptation to over-interpret the absence of a correlation is sometimes a little difficult, I thought I would gather them together here to emphasise the problems.

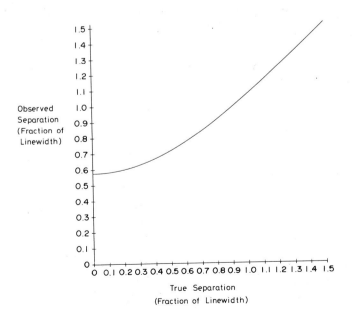

Figure 8.29 The relationship between the observed and true line separation for overlapping antiphase Lorentzian lines of equal width.

A first important point to clarify is the meaning of 'absent'. In practice this only implies that the cross peak of interest is below the lowest contour level plotted, or ultimately is below the noise in the spectrum. Thus there is no clear cut point at which a correlation will disappear; rather the weaker the peak the less likely we are to see it. Anything which serves to reduce the intensity of cross peaks may therefore assist in their disappearance. Four important contributions can be identified: the size of the coupling involved, cancellation of antiphase pairs due to inadequate effective resolution, mismatch of window function and FID envelope (which arises in cases of widely varying T_2) and non-optimum repetition rate (which arises in cases of widely varying T_1).

Notice that the size of the coupling crops up twice here. It will determine the degree of cancellation of antiphase pairs; it can also be shown that the amount of magnetisation transfer between nuclei with coupling J is proportional to:

$$e^{-t_1/T_2} \sin \pi J t_1 \qquad\qquad (8.3)$$

so for very small couplings the signal will be attenuated by transverse relaxation. This is a fundamental problem which means that correlations due to small couplings will be weak. What counts as a 'small coupling' evidently depends on T_2, or in other words on the linewidth. Optimisation of the detection of small couplings is discussed in section 8.4.2.

Problems due to misuse of window functions are greatest when the magnitude mode of display is used, because of the need for strong resolution enhancement. As this was discussed previously, I will simply stress once again that this is probably the most common reason for failure to observe cross peaks. Nuclei involved in chemical exchange, such as NH or OH protons, are most likely to fall foul of this pitfall; such resonances also often fail to show resolvable couplings in 1D spectra. With phase-sensitive spectra choice of window function can be approached in a similar manner to 1D spectra, depending on whether signal to noise improvement or resolution enhancement is required. A point to watch, though, is that more care is required to ensure proper apodisation. This is both because the FID is likely to be more truncated (particularly in ν_1), and because in a spectrum in which both positive and negative peaks are significant the negative going excursions of improperly apodised lines can be confusing, especially in a contour plot.

Wide T_1 variation is less common for protons in normal organic molecules, particularly when non-degassed solutions are in use, but it is still advisable to keep this possibility at the back of one's mind. The suggestion made in the section on digital resolution and acquisition times was for a repetition rate of 2 s; this can easily be a little on the fast side. Often you can get away with this and obtain the desired information quickly, but as soon as you start working with fairly dilute solutions it pays to examine the question of repetition rate more carefully (see Chapter 7).

Noise in two dimensions

Two dimensional spectra contain random noise, largely originating from thermal noise in the probe and early stages of the receiver. This is identical with noise encountered in 1D spectra, and for the COSY experiment with proton observation only becomes significant with quite dilute samples. Far more troublesome random interfering signals originate from the way in which the experiment is performed. As the interferograms which constitute the t_1 dimension are generated over a long series of experiments, a variety of instrumental instabilities can introduce spurious modulations of the signal. For example, imagine what would happen if the pulses used to excite the signal were not always exactly the same length or intensity. The amplitude of signals would vary for a reason other than the desired frequency labelling during t_1, resulting in the introduction of random frequency components in that dimension. Similarly, any instability in the field/frequency ratio, as may arise if the lock system is inefficient or if the instrument is subject to external disturbance, will introduce spurious frequency modulation. Such effects, together with a host of others[7] (so many, in fact, that it sometimes seems a bit surprising that the experiment works at all) cause the phenomenon known as t_1 *noise*.

The characteristics of t_1 noise are quite different from those of the usual thermal noise. As it originates from instabilities over the long building up of the t_1 interferograms, it is only present in the ν_1 dimension. Furthermore, since these instabilities need to modulate a signal in order to appear as noise, it occurs only along sections intersecting signals, and is proportional to the amplitude of the signal concerned. The practical result of this is that strong diagonal peaks, such as those of methyl groups or the solvent line, are accompanied by ridges of noise running along the ν_1 dimension. This is usually much the worst source of interfering signals in 2D spectra (Figure 8.30). Whether or not it is a problem depends on the relative sizes of interesting cross peaks and the strongest t_1 noise; if the cross peaks are large enough then it can be arranged that the lowest contour level lies above the t_1 noise.

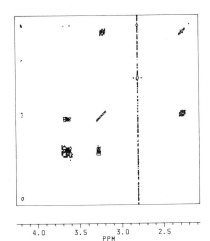
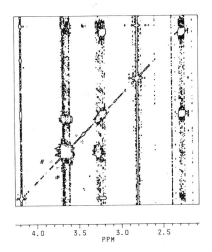

Figure 8.30 With a fairly high minimum contour level (left) a band of t_1 noise is only apparent for the strong signal at 2·8 p.p.m., but by plotting a lower contour (right) it is revealed as being associated with all signals.

Many of the sources of t_1 noise are intrinsic to the design of the spectrometer, so you will just be stuck with what you get. However there are certain precautions that can be taken to ensure that the noise is not further increased by incorrect operation. This is very similar to the optimisation of conditions for running nOe difference spectra, which has already been discussed in detail in Chapter 5; briefly the important factors are a sharp and strong lock signal and good environmental control. If many scans are being accumulated for each t_1 increment (relatively rare for COSY spectra), then it may be helpful to distribute them throughout the acquisition period by cycling through the incrementation of t_1 several times; this helps to average out long term instabilities.

Folding in two dimensions

When peaks lie at frequencies outside the spectral range determined by our sampling rate in either dimension, they appear folded as usual. There is no conceptual difference between one and two dimensional spectroscopy in this respect, but there *are* some practical differences. First, we are much more likely to have signals outside the spectral window in an average 2D experiment, because we are always striving to minimise the size of the dataset by cutting down the spectral width. Signals arising from non-transferred magnetisation give rise to out of position peaks on the diagonal if they are folded; these are not usually a problem provided you watch out for them. Rather more confusion arises for coherence transfer signals, because they may potentially be folded in either or both dimensions. The position to which peaks move when folded in one or two dimensions depends, of course, on the extent to which their true frequencies exceed the respective Nyquist frequencies, but also on the precise details of the method of quad detection in each dimension. Predicting the location of folded cross peaks is thus quite tricky, and experimentation with a particular spectrometer is usually the best way to get a feel for what is going on. Peaks folded in v_1 will be stronger than those folded in v_2, for reasons described below.

In COSY there is no reason why the spectral width should differ between v_1 and v_2 (unless you are trying to save space by *deliberately* folding in v_1 - an experiment called FOCSY[8] does this), so the partner (on the other side of the diagonal) of a cross peak folded in v_1 will be folded in v_2. However, because of the difference between the modes of detection during t_1 and t_2 the *intensities* of peaks folding in v_1 differ from those folding in v_2. In the v_2 dimension the bandpass filter of the spectrometer attenuates signals outside the spectral region, and if they are far outside they can be essentially eliminated. No such filtration occurs in v_1, however, so folded peaks occur with their full intensity in that dimension. This is illustrated for an AX system in Figure 8.31.

Figure 8.31 Peaks folded in v_1 are not attenuated in a 2D experiment, unlike those folded in v_2. As a free bonus, this figure also demonstrates very bad t_1 noise in a stack plot format.

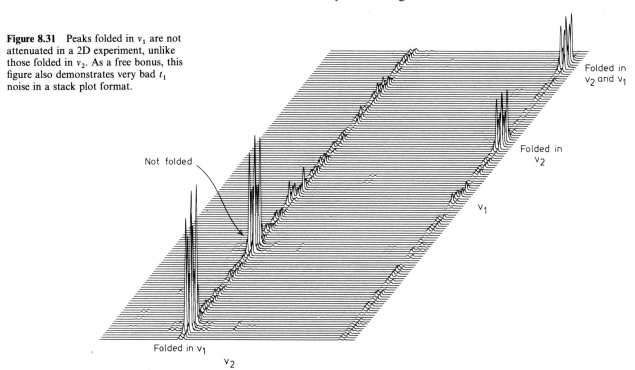

Not folded

Folded in v_2 and v_1

Folded in v_2

v_1

Folded in v_1

v_2

Symmetrisation

Both t_1 noise and folded cross peaks can be reduced in intensity by a software manipulation called *symmetrisation*. Most spectrometer software packages offer this facility, and its use has become very popular. This is not entirely a good thing; *extreme care must be used when applying symmetrisation*. It is only genuinely advantageous in certain specific cases, and even then it may give rise to deceptive effects in the spectrum, as can easily be seen by examining what is involved.

Symmetrisation takes advantage of the fact that in COSY (and certain other experiments) the real peaks have symmetry about the diagonal (i.e. a cross peak at (v_1, v_2) is matched by one at (v_2, v_1)). Artefacts such as t_1 noise or folded peaks lack this symmetry, because they are random (in the case of noise) or for the reasons discussed above (in the case of folded peaks). So reprocessing the spectrum in such a way as to *discard* any data which is not symmetrical about the diagonal would be expected to eliminate these features. Indeed it does, but there are two problems. If the experiment has substantially different digital resolution in each dimension *and* it contains antiphase peaks, then symmetrisation degrades the sensitivity by propagating the less intense signals along v_1 onto v_2. This rules out symmetrisation for COSY experiments run as suggested in this chapter; however it can be helpful when the digital resolution is similar in each dimension, or for experiments in which all the peaks have the same phase (such as NOESY described in section 8.5.2)

The second problem is a very important one. Figure 8.32 shows a COSY experiment run on a mixture of chloroform and dichloromethane, containing two singlets whose protons are not even in the same molecule, let alone coupled. In the unsymmetrised spectrum the usual ridges of t_1 noise are apparent; symmetrisation eliminates most of these *except where they appear symmetrical about the diagonal*. This, of course, occurs precisely where a cross peak would appear if the two signals were really coupled. Looking at the contour plot of the symmetrised spectrum, we would erroneously conclude that this represented an AX system. The fact that the diagonal and cross peaks appear to be singlets would be unlikely to alert us, since in a magnitude display doublets are often ill-resolved in this way. Intersection of t_1 noise ridges, or the wide wings of intense lines in magnitude spectra, can therefore lead to very dangerous results after symmetrisation. As a general principle, if you want to symmetrise a spectrum (which it has to be admitted improves its appearance greatly) it is *essential* also to examine the unsymmetrised version for problems of this kind.

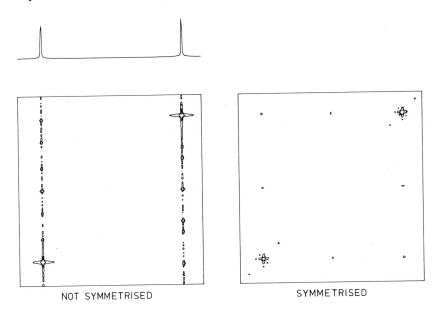

NOT SYMMETRISED   SYMMETRISED

Figure 8.32 The symmetrisation trap; beware of falling in to it.

Nuclei other than protons

The COSY experiment is restricted neither to proton observation, nor to spin-$\frac{1}{2}$ nuclei. The extension to other spin-$\frac{1}{2}$ nuclei which exhibit homonuclear coupling (such as ^{31}P or ^{19}F) is of course straightforward. The practical advantage here is likely to be considerable, because most spectrometers are not equipped to perform homonuclear decoupling of nuclei other than protons. Figure 8.33 illustrates a ^{31}P-^{31}P COSY experiment (for a compound of putative structure **5**), which as usual allowed the rapid elucidation of the coupling network. The only difficulty you are likely to encounter in extending the technique to various spin-$\frac{1}{2}$ nuclei is the wide spectral ranges that could arise, necessitating large datasets in some

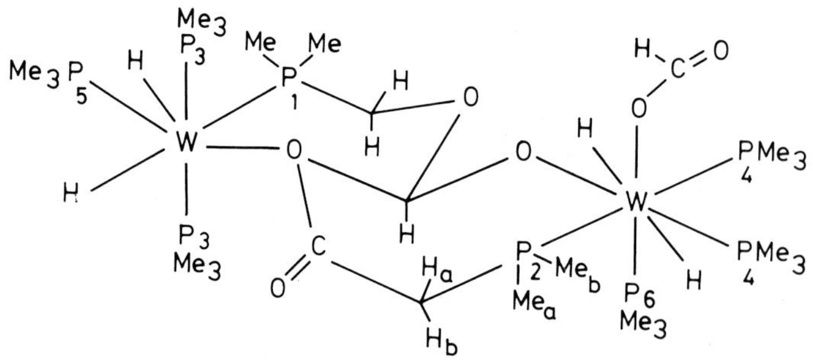

5

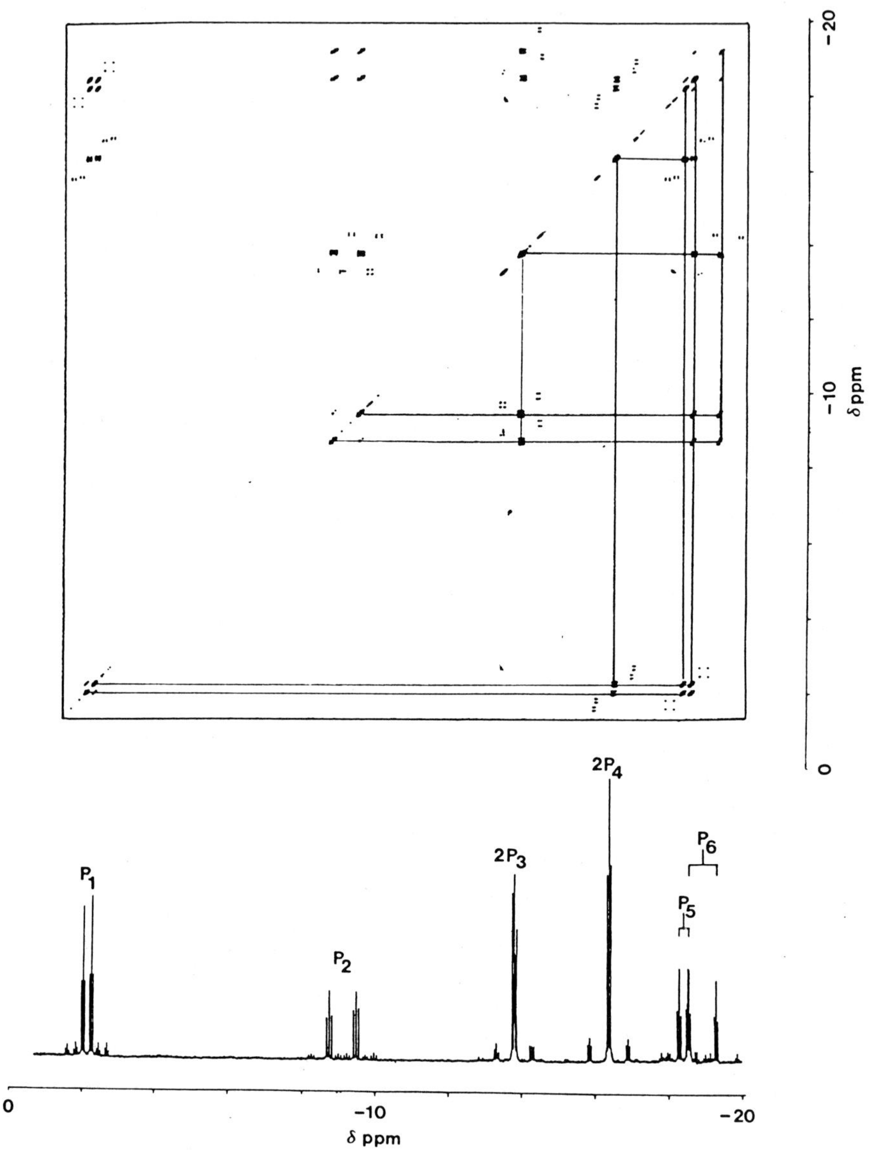

Figure 8.33 Tracing a ^{31}P coupling network for the compound tentatively assigned structure **5**.

cases. For quadrupolar nuclei, the relaxation times must not be so short that the coupling is completely eliminated, but couplings that are not fully resolved in the 1-D spectrum may still give rise to cross-peaks.

This experiment has found its most spectacular application in the study of elements which form polynuclear cluster compounds; COSY proves to be the ideal method for elucidating complex structures in this area. For example, COSY has been used to correlate ^{11}B spectra[26] and ^{183}W spectra[27], and some of the other shift correlation methods discussed in section 8.5 have also been applied[28] to ^{183}W.

8.4 EXPERIMENTS RELATED TO COSY

8.4.1 COSY-45

Varying the length of the second pulse in the COSY experiment, i.e. using the sequence:

$$\frac{\pi}{2} - t_1 - \alpha$$

with variable α, leads to variation in the relative intensities of cross peaks and peaks close to the diagonal. Specifically, it is possible to show (using arguments based on pulse cascades, or a formal quantum mechanical treatment) that the ratio of the intensities of coherence transfer between directly and indirectly connected transitions is given by $\cot^2(\alpha/2)$ (*directly* connected transitions are pairs such as A_1 and X_1 in the AX system which share a common energy level; *indirectly* connected transitions may arise in more complex spin systems and are non-parallel transitions which do not share a common energy level). Setting α to $\pi/4$ gives a ratio of about 6; the intensities of peaks arising by transfer between parallel transitions are also reduced by varying amounts.

This is the popular COSY-45 experiment, which has two advantages over basic COSY. Reducing the intensity of transfer between parallel transitions simplifies the appearance of the spectrum around the diagonal, by reducing cross peaks *within* multiplets. In a complex spectrum this can make it possible to identify correlations that would otherwise be hidden in the clutter of peaks close to the diagonal. Restricting between-multiplet transfer largely to directly connected transitions allows the determination of the relative signs of coupling constants in systems with three or more spins, in a manner somewhat analogous to the 1D spin-tickling experiment. The latter information is relatively specialised from the point of view of structure elucidation, and as it is also extensively discussed in Bax[5] (section 2.3.4) I will not pursue it further.

Figure 8.34 shows COSY and COSY-45 spectra of 2,3-dibromopropionic acid (both in the magnitude mode). The loss of some components of cross peaks and correlations within multiplets is clear in the latter. However, this illustration is perhaps a little deceptive; it is important to remember that these correlations are only reduced by a moderate factor, not entirely eliminated. In a more complex case it is not possible to set the lowest contour level so as to exclude all the attenuated correlations, so such perfect clarification of the diagonal region is not obtained. Contrast Figure 8.35 (COSY-45) with Figure 8.10 (COSY-90 of the same compound) for a slightly more realistic indication of what is to be expected. For routine spectra run with echo selection and magnitude display COSY-45 is generally preferable nonetheless. However, there is loss of sensitivity over COSY-90, and it *cannot* be used in phase-sensitive mode because the signals are not experiencing pure amplitude modulation. For a more recent experiment which can generate phase-sensitive data, see E. COSY[9].

8.4.2 Detection of Small Couplings

Small couplings significantly less than the linewidth still give rise to cross peaks in the COSY experiment, although with low intensity. For proton spectra, this means that it is often possible to identify 4 and 5 bond coupling, which is usually in the range of 0·1-0·5 Hz and hence unresolvable in average 1D experiments. Spectroscopists find this possibility

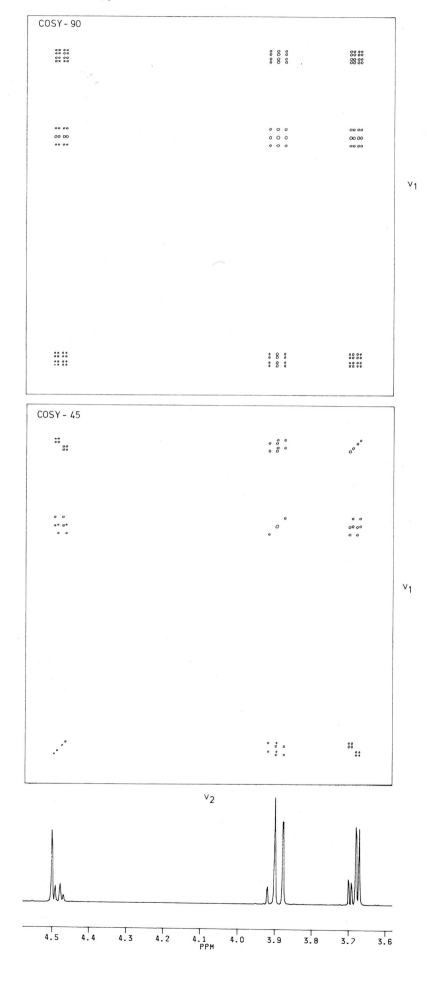

Figure 8.34 COSY-90 and COSY-45
spectra of 2,3-dibromopropionic acid,
and a cross section through the COSY-
45 spectrum.

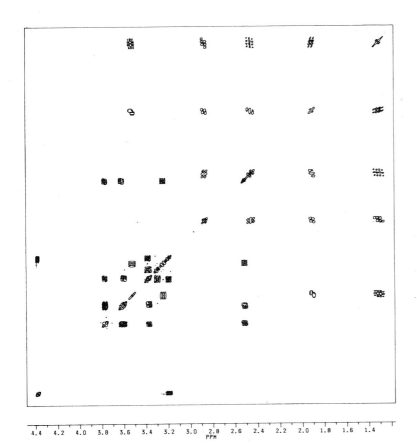

4.4 4.2 4.0 3.8 3.6 3.4 3.2 3.0 2.8 2.6 2.4 2.2 2.0 1.8 1.6 1.4
                                   PPM

Figure 8.35 COSY-45 spectrum of **1**.

exciting, as it seems to be extending the limits of measurement, but as chemists we may be a little more sceptical. Certainly there are times when it is helpful to pick up very small couplings, as for instance in the case of compound **1** discussed in section 8.3.4. There are also times when it becomes a tremendous nuisance, though, that all kinds of unlikely correlations occur. Particularly in molecules with rigid (or, especially, strained) carbon frameworks, or in polyunsaturated olefines, it can sometimes start to seem that everything correlates with everything else, which is hardly informative from a structural point of view. Presently the only way to attenuate correlations due to small couplings is to work with low digital resolution and short acquisition times. They may be emphasised in the manner described below.

We have already seen the important factors which influence the intensity of cross peaks due to small couplings; to remind you, briefly they are the degree of cancellation of antiphase peaks, and the balance of coherence transfer and transverse relaxation represented by equation 8.3. The latter describes, in effect, the envelope of the FID for the coherence transfer signal in *both* dimensions. It reaches a maximum at a time t_{max} given by:

$$t_{max} = \frac{1}{\pi J}\tan^{-1}\pi J T_2 \qquad (8.4)$$

For very small values of J ($JT_2 \ll 1$) this is well approximated by $t_{max} = T_2$. So to avoid throwing away signals arising from small couplings, we need acquisition times in both dimensions significantly greater than T_2, a condition not met in routine COSY experiments.

In experiments with echo selection and magnitude display, ideally we also want to arrange that the window function is echo-like (i.e. symmetrical about the mid-point of the acquisition) to give absorptive lineshapes. In order for the maximum of this window function to correspond with the maximum in the coherence transfer signal, we therefore need acquisition times of around $2T_2$ in both dimensions. There is no escape from this timing requirement, but it is possible to avoid the need to sample and store what may otherwise become a very large number of data points. This is achieved by introducing fixed delays into the t_1 and t_2 intervals, so the sequence becomes:

$$\frac{\pi}{2} - \Delta - t_1 - \alpha - \Delta - t_2$$

The requirement for the acquisition times is now reduced to $A_{t_1,t_2} \simeq 2T_2 - 2\Delta$, so although the same total time is spent acquiring the data, less points need to be digitised. α can be reduced from $\pi/2$ as in COSY-45 if required, but as sensitivity is likely to be a problem anyway this may be inadvisable. We already saw an example of a spectrum run in this way (Figure 8.13); in this case Δ was chosen to be 0·25 s. For a typical range of proton T_2's of 0·2-0·6 s, suitable Δ values would be about 0·05-0·45 s, assuming one is aiming for 0·3 s acquisition times. In experiments of this type it is possible for correlations due to larger couplings to be reduced in intensity if it happens that $J(T_2 - \Delta)$ is close to an integral value (e.g. for $J \simeq 7$ or 14 Hz for the above values), so a normal COSY spectrum should also be available to avoid confusion.

8.4.3 Multiple Quantum Filtration

If we add an extra $\pi/2$ pulse immediately after the end of the COSY sequence, thus:

$$\left(\frac{\pi}{2}\right)_\varphi - t_1 - \left(\frac{\pi}{2}\right)_\varphi \left(\frac{\pi}{2}\right)_x - t_2$$

where the third pulse should ideally follow on instantaneously, but in practice may be delayed by a few μs to allow phase shifting to occur, then multiple quantum coherences that happen to exist before the third pulse may be converted back to observable magnetisation (see 'Another look at magnetisation transfer' in section 8.3.5). As discussed previously, suitable choice of the phase cycle can then separate out signals which have arisen from different orders of multiple quantum coherence. For instance, double quantum coherence is twice as sensitive to phase changes in its excitation sequence as is single quantum coherence. Therefore if we shift the phase φ of the first two pulses (in general, *all* pulses before the coherence is created) by 90°, the phase of the detected signal which came via double quantum coherence is inverted. Inverting the receiver phase (i.e. subtracting the 90° from the 0° experiment) then selects the component which passed through double quantum coherence.

To select p-quantum coherence, the rule is to step the phases of the excitation pulses through the sequence 0, $180/p$, $2 \times 180/p$ $\cdots$ $(2p-1) \times 180/p$, alternating the receiver phase as you go[10] (this procedure also detects coherences of order $p(2m+1)$, $m = 0,1,2 \cdots$, but in general higher order coherences are excited with low intensity). The multiple quantum coherences are modulated as a function of t_1 in the usual way, so the resulting spectrum appears similar to normal COSY, but has been subject to *multiple quantum filtration*. Quad detection in v_1 in these experiments can be achieved by any of the methods discussed previously, provided you remember that to obtain the required 90° phase change in a p-quantum coherence, you only need to shift its excitation sequence by $90/p°$. Note that multiple quantum filtration may require phase shifts of other than integral multiples of 90°; spectrometers built after about 1983-4 generally provide these.

There are two distinct reasons why one might want to perform multiple quantum filtration: spectral simplification and improvement of the phase sensitive COSY experiment. Spectral simplification arises because multiple quantum coherences can only be generated in coupled systems; p-quantum coherence can only arise between at least p spins-$\frac{1}{2}$. So for the double quantum experiment singlets (for instance the solvent line) are eliminated, while the usual correlations remain. A triple quantum experiment eliminates in addition AX and AB systems, and so on for higher orders (sensitivity falls off rapidly as higher orders are selected, though). The elimination of peaks is occurring only by subtraction of signals, so in practice the best that can be expected is perhaps a few hundredfold reduction in their intensity; there is also no help with regard to dynamic range - the ADC must still be able to digitise the full range of signals present. Figure 8.36 compares double and triple quantum filtered spectra of the tripeptide Gly-Tyr-Gly. The triple quantum and higher experiments are obviously less general than basic COSY; it is necessary to anticipate the sorts of spin systems that will be present and choose the experiment accordingly.

DOUBLE-QUANTUM FILTERED COSY **TRIPLE-QUANTUM FILTERED COSY**

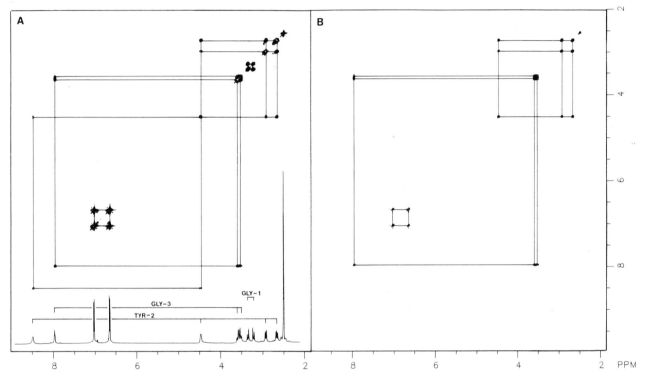

Figure 8.36 Double- and triple-quantum filtered COSY spectra of the tripeptide Gly-Tyr-Gly.

A second important feature of the double quantum filtered experiment, in its phase sensitive version, is that in contrast with phase sensitive COSY both diagonal and cross peaks can be adjusted into pure absorption phase. (In fact for systems with more than two spins there will still be some dispersive contribution to the diagonal; in practice the improvement is considerable nonetheless). This eliminates the sole disadvantage of phase sensitive COSY. The price to be paid is a theoretical reduction in sensitivity by a factor of 2; however this is sensitivity defined as signal-to-thermal-noise ratio. We already saw in section 8.3.5 that, for proton observation at least, it is more often the level of t_1 noise rather than thermal noise which interferes with detection, and as the singlets which cause the worst t_1 noise are eliminated from the double quantum filtered experiment it may show an effective *improvement* over an equivalent unfiltered spectrum. At present phase sensitive COSY with a double quantum filter (DQF-COSY) would appear to be the method of choice for assigning coupling networks; compare the two spectra in Figure 8.37 and you will see how much the diagonal region is clarified in the DQF-COSY spectrum.

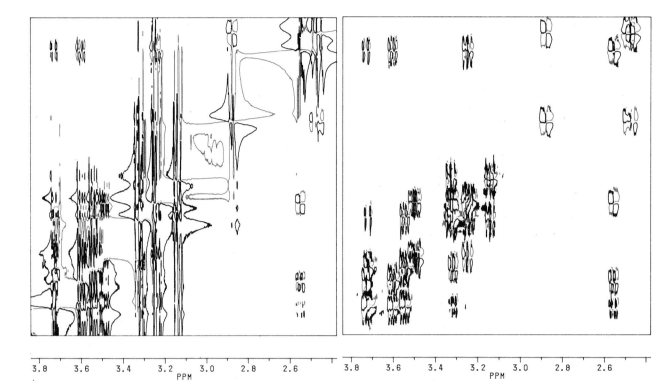

Figure 8.37 A comparison of normal phase-sensitive COSY (left, same data as Figure 8.27) with DQF-COSY (right). The tremendous clarification of the diagonal region of the latter spectrum more than compensates for the theoretical loss in sensitivity in this experiment.

8.4.4 Relayed Coherence Transfer

A quite common problem when assigning proton spectra is identifying linear chains of spins, i.e. a sequence CH-CH-CH-CH ⋯ such as may arise in polyfunctionalised compounds or olefines. Using COSY the assignment should be straightforward, but there may sometimes be ambiguity. Consider, for instance, just three spins in a line. This would constitute an AMX system, in which J_{AM} and J_{MX} would be vicinal couplings and J_{AX} would be zero (for a saturated compound). The corresponding COSY spectrum would show cross-peaks between A and M and M and X, but not between A and X. This pattern lacks any direct evidence that A and X are part of the same spin system, since the same result would be obtained from two two-spin systems AM and M'X, if it happened that M and M' had the same shift. In a complex spectrum (particularly protein spectra, in which this problem frequently arises) it may not be possible to tell from examination of the 1D data whether or not several signals seem to overlap at any given point.

The *relayed coherence transfer* (RCT) experiment attempts to solve this problem, by propagating the magnetisation transferred from A to M in a normal COSY experiment on through further couplings experienced by M. The principle of this is simply to add a third $\pi/2$ pulse to cause another redistribution of coherence, so the basic sequence would be:

$$\frac{\pi}{2} - t_1 - \frac{\pi}{2} - t_? - \frac{\pi}{2} - t_2$$

The problem here is to decide what to do with the interval $t_?$. Ideally one could imagine performing a *three*-dimensional experiment in which this interval was also incremented independently of t_1, leading after a 3D transform to data representing amplitude as a function of three frequencies. This is not a very practical idea, though, both because the experiment would take a long time to perform and because the data would be difficult to present in an informative manner. The alternative usually adopted is to fix $t_?$ at a value calculated to give the maximum propagation of coherence from A to X via M (a second alternative known as 'accordion spectroscopy', which essentially scales the 3D experiment down to two dimensions by making $t_?$ a function of t_1, will not be discussed here; see reference 11). It is also necessary to insert a π pulse in the middle of the $t_?$ interval, so as to make the result independent of chemical shift differences. The sequence therefore becomes:

$$\frac{\pi}{2} - t_1 - \frac{\pi}{2} - \tau - \pi - \tau - \frac{\pi}{2} - t_2$$

For an AMX system in which $J_{AX} = 0$, the amount of magnetisation propagated from A to X by this sequence is proportional to:

$$\sin 2\pi J_{AM}\tau \sin 2\pi J_{MX}\tau \qquad (8.5)$$

so an optimum value for τ can be chosen if the coupling constants can be estimated. For a typical vicinal proton-proton coupling in freely rotating systems of 7 Hz, the correct value for τ is about 35 ms. For other systems τ depends in a complex way on the pattern of couplings and J values; reference 29 contains extensive tabulations of optimum τ values to suit various circumstances.

Obviously this is not a general experiment, because it requires some foreknowledge of the spin systems present and the likely coupling constants involved. However, in combination with COSY it can often be used to clear up ambiguities. Figure 8.38 compares COSY and RCT spectra of compound **6**. The 1D proton spectrum together with COSY appear in Figure 8.38a, while the RCT spectrum, with additional correlations marked, is Figure 8.38b. Although τ was chosen for the small couplings round the carbohydrate ring, relayed correlations can still be seen for some other resonances. This approach has been widely used in protein NMR, where spin systems likely to be present are known (because they must arise from amino acids), and where the complexity of the spectrum is such that even cross-peaks may have insufficient dispersion. Use of the RCT experiment can relate amino acid β protons with their corresponding NH's even when the relevant α protons are unresolved[12]. For small molecule work probably the heteronuclear version of RCT (Chapter 9) is more useful.

Figure 8.38a Proton and COSY spectra of **6**.

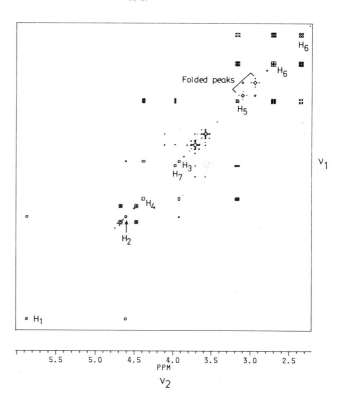

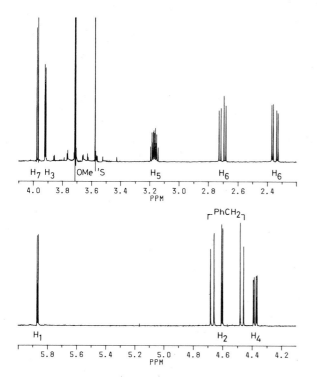

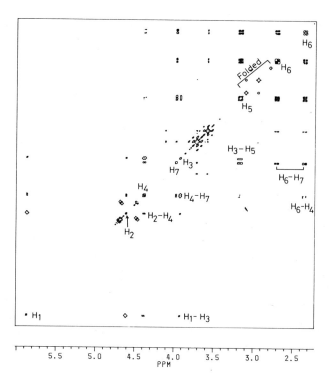

Figure 8.38b   RCT spectrum of **6**. Extra relay peaks have been marked in some cases.

8.5   OTHER HOMONUCLEAR SHIFT CORRELATION EXPERIMENTS

8.5.1   INADEQUATE

Introduction

*I*ncredible *N*atural *A*bundance *D*oubl*E* *QUA*ntum *T*ransfer *E*xperiment must be the most outrageous acronym in this book; appropriately one form of this experiment is virtually the ultimate weapon so far as tracing the skeletons of organic molecules is concerned. Unfortunately it is also just about the least sensitive experiment available; for this reason we will only look very briefly at this aspect of the technique. Historically this experiment evolved out of ideas concerned with measuring carbon-carbon coupling constants at natural abundance; however the concepts involved are much more general than this. I am going to approach it in reverse, starting from what we already know about COSY and working back to the original INADEQUATE experiment (in fact, we will discover that we have already discussed the principle involved).

So far the most we have done with the double quantum coherences which may be present at the end of the basic COSY sequence is convert them straight back to single quantum coherence. This led to the DQF-COSY experiment, in which the only role for the double quantum coherence is to distinguish nuclei without couplings from those which have them. In DQF-COSY all the interesting things, that is the evolutions during t_1, are still happening to single quantum coherence, so in that sense it is only marginally a multiple quantum experiment. Using exactly the same sequence of pulses, we could perform a *real* double quantum experiment by fixing the interval between the first two pulses and placing the t_1 interval between the second two, thus:

$$\frac{\pi}{2} - 2\tau - \frac{\pi}{2} - t_1 - \frac{\pi}{2} - t_2$$

With proper phase cycling to select signals which have travelled via double quantum coherence, the things which evolve during t_1 will be the frequencies of double quantum transitions. For an optimum experiment we need to choose the value of τ so as to create maximum double quantum coherence, and also put a π pulse in the middle of the 2τ interval to make the excitation independent of chemical shifts:

$$\frac{\pi}{2} - \tau - \pi - \tau - \frac{\pi}{2} - t_1 - \frac{\pi}{2} - t_2$$

This is the basic INADEQUATE-2D sequence. As we did for COSY, I am going to discuss the results obtainable first, then look fairly briefly at the technical details of performing the experiment.

INADEQUATE-2D spectra

The double quantum frequencies which evolve during t_1 in this experiment are simply the sums of the chemical shifts (relative to the transmitter frequency) of the pairs of nuclei which enter into the double quantum coherence. Peaks in the 2D spectrum arising from two nuclei with shifts v_A and v_B have coordinates $((v_A + v_B),X)$, where X is the frequency of the single quantum coherence to which the magnetisation was transferred by the last pulse. For an abundant nucleus such as ^1H this can lead to some complexity as we shall see shortly; this experiment was first devised and demonstrated[13] for ^{13}C. The advantage of a rare nucleus like carbon is that only two-spin AB and AX systems (which arise from the 0.01% of molecules containing *two* ^{13}C atoms) will contribute to the eventual spectrum with significant intensity. The much more intense signals from lone ^{13}C atoms are eliminated by the double quantum filtration, while three spin systems occur with negligible probability. With suitable choice of τ it can be arranged that only the larger one-bond couplings generate observable signals. Thus at each occupied frequency in v_1, we can extract a row parallel with v_2 which will be the AB or AX pattern of a pair of adjacent ^{13}C atoms, i.e. doublets occur at $((v_A + v_B),v_A)$ and $((v_A + v_B),v_B)$ in the 2D spectrum. The midpoint between each pair of signals (in the v_2 direction) therefore falls at $((v_A + v_B),(v_A + v_B)/2)$, so these points lie on a line of slope 2 (Figure 8.39). This fact can be used to distinguish genuine pairs of peaks from artefacts.

This is clarified by an example (Figure 8.40). As each non-terminal carbon is coupled to two or more others, in principle the whole skeleton of the molecule can be traced out directly (provided there are no breaks in the sequence of carbons, such as C-X-C linkages). Starting as usual from some clearly assignable signal, we run from its chemical shift in v_2 along the v_1 direction until we strike a peak; moving parallel with v_2 we then locate one of its coupling partners. Starting again from this signal and doing the same thing allows the pattern of couplings to be traced in an obvious way. This is a technique of phenomenal power, but also abysmal sensitivity. The snag is that we are trying to observe signals around the size of the ^{13}C satellites of ^{13}C signals, which have only a 1 in 10,000 chance of occurring. This rules out the experiment for most realistic problems; it may however find application in biochemical studies using ^{13}C labels.

If we try the same thing for observation of an abundant nucleus like ^1H the sensitivity problem goes away, but some other complications replace it[14]. Selection of the τ delay in the ^{13}C case is straightforward: for AX systems the optimum value is $1/4J$ (for strongly coupled systems different values are needed, see Bax[5]), and the range of J values for one-bond carbon-carbon couplings is relatively small (about 35-55 Hz). For protons, in contrast, the dependence of τ on J is liable to be complex, because we may often encounter complex spin systems, and also the range of coupling constants is larger (relatively speaking: say 2-20 Hz). A further problem in systems of more than two spins is that the double quantum coherence may be redistributed over all the transitions in the system by the last pulse, removing the simple interpretation of each row in v_1 representing the

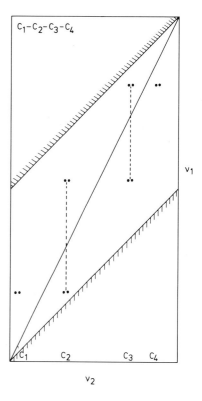

Figure 8.39 Schematic INADEQUATE-2D spectrum for a chain of AX systems. The regions outside the hatched lines can never contain signals.

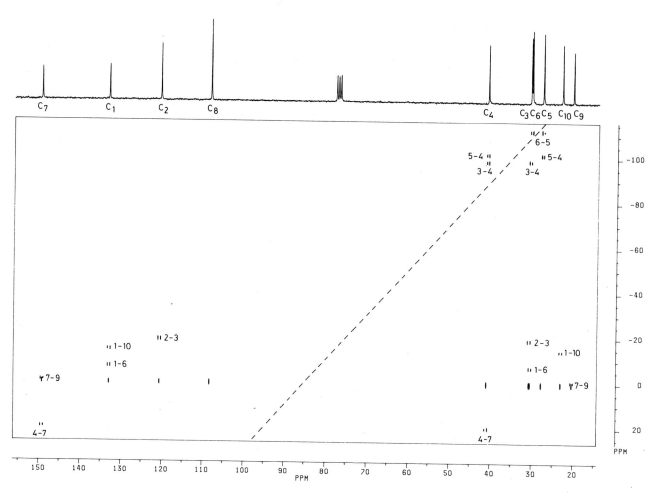

Figure 8.40 INADEQUATE-2D ^{13}C spectrum of limonene (7). The acquisition time in t_1 was about 30 ms (256 increments with a spectral width of 9 kHz; 128 scans per increment). Residual single-quantum signals are visible along the line $v_1 = 0$. Since τ was selected for the average aliphatic C-C coupling, the two correlations across double bonds were of low intensity, and have not been included in the plot.

signals from a pair of coupled nuclei. Fortunately this can be partially circumvented by setting the last pulse to $3\pi/4$ rather than $\pi/2$, which in a fashion akin to COSY-45 restricts the major part of the redistribution to certain transitions of the nuclei participating directly in the double quantum coherence[14] (I am using 'directly' here in its English sense, without meaning to imply anything about transition connectivities). Figure 8.41 shows INADEQUATE-2D proton spectra of 2,3-dibromopropionic acid obtained with $\pi/2$ and $3\pi/4$ final pulses.

The advantage of this experiment relative to COSY (which contains essentially the same information) is that the spectrum lacks diagonal peaks. Thus, for very complex problems, it may be possible to trace correlations in the INADEQUATE-2D spectrum which would be lost close to the diagonal of COSY. The disadvantage is less generality, for it will not be possible to select an optimum τ value to encompass all conceivable arrangements of spins and couplings. At present there have been some demonstrations of its usefulness in protein NMR[15], but it is still too early to assess the balance of advantages and disadvantages. For small molecule NMR this is definitely not the experiment of first choice, but can be considered as an alternative in circumstances where COSY or its variants do not provide the required answer.

Details of INADEQUATE-2D

In setting up the experiment we are faced with the normal 2D problems of selecting spectral widths, digitisation parameters and phase cycle. There are quite a number of approaches to this in the literature, because the experiment has been evolved in a circuitous way by several research groups, and for historical reasons related to availability of spectrometer

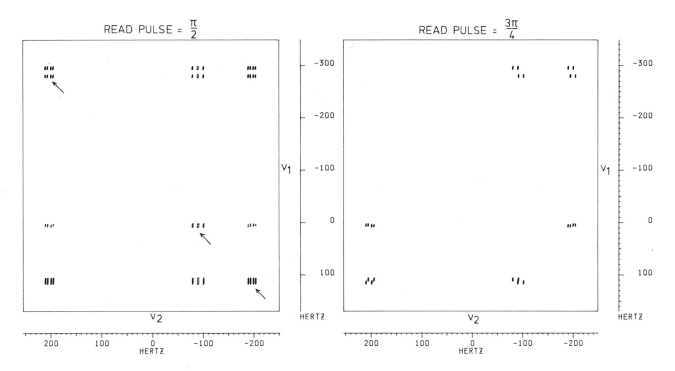

READ PULSE = $\frac{\pi}{2}$ READ PULSE = $\frac{3\pi}{4}$

Figure 8.41 INADEQUATE-2D ^{1}H spectra of 2,3-dibromopropionic acid with variation of the final (or 'read') pulse angle. Peaks arising from the undesirable transfer of magnetisation to the 'passive' spins are indicated by arrows in the left hand spectrum.

hardware. I am going to proceed by analogy with what we already know about DQF-COSY and COSY, but for completeness I will mention some other possibilities you may encounter.

In building up the phase cycle, we must account for double quantum filtration, suppression of artefacts and quad detection in v_1. The double quantum filtration can be achieved just as for DQF-COSY, exploiting the fact that shifting the excitation sequence (i.e. the first three pulses) 90° inverts the signals which have passed via double quantum coherence. Literature reports of this experiment, however, more often describe cycling the phase of the last pulse, presumably for reasons of programming convenience. A shift in phase of this pulse by 90° shifts the phase of the signal which has passed through double quantum coherence *270°*, which can be understood by imagining that we have a 90° shift relative to the excitation sequence (causing a 180° phase change) *plus* a direct 90° shift in the created single quantum coherence. If we step the last pulse through the sequence $x, y, -x, -y$ the receiver phase must therefore step the opposite way (i.e. $x, -y, -x, y$) to follow the double quantum signal. Suppression of artefacts can be added to this four step cycle in the form of CYCLOPS, and it may also be profitable to phase alternate the π pulse. There are further possibilities too, for which see reference 16. Generally it is necessary to find out experimentally how much phase cycling is really necessary on a given spectrometer, because one does not want to be committed without good cause to a large number of scans per t_1 increment.

Quad detection in v_1 requires as usual a shift by 90° in the received signal, which for signals passing through double quantum coherence means a 45° shift in the excitation pulses. On recent spectrometers this can be achieved directly; however when INADEQUATE was being developed the hardware to do this was uncommon and two alternatives were proposed. The first is the use of a so-called *composite z pulse*[17]. This consists of the sequence:

$$\left(\frac{\pi}{2}\right)_{-x} - \varphi_{-y} - \left(\frac{\pi}{2}\right)_{x}$$

where there is no delay between pulses, and φ is a pulse angle corresponding with the desired phase shift (i.e. $\pi/4$ in this case). It is easy to see for x-y magnetisation how, in a perfect world, this sequence would rotate the

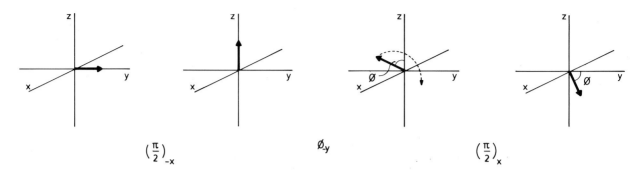

$$\left(\frac{\pi}{2}\right)_{-x} \qquad\qquad \emptyset_y \qquad\qquad \left(\frac{\pi}{2}\right)_x$$

Figure 8.42   The 'composite z pulse'.

magnetisation through an angle φ about the z axis (Figure 8.42). What happens on a real spectrometer will be highly dependent on the quality of the hardware. For the imaginary part of a quad detection experiment we put this group of pulses immediately after the third pulse of the INADEQUATE sequence[18], and noting that this leads to the pair $(\pi/2)_x$ $(\pi/2)_{-x}$ we cancel these two pulses, giving:

$$\left(\frac{\pi}{2}\right)_x - \tau - \pi - \tau - \left(\frac{\pi}{4}\right)_{-y}\left(\frac{\pi}{2}\right)_x - t_1 - \left(\frac{\pi}{2}\right)_x - t_2$$

A third possibility for achieving echo selection also exists in this experiment: the dependence of the echo and antiecho signal intensities on the length of the final pulse. The basis for this is too complex to discuss here, but setting this pulse to $2\pi/3$ essentially eliminates the antiecho component without any need for extra phase shifted experiments[19].

Selection of digital resolution for the experiment can be made in the usual way, aiming to keep the t_2 acquisition time at least equal to T_2^* to avoid loss of sensitivity. The t_1 acquisition time can be very short. It appears at first sight as if the spectral width in v_1 must be twice that in v_2, because the double quantum frequencies are the sums of chemical shifts. In fact, if we set both spectral widths the same, then although there *is* folding of double quantum frequencies, no ambiguity is introduced (Figure 8.43). Connectivities can be mapped out in the manner already described, and in certain circumstances sensitivity is improved by $\sqrt{2}$ for equal acquisition time[20].

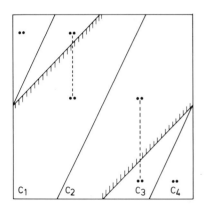

Figure 8.43   Schematic INADEQUATE-2D spectrum as for Figure 8.39, but with deliberate folding in v_1. No ambiguity is introduced.

The original INADEQUATE experiment

Finally we arrive where the whole thing started, which was a method for measuring carbon–carbon coupling constants. Although this does not strictly fit into a chapter on homonuclear shift correlation, it is so closely related to the above experiments that it would be silly to place it anywhere else. Carbon–carbon couplings are manifested in ^{13}C spectra as ^{13}C satellites of the lines. The satellites due to one-bond couplings are about 15–25 Hz from the lines, and of course have an amplitude 0·55% of the main line. Two- and three-bond couplings are in the range 0–15 Hz and the corresponding satellites are therefore 0–7 Hz from the main line. For practical lineshapes obtained on superconducting magnets this means that the one-bond satellites should easily be observed directly, while the others may be lost in the wings of the line. Lineshape may not be the only problem, however, because spinning sidebands around the 0·1–1% level are common, as of course are impurities. In order to detect the satellite lines unambiguously, it is helpful if the main signal and its sidebands (and in general any non-coupled signals) can be eliminated.

As we now know this can be achieved by double quantum filtration. If we eliminate the incrementation of t_1 from INADEQUATE-2D (i.e. fix t_1 at 0) we have the sequence:

$$\left(\frac{\pi}{2}\right)_x - \tau - \pi - \tau - \left(\frac{\pi}{2}\right)_x\left(\frac{\pi}{2}\right)_\varphi$$

which is a 1D experiment with double quantum filtration. Using the phase cycling already described (without the t_1 quad detection component, of course) leads to a spectrum in which the signals due to non-coupled nuclei are suppressed, while those due to AX and AB systems remain (without any enhancement in sensitivity, though; we are still looking at 0·01% of the nuclei). Components of doublets appear in a characteristic antiphase pattern (Figure 8.44). Success in this experiment requires a large amount of sample and careful attention to optimisation of spectrometer stability (as for nOe's), scan repetition rate and choice of τ. As the latter two points are extensively discussed in Bax[5] (section 5.2) I will not pursue them further here.

Because of the sensitivity limitation, and because the information available from an INADEQUATE spectrum is of a specialised sort, this is not likely to become a widely used experiment in its original form. However, the idea of using a double (or higher order) quantum filter in a 1D spectrum to separate out signals of interest from an intense background has many possible applications, particularly in biochemistry. One obvious example would be following ^{13}C double-labelled metabolites directly in a biological system. Such an experiment would have to be compared with the reverse polarisation transfer method already mentioned in Chapter 6, and the spin echo difference techniques of Chapter 10.

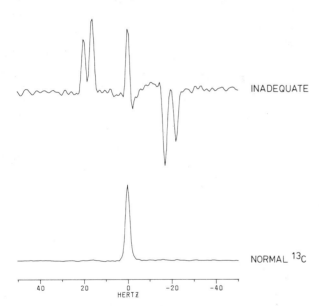

INADEQUATE

NORMAL ^{13}C

40   20   0   −20   −40
         HERTZ

Figure 8.44 1-D INADEQUATE ^{13}C pattern for a single carbon coupled to two others; the two different one-bond couplings can be distinguished. Some residual signal from the main (uncoupled) line is apparent; this does not quite fall at the centre of the doublets because of the isotope shift when ^{12}C is replaced by ^{13}C.

8.5.2 NOESY or Exchange Spectroscopy

Introduction

If you look all the way back to Figure 8.1, you will discover that there is a component of magnetisation created by the COSY sequence which so far we have ignored completely. This is the part of the transverse magnetisation returned to the z axis by the second pulse; notice that this magnetisation is also modulated as a function of t_1, i.e. it is frequency labelled just as is the x-y magnetisation. Of course, in COSY this component generates no signals, so we do not even have to worry about eliminating it by phase cycling. However, if it were involved in some physical process, we might expect to be able to make a 2D experiment to monitor that process by returning it later to the x-y plane and detecting the resulting signal. It is possible to imagine at least two informative interactions in which a frequency labelled z component might be involved: chemical exchange with another nucleus, and the nuclear Overhauser effect.

The result of chemical exchange is perhaps a little easier to visualise. If we wait a while after the second pulse, using the sequence:

$$\frac{\pi}{2} - t_1 - \frac{\pi}{2} - \tau_m - \frac{\pi}{2} - t_2$$

(referred to as NOESY), then in the presence of exchange a nucleus whose z magnetisation was modulated by one chemical shift during t_1 may have the opportunity to migrate to another site during τ_m (referred to as the *mixing time*). Thus it has a different chemical shift during t_2, and the resulting spectrum will have a COSY-like appearance, with cross-peaks between the shifts of exchanging sites. This is a 2D equivalent of the common magnetisation transfer experiment for detecting chemical exchange, and should show the usual 2D advantages for complex systems. There are, however, quite a number of technical disadvantages as we shall see shortly.

The nuclear Overhauser effect also allows changes in the z magnetisation of one nucleus to lead to variation in the z magnetisation of another, and so pairs of nuclei which would show an nOe in a 1D experiment may show cross peaks in this 2D experiment. In the common equilibrium nOe method discussed in Chapter 5, the variation in z magnetisation we use is *saturation* (i.e. elimination) of it, but any other disturbance from thermal equilibrium may also affect the populations of neighbouring nuclei. NOESY is the 2D equivalent of the transient nOe of section 5.2.4.

This experiment appears, therefore, to offer a means of determining either chemical exchange pathways or spatial relationships with 2D speed and resolution, a very attractive possibility. However, without wishing to seem over-pessimistic, I now have to present a number of major problems with the idea. Already evident is the fact that exchange and nOe interactions are not distinguishable, but this is shared with the equivalent 1D methods and so is no reason to avoid using NOESY. It is normally clear from the chemical circumstances whether exchange processes are likely to be present, so the appropriate interpretation can be made of the NOESY spectrum. (There are experiments under development, e.g. the so-called *zz*-spectroscopy[21], which should allow exchange to be distinguished from the nOe, but these are not yet in routine use). Two much more serious problems are choice of the delay τ_m, and the possible presence of interfering cross-peaks due to J-coupling.

Choosing τ_m

τ_m has to be selected so that the maximum amount of exchange takes place, or the maximum nOe builds up, before the final pulse samples the z magnetisation. This would require knowledge of the T_1's of the nuclei involved, and of the rate of exchange or nOe build up. If we knew either of the latter two parameters then there would hardly be any point doing the experiment in the first place, so guesswork will be needed. There is a distinct difference here between the possible range of τ_m for detection of exchange and nOe's (I assume in the following some familiarity with dynamic effects in NMR spectra; see for instance Günther[22], Chapter 8).

Chemical exchange can, of course, be occurring at virtually any rate. Evidently if we are trying to detect exchange by NOESY the nuclei involved are not in the fast exchange regime, though, so we can put an upper limit on the likely exchange rate according to the shift differences involved. Remembering that for two-site exchange between nuclei with equal populations and shift difference Δv the coalescence rate k_c is $\pi \Delta v / \sqrt{2}$, we certainly know that the exchange rate k must be less than k_c. At the other extreme, k must not be much less than $1/T_1$ or frequency-labelled z magnetisation will disappear before it has a chance to migrate; the range of k values over which the NOESY experiment is useful lies between these two limits. We are most likely to want to use 1D magnetisation transfer or NOESY in the rather slow exchange area, where k is not too different from $1/T_1$, as there is no other evidence here for the exchange pathway. In this region, the optimum τ_m value for detecting exchange at rate k (between equally populated sites) varies slowly as a function of kT_1, from about $0.5/k$ for $k = 1/T_1$ to about $1.5/k$ for $k = 10/T_1$. Given that k will not be known with any confidence, it seems reasonable therefore to pick a k and use $1/k$ for τ_m.

In practice it will usually be necessary to perform several experiments using different estimates for k. Even with this potentially time consuming approach, the experiment is still only providing a qualitative indication of exchange routes; there is no straightforward way to extract quantitative rate-constant information. Both the problem of selecting τ_m, and the lack of quantitative results, can be tackled using the 'accordion spectroscopy' approach[11], where τ_m is made proportional to t_1. This effectively condenses a 3D spectrum down to two dimensions, the v_1 dimension consisting of lines whose shape shows a dependence on k. Extracting rate constants from such a spectrum requires numerical lineshape analysis of these signals, and as the software for this is not yet widely available I will not pursue it further. It seems possible that this may become a useful experiment in the future.

In contrast with chemical exchange, build up of the nuclear Overhauser effect is constrained by the rate of longitudinal relaxation. Thus, although many circumstances might make the nOe build up more slowly than $1/T_1$, nothing can make it happen faster. This means that choosing τ_m in the region of $1/T_1$ gives a reasonable chance of observing nOe cross peaks between neighbouring nuclei. NOe's between more distant nuclei build up more slowly, so that longer mixing times are needed, but the competitive longitudinal relaxation ensures that cross peaks in this case will be very weak. Of course, such nOe's are weak in 1D experiments too, but because NOESY is a transient rather than an equilibrium experiment this problem is worse. Also as discussed below the presence of J-coupling can cause very confusing artefacts in the 2D spectrum which may make identification of true nOe cross peaks difficult. Finally, once again there is no simple way of getting quantitative measurements from the experiment; we only have an indication of the presence of an nOe, not a reliable measure of its size. These factors have meant that NOESY has been little used for small-molecule NMR to date; however in protein studies it has been of great importance. This is because proteins are generally of such a size that they tumble relatively slowly in solution, which puts their dipole-dipole relaxation into the W_0 dominated region (see Chapter 5), where nOe's are large, negative and fast-growing. For examples of the application of NOESY to assigning signals in proteins, see for instance references 23,24.

Problems due to J-coupling

It may perhaps have struck you that, aside from the introduction of τ_m, the NOESY sequence is the same as that used for DQF-COSY. This means that various coherence transfer signals may be present, and as usual the desirable component is selected by the phase cycle. This is a particularly easy case to understand: obviously the phase of signals arising from magnetisation which was along the z axis during τ_m is independent of the phase of the first two pulses, but follows the phase of the third. Thus either cycling the first pair of pulses together through the sequence $x, y, -x, -y$ with constant receiver phase, or cycling the last pulse and the receiver together, eliminates most undesired signals. Quad detection can be achieved either phase-sensitive or by echo selection in the usual ways.

Unfortunately, there is a class of coherence transfer signals which has exactly the same phase behaviour as the desired z component, and that is the part which has passed through zero-quantum coherence. This means that, superimposed on the cross peaks due to nOe or exchange, we have a zero-quantum filtered COSY spectrum of the compound; phase cycling cannot eliminate this. Obviously for coupled systems this is an almost fatal problem; it is akin to the SPT effects which interfere with 1D nOe experiments, but it is difficult or impossible in the 2D experiment to distinguish real and zero-quantum cross peaks.

Several solutions to this have been proposed, none of which can be considered completely satisfactory. The commonest approach is to intro-duce a random variation between scans in the interval τ_m. The intensity of nOe or exchange cross peaks varies only slowly with changes in τ_m, whereas the coherence transfer peaks are modulated with the zero-quantum frequencies (which are the chemical shift differences between the coupled nuclei). At least some of the zero quantum signals therefore change rapidly as τ_m is varied, and so with luck should be cancelled on average over a number of scans. A better way to implement the same idea is to introduce a π pulse into τ_m:

$$\frac{\pi}{2} - t_1 - \frac{\pi}{2} - \tau_i - \tau_i^r - \pi - \tau_i^r - \frac{\pi}{2} - t_2$$

so that τ_m now equals $\tau_i + 2\tau_i^r$ (the notation used here is that of reference 25). τ_i and τ_i^r can be varied while keeping τ_m constant, so the exchange or nOe peaks do not change from scan to scan, while the zero-quantum peaks are modulated as a function of τ_i only (because effects during the remainder of τ_m are refocused). τ_i can be varied randomly, but this still relies on luck and is only likely to give adequate suppression over a large number of scans. Alternatively, it is possible to calculate an optimum set of values for τ_i as a function of the number of scans available for averaging and the range of zero-quantum frequencies (i.e. chemical shift differences between pairs of coupled nuclei) to be eliminated. The best procedure appears to be combining calculated and random variations with incrementation of τ_i as a function of t_1, so that τ_i is expressed as:

$$\tau_i = \tau_i^0 + \chi t_1 + \tau_{random} \qquad (8.6)$$

Reference 25 contains tables and formulae for determining the best τ_i^0 values for given combinations of chemical shift ranges and number of scans per t_1 increment.

Even with this elaborate procedure, suppression of zero-quantum com-ponents is only moderate for small numbers of scans per t_1 increment, particularly if there is a large range of chemical shift differences present. However, as the NOESY experiment has low sensitivity anyway, it is likely that large numbers of scans will be in use for purposes of signal averaging; with more than ten scans per increment zero quantum suppression becomes adequate for typical proton shift ranges. Naturally this, together. with the need for τ_m to be of the order of T_1, means that NOESY will be a slow experiment to perform.

Examples

Despite the comments of the previous two sections, it *is* actually possible to perform these experiments! I am inclined to think that it is the detection of chemical exchange that is the most practical use of this technique for small molecules; Figure 8.45 illustrates the tracing of an exchange network in a fluxional organometallic compound. As an example of the detection of nOe's, Figure 8.46 is the NOESY spectrum of a penicillin derivative **8**. In this compound the T_1's (in degassed D_2O) varied quite widely, from 0·4 s

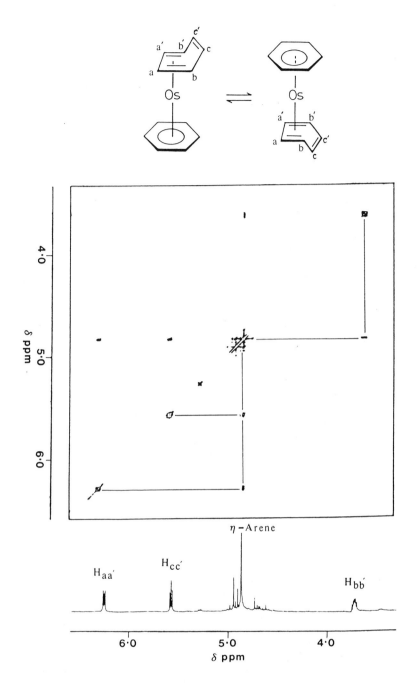

Figure 8.45 Tracing a chemical exchange pathway in a fluxional organometallic compound.

to 2 s, so τ_m was assigned the rather arbitrary value of 1 s. Nevertheless, cross peaks are evident in the places expected from equilibrium nOe measurements, and as there is no problem with coupling in this case we can be quite confident that they arise from the Overhauser effect.

REFERENCES

1. S. R. Evans, A. R. Hayman, L. E. Fellows, T. K. M. Shing, A. E. Derome and G. W. J. Fleet, *Tet. Lett.*, **76**, 1465-1468, (1985).
2. D. J. States, R. A. Habekorn and D. J. Ruben, *J. Mag. Res.*, **48**, 286-292, (1982).
3. J. Keeler and D. Neuhaus, *J. Mag. Res.*, **63**, 454-472, (1985).
4. D. Marion and K. Wüthrich, *Biochem. Biophys. Res. Comm.*, **113**, 967-974, (1983).
5. A. Bax, *Two Dimensional Nuclear Magnetic Resonance in Liquids*, Delft University Press, Dordrecht, (1982).
6. M. H. Levitt, G. Bodenhausen and R. R. Ernst, *J. Mag. Res.*, **58**, 462-474, (1984).
7. A. F. Mehlkopf, D. Korbee, T. A. Tiggelman and R. Freeman, *J. Mag. Res.*, **58**, 315-323, (1984).

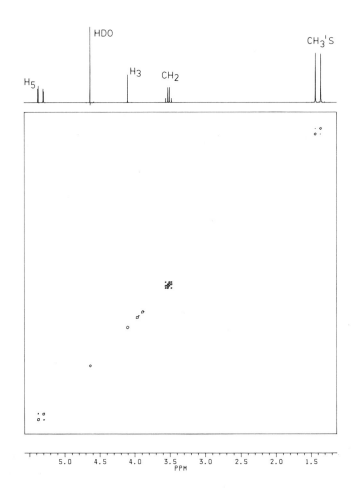

Figure 8.46 NOESY spectrum of a penicillin derivative (**8**). The methyl group on the lower face of the molecule shows a cross peak to one of the β-lactam protons (H₅); in equilibrium experiments irradiating this methyl generates a 15% nOe at the β-lactam. Conversely, the methyl on the upper face is correlated with H₃.

8. K. Nagayama, A. Kumar, K. Wüthrich and R. R. Ernst, *J. Mag. Res.*, **40**, 321, (1980).
9. C. W. Griesinger, O. W. Sørensen and R. R. Ernst, *J. Amer. Chem. Soc.*, **107**, 6394-6396, (1985).
10. U. Piantini, O. W. Sørensen and R. R. Ernst, *J. Amer. Chem. Soc.*, **104**, 6800-6801, (1982).
11. G. Bodenhausen and R. R. Ernst, *J. Amer. Chem. Soc.*, **104**, 1304-1309, (1982).
12. G. Wagner, *J. Mag. Res.*, **55**, 151-156, (1983).
13. A. Bax, R. Freeman and T. A. Frenkiel, *J. Amer. Chem. Soc.*, **103**, 2102-2104, (1981).
14. T. H. Mareci and R. Freeman, *J. Mag. Res.*, **51**, 531-535, (1983).
15. J. Boyd, C. M. Dobson and C. Redfield, *J. Mag. Res.*, **55**, 170-176, (1983).
16. A. Bax, R. Freeman and S. P. Kempsell, *J. Amer. Chem. Soc.*, **102**, 4849-4851, (1980).
17. R. Freeman, T. A. Frenkiel and M. H. Levitt, *J. Mag. Res.*, **44**, 409, (1981).
18. A. Bax, R. Freeman, T. A. Frenkiel and M. H. Levitt, *J. Mag. Res.*, **43**, 478, (1981).
19. T. H. Mareci and R. Freeman, *J. Mag. Res.*, **48**, 158, (1982).
20. D. L. Turner, *J. Mag. Res.*, **58**, 500-501, (1984).
21. G. Wagner, G. Bodenhausen, N. Müller, M. Rance, O. W. Sørensen, R. R. Ernst and K. Wüthrich, *J. Amer. Chem. Soc.*, **107**, 6440-6446, (1985).
22. H. Günther, *NMR Spectroscopy*, Wiley, London, (1980).
23. G. Wagner, A. Kumar and K. Wüthrich, *Eur. J. Biochem.*, **114**, 375, (1981).
24. K. Wüthrich, M. Billeter and W. Braun, *J. Mol. Biol.*, **180**, 715-740, (1984).
25. M. Rance, G. Bodenhausen, G. Wagner, K. Wüthrich and R. R. Ernst, *J. Mag. Res.*, **62**, 497-510, (1985).
26. T. L. Venable, W. C. Hutton and R. N. Grimes, *J. Amer. Chem. Soc.*, **106**, 29-37, (1984).
27. C. Brevard, R. Schimpe, G. Tourné and C. M. Tourné, *J. Amer. Chem. Soc.*, **105**, 7059-7063, (1983).
28. R. G. Finke, B. Rapko, R. J. Saxton and P. J. Domaille, *J. Amer. Chem. Soc.*, **108**, 2947-2960, (1986).
29. A. Bax and G. Drobny, *J. Mag. Res.*, **61**, 306-320, (1985).

9

Heteronuclear Shift Correlation

9.1 INTRODUCTION

In building up the INEPT experiment from SPI in Chapter 6, we were careful to exclude the effects of the chemical shifts of the S spins by using a spin echo. If we take away the echo, and make the interval between the remaining pulses variable, we are left with:

S: $\qquad \dfrac{\pi}{2} - t_1 - \dfrac{\pi}{2}$

I: $\qquad\qquad\qquad \dfrac{\pi}{2} \quad Acquire \ (t_2) \ \cdots$

Clearly this experiment might still bring about polarisation transfer, but its extent will depend on the particular disposition of multiplet components at the time of the second pulse, which in turn depends on the offsets of S resonances and the length of t_1. Thus we have the basis for a 2D experiment: the amplitude of the I signal detected during t_2 will be modulated as a function of t_1 by the frequencies of the S spins. This is the fundamental scheme for *heteronuclear shift correlation*. Another way of looking at this sequence is by comparison with COSY, when the only difference found is that the coherence transfer processes of the second pulse have been extended to another nucleus, by providing a simultaneous pulse at another frequency. All the experiments in Chapters 6, 8 (excluding NOESY) and 9 are therefore seen to involve the *same* underlying phenomenon: the transfer of coherence amongst coupled spins, which can be most simply understood in the context of SPI.

As all the groundwork for understandng 2D NMR has been laid in the previous chapter, we are left here with the relatively simple task of examining a few technical details of the heteronuclear experiment, and reviewing some of its variants. The simplest form of heteronuclear correlation, described in the following section, and COSY (preferably in the form of phase-sensitive DQF-COSY), are the two central methods of 2D NMR, to which you should turn first when tackling a problem. All the other experiments are more specifically tailored to particular situations, and are best held in reserve until the analysis has advanced somewhat.

Rather strangely, considering how popular a sport is the invention of acronyms, no consensus has emerged regarding what to call heteronuclear shift correlation. As this full name is such a mouthful, authors usually arrive at some abbreviated version, and it is common to see it referred to as H-X COSY (but what if the other nucleus isn't H?) or, confusingly, just COSY. For clarity and compactness, in what follows I will call this experiment in its original form HSC; no doubt readers with greater creative flair will be able to devise their own more memorable designation.

9.2   DETAILS OF HSC

9.2.1   Eliminating Various Couplings

Introduction

What you get by performing the basic HSC sequence is Figure 9.1 (for an AX system). The signals in each dimension are connected by cross peaks; just as in phase-sensitive COSY, multiplet components appear in antiphase. Very often, the v_1 dimension of a spectrum like this will represent proton chemical shifts, and v_2 will contain the shifts of a heteronucleus like carbon. Clearly the experiment as it stands does not give us quite what we expect for spectra of such nuclei in either dimension. For instance, broadband proton decoupling is usually applied during the acquisition of heteronuclei, so we do not see multiplet structures due to proton coupling. In addition, at the moment the v_1 dimension contains the frequencies of proton lines coupled to the heteronucleus (i.e. the ^{13}C satellites of the proton lines, if carbon is being observed). For simplicity of interpretation, and optimum sensitivity, it is desirable to eliminate the heteronuclear coupling in *both* dimensions of the experiment. However, in trying to do this, we have to be careful not to eliminate the experiment as well, because it depends on the presence of the coupling for its operation.

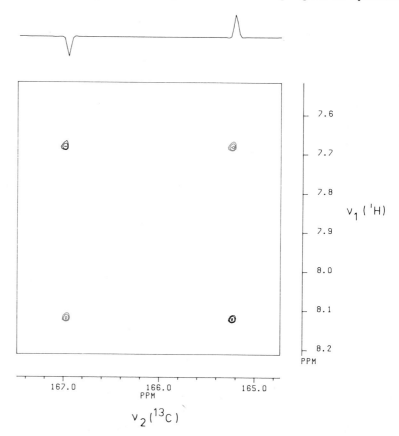

Figure 9.1 Basic ^1H-^{13}C heteronuclear shift correlation experiment for an AX system (formic acid, v_1 quad detection by TPPI). Both the ^1H and ^{13}C dimensions contain antiphase doublets (positive and negative contours red and black). A slice taken parallel with v_2 is shown above the contour plot.

Eliminating the coupling in v_2

In order to be able to turn on broadband proton decoupling during t_2, we have to get the multiplet components of the detected nucleus back in phase. This is exactly the same as the problem of decoupling in INEPT, and the solution is to wait a time after the final pulses as usual. This delay, usually referred to as Δ_2 in the context of HSC, has precisely the same role as the INEPT delay Δ (indeed, it *is* the same delay), and so can be chosen using the criteria of Chapter 6. The only difference is that, if the 2D experiment is destined to be displayed in magnitude mode, we need not be concerned with phase changes due to chemical shifts during this interval. Thus there is no need to place π pulses at the centre of Δ_2 to create a spin echo for magnitude-mode experiments, but they must be included if the preferred phase-sensitive form of detection is in use.

Eliminating the coupling in v_1

To give the v_1 dimension the appearance of not being coupled to the heteronucleus, a rather subtle approach is required. To see why, imagine first that true broadband decoupling of the I nucleus is available. If this was applied throughout t_1, the experiment would not work at all, because multiplet components due to the heteronuclear coupling would never have the antiphase disposition required for polarisation transfer. What we really want to do is stop the heteronuclear coupling modulating the signal during t_1, while still allowing it to generate antiphase lines ready for polarisation transfer. This can be achieved by decoupling during t_1, but then inserting another delay Δ_1 between the end of t_1 and the polarisation transfer step during which decoupling is removed, which makes the complete HSC sequence (for a magnitude-mode experiment) Figure 9.2, sequence A.

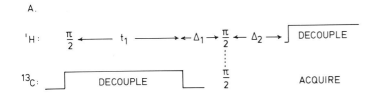

Figure 9.2 Improved HSC schemes: A, with heteronuclear decoupling in v_1 and v_2; B, the same with refocusing in the Δ delays, for phase-sensitive experiments; C, avoiding the need for true broadband carbon decoupling by use of a ^{13}C π pulse at the centre of t_1.

This may seem to contain an element of having your cake and eating it, but it must be remembered that it is only *changes* during t_1 that count. Constant delays such as Δ_1 do not affect the amplitude modulation of the signal, but only introduce frequency-dependent phase shifts in v_1. Once again, this does not matter for magnitude-mode experiments, but should be eliminated with a spin echo if the phase-sensitive variation is chosen (Figure 9.2, sequence B).

A practical problem with this scheme is the need for broadband decoupling of an X-nucleus, because this is likely to be unavailable or difficult. Fortunately, since the t_1 dimension is built up in discrete steps, a convenient alternative is to apply a single π pulse to I halfway through t_1. Just as in the hypothetical broadband decoupling scheme discussed in section 7.4 of Chapter 7, this reverses the direction of precession of multiplet components, causing the effect of the coupling to disappear by the end of t_1. Since we only have to apply one pulse, this is not subject to the accumulating errors that may sabotage such an approach to ordinary broadband decoupling. Nevertheless, it is often helpful to use a composite pulse here. The optimum scheme for phase-sensitive HSC with decoupling in v_1 and v_2 would then be Figure 9.2, sequence C.

The delay $\frac{1}{2}\Delta_1$ in this sequence is equivalent to τ in INEPT, and should therefore be set to $1/4J$ (i.e. $\Delta_1 = 1/2J$). For correlation of protons and carbon, the full range of coupling constants encountered is 125-210 Hz. However in most cases assuming a value around 130-150 Hz, and therefore setting Δ_1 to 3-4 ms, will be suitable. Δ_2 can be used for editing if required, but commonly is set to $0{\cdot}3/J$ (2-2·5 ms for H-C correlation) to give reasonable intensity for IS, IS_2 and IS_3 groups. The resulting spectrum for an AX system appears as in Figure 9.3.

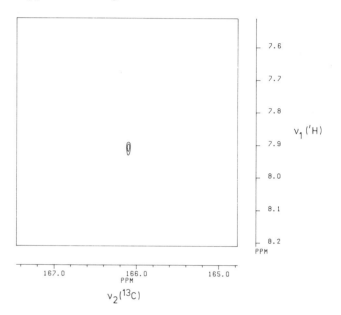

Figure 9.3 HSC with heteronuclear decoupling in both dimensions (formic acid again).

9.2.2   Other Experimental Aspects

Phase cycle

As usual, the most important considerations in building the phase cycle for HSC are elimination of axial peaks and v_1 quad detection. Phase alternating the final S pulse and the receiver eliminates axial peaks (note that this is the same procedure as that for eliminating the natural I magnetisation in INEPT). Quad detection should preferably be brought about in phase sensitive mode, by the RuSH procedure or TPPI as appropriate for your spectrometer, but echo selection/magnitude display is very commonly encountered in literature examples of the experiment. For echo selection, the final S spin $\pi/2$ pulse is shifted 90° together with a $-90°$ shift of the receiver. The basic phase cycle thus has four steps (or two pairs of two steps each in phase-sensitive form).

If necessary this four-step cycle can be extended with CYCLOPS, by cycling the phases of all the pulses and the receiver in 90° steps, and by phase alternations of any π pulses present. In practice, though, the experiment is not too sensitive to pulse defects.

Acquisition times and repetition rate

Since this experiment involves heteronuclear detection, careful optimisation of the various timings may be required to ensure acceptable sensitivity. The principles involved here have been discussed in Chapter 7, section

7.5.2 and Chapter 8, section 8.3.5. Since polarisation transfer is involved, it is the T_1's of the S nuclei (usually protons) which determine the repetition rate; a period around $1 \cdot 3 T_1$ should be allowed between scans for relaxation. Note that this is *between* scans, rather than being the total repetition rate including the acquisition time, because broadband decoupling of S is in effect during t_2.

The acquisition time in t_2 should be at least equal to the average T_2^* for I; this usually presents no problem, as this time is not critical to the total experiment time. The t_1 acquisition time, on the other hand, can be very short. Unlike COSY, where we have to avoid reducing the signal by cancelling antiphase peaks, the signals in HSC are in phase. Thus A_{t_1} need only be long enough to resolve the features of interest in that dimension. In a routine application of HSC, we simply want to correlate approximate proton shifts with their X-nucleus partners, so there is no objection to a digital resolution in v_1 of 10 or more Hz/point, corresponding with $A_{t_1} =$ 100 ms or less. If it is desirable to resolve fine structure in v_1 then A_{t_1} must be increased, but the sensitivity of the experiment is rapidly reduced (see section 10.4.3 of Chapter 10 for an alternative approach in this case). The ideal, but technically difficult, way to achieve high resolution of the proton dimension is to make it v_2, i.e. to perform the experiment 'backwards'[1].

With a small number of t_1 increments and a repetition rate optimised for the proton T_1's, the sensitivity of this experiment is very high. While it is quite hard to predict the sensitivity exactly, because it depends on a large number of factors, I find the following rule of thumb useful. Determine the number of scans per t_1 increment that will be possible, given the amount of time available and the number of increments required. Run an INEPT spectrum with this number of scans (i.e. perform the $t_1 = 0$ experiment). If most of the expected resonances are visible in this spectrum, however bad their signal-to-noise ratio, the 2D shift-correlation will certainly have adequate sensitivity (assuming correct choice of A_{t_1} and proper selection of window functions). Even with less material than this, such that no signals are visible in the $t_1 = 0$ experiment, the shift correlation has a chance of success; we must not forget that *all* the scans contribute to the eventual signals.

9.2.3 Using HSC

The method of applying HSC to a problem is largely self-evident; it allows the identification of coupling relationships in heteronuclear systems. Here I would like to demonstrate this, and point out some of the remarkable advantages of the 2D mode of operation. The first advantage is *speed*. Figure 9.4 shows the proton and carbon spectra of compound **1**, which we have already investigated extensively in Chapter 8, while Figure 9.5 is the HSC spectrum. With the assignments of the proton spectrum to hand, it is a trivial exercise to assign the carbon resonances. Obtaining the same information by selective heteronuclear decoupling would quite likely be impossible in this case, because of excessive overlap in the proton spectrum. Even when such decouplings are possible, performing more than two or three is extremely tedious in comparison with running HSC.

HSC is a very straightforward experiment to apply in this fashion, the only potential problem being strong coupling amongst protons[7]. In weakly coupled systems, cross peaks only arise between protons and their directly bonded heteronuclear partners. In the presence of strong coupling, however, magnetisation may be transferred throughout the system, leading to *indirect* correlations. That is, in a system such as H_A-C_A-C_B-H_B, H_A may show a cross peak to C_B if it is strongly coupled to H_B. This is not difficult to spot in practice, but a subtle point to watch for is that it is the ^{13}C satellites which need to be strongly coupled. A system which appears first order (from the ^{12}C lines) may still have strongly coupled satellites, because of course the relative chemical shifts of the lines are significantly

1

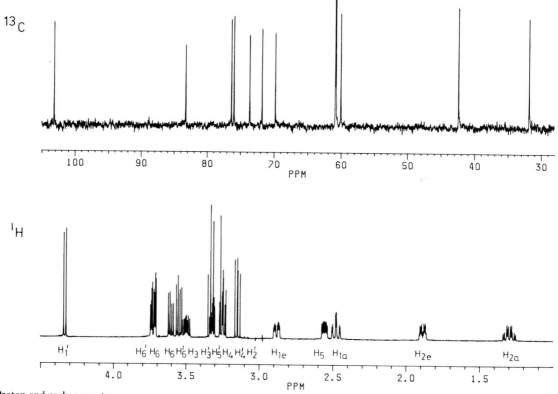

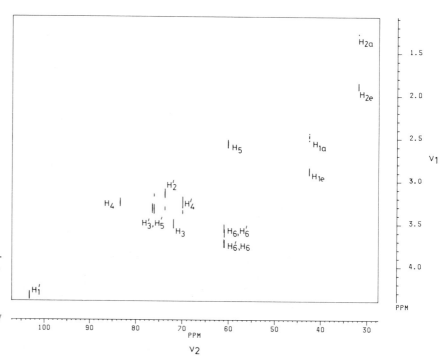

Figure 9.4 Proton and carbon spectra of compound **1**. The carbon spectrum was acquired using DEPT (400 scans), on the same sample used for the HSC experiment in the following figure.

different. The opposite (i.e. weakly coupled ^{13}C satellites even when the ^{12}C system is strongly coupled) may also sometimes be true.

The second advantage is *dispersion*. Common heteronuclei like ^{13}C and ^{31}P have much wider chemical shift ranges than protons (in Hz as well as p.p.m.). It also often happens that their spectra consist of collections of singlets, because of an absence of homonuclear coupling (due either to low isotopic abundance or the presence of only one abundant heteronucleus per molecule), and broadband decoupling of protons. These two facts

Figure 9.5 HSC spectrum of **1** (128 t_1 increments for an acquisition time in that dimension of 76 ms; 192 scans per increment). The proton assignments (derived using COSY in the previous chapter) are marked. In the crowded central region it is a little difficult to identify relative proton shifts from the contour plot, but these were readily measured by examining vertical slices through the spectrum. The assignments of H_4 and H_4' may be interchanged.

together mean that the effective dispersion of heteronuclear spectra is much greater than that of proton spectra; overlap of resonances is rare even for quite large molecules. In the HSC spectrum, the information about proton frequencies is carried on the chemical shifts of the heteronuclei, so this offers the possibility of analysing complex proton spectra with the dispersion of a heteronucleus. In the absence of strong proton-proton coupling, and provided sufficient digital resolution can be achieved in v_1, this idea can be applied in a straightforward way. Taking a slice through the HSC spectrum (parallel with v_1) at the chemical shift of an X nucleus yields the ^1H signals of its attached protons. In fact, it is often more useful to plot selected slices in this fashion, rather than a complete contour plot.

Figures 9.6 and 9.7 illustrate the effectiveness of this technique. Figure 9.6 is the aromatic region of the 500 MHz proton spectrum of an iridium complex, as indicated (this is an intermediate in catalytic hydrogenation). Our aim was to assign all these resonances, so that by means of nOe experiments we could determine the arrangement of ligands around the iridium[2]. The connections amongst the protons were readily determined using COSY, but it remained to identify starting points for the assignment. Figure 9.7 shows slices taken through the ^1H-^{31}P HSC spectrum at the chemical shifts of each phosphorus; the *ortho* protons of the phenyl rings are extracted neatly by this means. Note in particular H_1, which has been identified even though its region of the 1D proton spectrum looks fairly intractable. Small responses are also evident at the shifts of some of the other protons on the phosphorus bearing rings. These arise, albeit with low intensity, by polarisation transfer through the long range proton-phosphorus couplings, even though Δ_1 and Δ_2 were chosen according to the estimated *ortho* (3-bond) coupling.

This example also illustrates the third advantage: *power*. It is clear that, by the standards of 2D NMR, this was a 'simple' compound to deal with. Once the experiments had been performed, the assignment followed in a matter of minutes. There is little difficulty extending the technique to much more complex problems than this. A particularly important feature is the advantage of using both COSY and HSC spectra together, because this may allow the complete framework of a molecule to be traced. Starting from some readily assigned resonance in, say, a carbon spectrum, the corresponding proton shift(s) can be identified from HSC. COSY then allows the neighbouring protons to be located, which in turn allows the

Figure 9.6 Proton spectrum of an iridium complex.

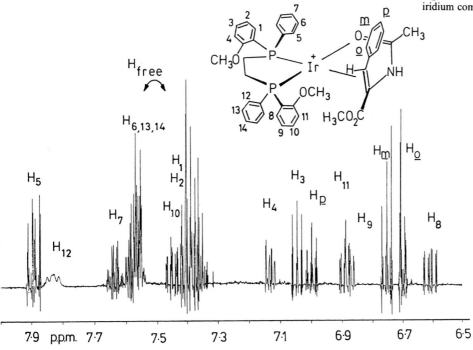

Figure 9.7 Slices parallel with v_1 through the ^1H-^{31}P HSC spectrum of the iridium complex of Figure 9.6. Although this was not a phase-sensitive experiment, the individual slices have been adjusted into absorption mode.

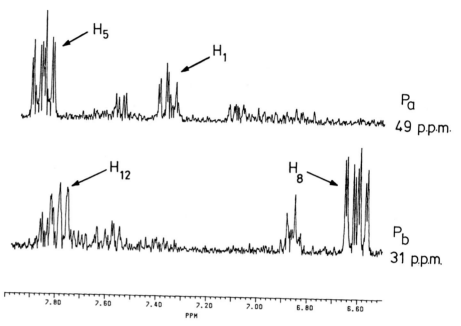

identification of further carbon resonances. There is no need for the proton spectrum to be well dispersed for this analysis to be feasible, provided cross peaks can be located. For an example of this approach to the assignment of a hopelessly unresolved proton spectrum, see reference 3. Several other techniques offer the possibility of tracing the carbon skeleton of a molecule (for instance, INADEQUATE and RCT), but the combined use of COSY and HSC is most practical, and, for realisitic situations where only limited amounts of material are available, is the method of choice.

9.3 EXPERIMENTS RELATED TO HSC

9.3.1 HSC + Broadband Homonuclear v_1 Decoupling

In a typical HSC spectrum run with rather low v_1 resolution, the presence of homonuclear proton-proton coupling is a nuisance. Probably individual lines will not be resolved because of the limited digitisation, and so homonuclear coupling simply leads to broadening of the resonances and a reduction in sensitivity. A variation of the experiment[4] allows most homonuclear couplings to be eliminated from v_1, and, subject to certain limitations, is to be preferred for low-resolution applications.

The manner in which this can be achieved can be understood as follows. Suppose in some way a π pulse could be applied to most of the proton resonances, while leaving the ^{13}C satellites of lines unaffected. A typical proton which is attached to ^{13}C will also be coupled to other protons *on adjacent ^{12}C atoms*, so by use of such a 'semi-selective' pulse it should be possible to invert the state of the ^{12}C protons without affecting their coupling partner. Now, we can recall from Chapter 4 that the reason homonuclear coupling is not usually refocused by π pulses is that both nuclei involved have their spin states inverted, reversing the direction of precession of multiplet components after the pulse. This 'semi-selective' pulse, however, does not invert the nuclei attached to ^{13}C, and so placing it at the centre of t_1 during an HSC experiment should refocus homonuclear coupling. Excluded from this, of course, would be *geminal* couplings, because then both protons are attached to ^{13}C.

It remains to find a way of inverting resonances without affecting their ^{13}C satellites. A sequence with this property is the so-called 'bilinear rotation operator'[5], which consists of:

^{1}H: $\quad \left(\dfrac{\pi}{2}\right)_x - \dfrac{1}{2J_{CH}} - \pi_y - \dfrac{1}{2J_{CH}} - \left(\dfrac{\pi}{2}\right)_x$

^{13}C: $\quad\quad\quad\quad\quad\quad\quad\quad \pi$

The effect of this on protons attached to ^{13}C, and other ('distant') protons is illustrated in Figure 9.8. This should be compared with the TANGO sequence described in Chapter 10, section 10.2.2, which in an analogous way acts as a $\pi/2$ pulse for attached nuclei and a π pulse for distant ones. Inserting the bilinear rotation operator in place of the ^{13}C π pulse half way through t_1 of HSC (Figure 9.9), we get a spectrum without homonuclear coupling in v_1 (Figure 9.10). In this figure, notice how the *geminal* couplings are still present, while all other resonances have been reduced to singlets in v_1.

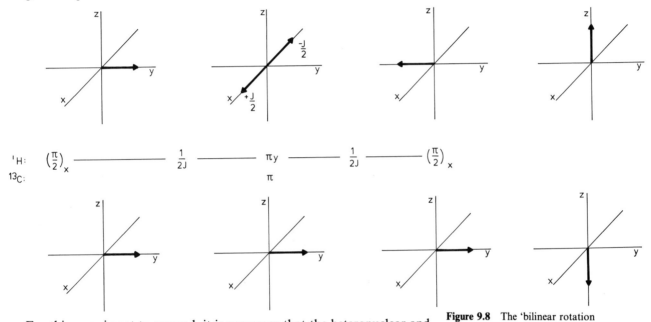

^{1}H: $\quad \left(\dfrac{\pi}{2}\right)_x \rule{3cm}{0.4pt} \dfrac{1}{2J} \rule{2cm}{0.4pt} \pi_y \rule{2cm}{0.4pt} \dfrac{1}{2J} \rule{3cm}{0.4pt} \left(\dfrac{\pi}{2}\right)_x$

^{13}C: $\quad\quad\quad\quad\quad\quad\quad\quad\quad\quad\quad\quad \pi$

Figure 9.8 The 'bilinear rotation operator', here being used to invert protons attached to ^{12}C while leaving those attached to ^{13}C unaffected.

For this experiment to succeed, it is necessary that the heteronuclear and homonuclear couplings have substantially different values. A suggested guideline[4] is that the total width of the proton multiplets should be at least five times smaller than the heteronuclear coupling; for H-C one-bond couplings this means that multiplets should be no more than about 25-30 Hz wide. The range of heteronuclear couplings should also be small, so that the interpulse delay of the bilinear rotation operator does not deviate too much from $1/2J$. Artefacts arising from errors in this delay can be reduced by cycling the phases of all the proton pulses of the bilinear rotation sequence together in 90° steps, with constant receiver phase[6]. A final restriction is that transverse relaxation times for the protons should not be too short, so as to avoid signal loss during the extra $1/J$ interval inserted in t_1.

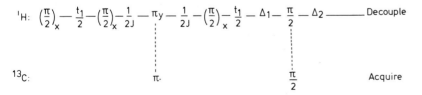

^{1}H: $\left(\dfrac{\pi}{2}\right)_x - \dfrac{t_1}{2} - \left(\dfrac{\pi}{2}\right)_x - \dfrac{1}{2J} - \pi_y - \dfrac{1}{2J} - \left(\dfrac{\pi}{2}\right)_x \dfrac{t_1}{2} - \Delta_1 - \dfrac{\pi}{2} - \Delta_2 \rule{1.5cm}{0.4pt}$ Decouple

^{13}C: $\quad\quad\quad\quad\quad\quad\quad\quad\quad\quad\quad \pi \quad\quad\quad\quad\quad\quad\quad\quad \dfrac{\pi}{2}$ $\quad$ Acquire

Figure 9.9 HSC with broadband homonuclear decoupling in v_1.

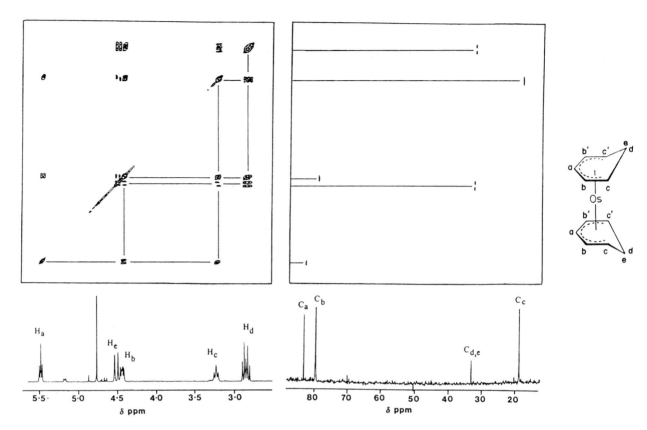

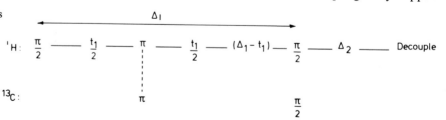

Figure 9.10 HSC spectrum (top right) obtained with the sequence of Figure 9.9, together with proton and COSY spectra of the same compound.

9.3.2 Small Couplings - COLOC

Because the HSC sequence contains fixed delays related to the heteronuclear coupling, sensitivity is reduced by transverse relaxation when coupling constants are small. This might arise when trying to correlate protons and carbon through two- or three-bond couplings, for instance, or in cases where no one-bond couplings are present (e.g. the ^{1}H-^{31}P example of section 9.2.3.). Once the point is reached where Δ_1 is greater than $\frac{1}{2}A_{t_1}$, a modified sequence known as COLOC[8] becomes attractive. For a typical A_{t_1} of 100 ms, this point would be reached when J_{HX} fell below 20 Hz. The delays Δ_1 and Δ_2 are unavoidable, but COLOC economises on total sequence duration by incorporating t_1 *inside* Δ_1 (Figure 9.11).

An interesting side effect of the COLOC sequence is that it generates a spectrum with broadband homonuclear decoupling in ν_1 without further modification. Since the interval between the first pulse and the polarisation transfer step is fixed, homonuclear couplings (which are not affected by the mobile π pulse during this interval) do not modulate the signal as a function of t_1. Chemical shifts, on the other hand, have an effect which does depend on the position of the π pulse, because they are refocused at time t_1, and then evolve during the remainder of Δ_1. The ν_1 dimension of a COLOC spectrum thus contains only proton chemical shifts. Correlations occur between coupled nuclei as for HSC, and it should be remembered that, even though Δ_1 and Δ_2 will be selected according to the small couplings of interest, correlations due to large couplings will not necessarily be eliminated. The Δ values selected for the small couplings may happen

Figure 9.11 The COLOC sequence. t_1 is defined by moving a π pulse within the Δ_1 interval.

to be suitable multiples of those appropriate for larger couplings, permitting both types of correlation to be observed. If this is troublesome, the low-pass *J*-filter mentioned in the following section might be incorporated in the experiment.

9.4 RELAYED COHERENCE TRANSFER

9.4.1 Introduction

A constant goal of NMR research is the discovery of a method which will allow the direct determination of the skeleton of an organic molecule in a single experiment. ^{13}C-^{13}C INADEQUATE-2D, which we saw in Chapter 8, comes close to providing this, but at the cost of intolerably poor sensitivity. At the more practical end of the scale we find the combined use of COSY and HSC, but of course this is two experiments, not one. Somewhere between these extremes fall the various relayed coherence transfer (RCT) methods, which, although of substantially lower sensitivity than HSC, are better than INADEQUATE-2D and convey similar (but not identical) information.

Relayed techniques generally combine two coherence transfer steps, one of which is used to modulate the signals and generate the t_1 dimension, while the other remains fixed and serves to pass on the signal to a more interesting destination. For instance, the most common path is to transfer magnetisation from one proton to another, and then on to a heteronucleus coupled to the second proton. Thus, the original proton can be correlated not only with the heteronucleus to which it is directly attached, but also with another one nearby, providing the required information about the molecular skeleton. The essential problem with techniques of this type is lack of generality, because it proves impossible to optimise the scheme in the presence of diverse spin systems and a wide range of proton-proton couplings. Thus these are not experiments of first choice, but may be resorted to when other approaches have proved fruitless.

9.4.2 H-H-C Relay

The H-H-C relay experiment can be built up[9] from HSC as indicated in Figure 9.12. The first step is to remove the pulse on carbon at the end of

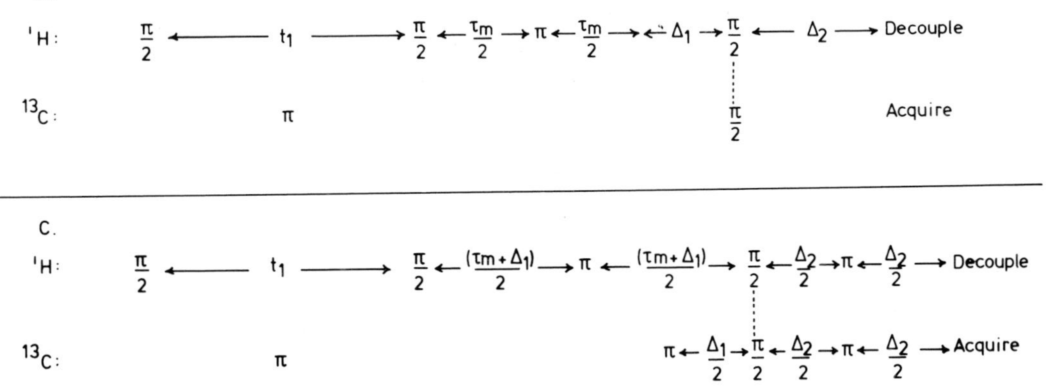

Figure 9.12 Building up the H-H-C relay experiment from HSC: A, normal HSC; B, addition of a second proton-proton transfer step before the heteronuclear transfer; C, the optimum sequence (see text).

t_1, thus restricting magnetisation transfer to the protons. To make it possible for magnetisation which has been transferred between protons to be passed on to carbon, it is then necessary to wait for multiplet components due to the homonuclear coupling (which at this point are in antiphase, as in COSY) to get back into phase. This occurs during the period τ_m, in which a spin echo is created to eliminate effects due to chemical shifts (Figure 9.12, sequence B). We are then back in a state similar to normal HSC, and have to wait a time Δ_1 as usual for the multiplet components due to *heteronuclear* coupling to get into antiphase, before completing the magnetisation transfer with pulses on protons and carbon. Δ_2 also has the same role as in HSC. A final modification is to insert proton and carbon π pulses at the centres of the Δ_1 and Δ_2 delays; this is necessary as before for phase-sensitive spectra, and in the relay experiment also optimises the magnetisation transfer process[9]. The protons are now found to be subject to two π pulses in succession, and it proves possible to merge them into one, which is placed halfway between the second two proton $\pi/2$ pulses (I would recommend not worrying about the details of this!). Figure 9.12 sequence C is then the optimum H-H-C relay experiment.

The stumbling block in this experiment is the interval τ_m, which has to be adjusted according to the proton spin system and coupling constants. This problem has been analysed in some detail, and reference 10 includes graphs of magnetisation transfer efficiency against τ for various spin systems (note that this paper uses the symbol τ to refer to the total mixing time, which would be $\tau_m + \Delta_1$ in our notation). In the absence of a proper calculation of τ_m (which, for an unknown structure, is clearly impossible), a value of about $1/5J_{HH}$ is chosen. Typical vicinal proton-proton couplings of 7 Hz therefore require τ_m to be set to about 28 ms. Mismatch of τ_m with the requirements of the spin system can cause very strong attenuation of the relayed coherence transfer signals; *at best* the experiment is likely to be four times less sensitive than HSC.

9.4.3 Using RCT

The manner in which H-H-C relay is to be applied can be understood by reference to Figure 9.13. This compares schematic HSC and relay spectra for a fragment of the type H_A-C_A-C_B-H_B. In the HSC spectrum, cross peaks occur at the chemical shifts (H_A, C_A) and (H_B, C_B). In the H-H-C relay spectrum these peaks are still present, but they are joined by others at (H_A, C_B) and (H_B, C_A), arising from the indirectly transferred magnetisation. Thus, the carbon skeleton of the molecule can be traced out through vicinal proton-proton couplings. Figure 9.14 illustrates an actual H-H-C relay spectrum of compound **2**.

The fact that directly bonded atoms still give cross peaks in a relayed coherence transfer spectrum can be confusing. A method has been proposed[11] which attenuates these cross peaks between 'neighbours', by ex-

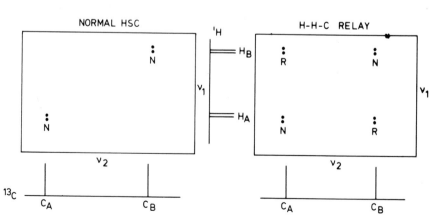

Figure 9.13 Schematic HSC (left) and H-H-C relay (right) spectra.

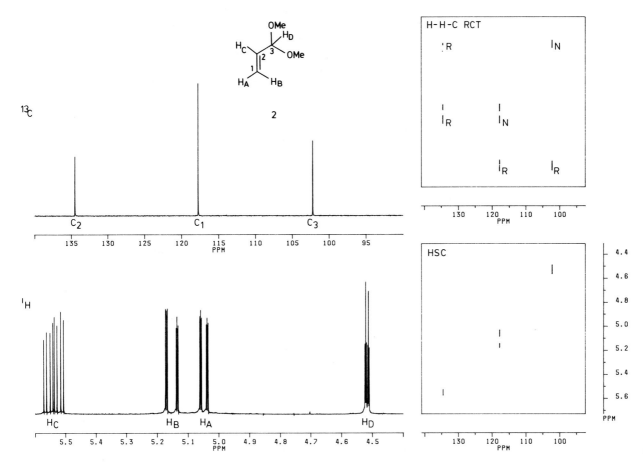

ploiting the difference between 1-bond and long range proton-carbon coupling constants. This 'low-pass J filter', although attractive in principle, suffers from the practical disadvantage of a long phase cycle, which increases the minimum number of scans per t_1 increment to a level that may be unacceptable in many cases.

A variety of other relayed coherence transfer experiments have been proposed, for instance: H-C-C relay[12], H-X-H relay[13] (regrettably designated HERPECS) and C-H-H relay[14]. The applicability of these methods obviously depends on circumstances. For instance, the H-C-C relay, like INADEQUATE, requires adjacent ^{13}C atoms, so in unlabelled systems will show very low sensitivity. However, it is slightly better in this respect than INADEQUATE, and can be optimised for the detection of quaternary carbons. H-X-H relay, involving proton detection, is a potentially sensitive experiment which can bridge gaps in a coupling network by correlating protons through a mutual heteronuclear partner. This has the advantage that the heteronucleus never needs to be measured at all, and at least for abundant nuclei like ^{31}P should be quite practical.

Figure 9.14 H-H-C relay experiment on compound **2**. Direct correlations are labelled N (for 'neighbour') and relayed correlations R (for 'remote'). Note how the intensity of peaks varies considerably, to the extent that in one case the 'N' peak is completely absent. This variation in signal strength is the main flaw in the experiment.

REFERENCES

1. D. Neuhaus, J. Keeler and R. Freeman, *J. Mag. Res.*, **61**, 553-558, (1985).
2. N. W. Alcock, J. M. Brown, A. E. Derome and A. R. Lucy, *J. Chem. Soc. Chem. Commun.*, 575-578, (1985).
3. N. S. Bhacca, M. F. Balandrin, A. D. Kinghorn, T. A. Frenkiel, R. Freeman and G. A. Morris, *J. Amer. Chem. Soc.*, **105**, 2538-2544, (1983).
4. A. Bax, *J. Mag. Res.*, **53**, 517-520, (1983).
5. J. R. Garbow, D. P. Weitekamp and A. Pines, *Chem. Phys. Lett.*, **93**, 504, (1982).
6. P. H. Bolton and J. A. Wilde, *J. Mag. Res.*, **59**, 343-346, (1984).
7. G. A. Morris and K. I. Smith, *J. Mag. Res.*, **65**, 506-509, (1985).
8. H. Kessler, C. Griesinger, J. Zarbock and H. R. Loosli, *J. Mag. Res.*, **57**, 331-336, (1984).

9.   H. Kessler, M. Bernd, H. Kogler, J. Zarbock, O. W. Sørensen, G. Bodenhausen and R. R. Ernst, *J. Amer. Chem. Soc.*, **105**, 6944-6952, (1983).
10.   S. K. Sarkar and A. Bax, *J. Mag. Res.*, **63**, 512-523, (1985).
11.   H. Kogler, O. W. Sørensen, G. Bodenhausen and R. R. Ernst, *J. Mag. Res.*, **55**, 157-163 (1983).
12.   H. Kessler, W. Bermel and C. Griesinger, *J. Mag. Res.*, **62**, 573-579, (1985).
13.   M. A. Delsuc, E. Guittet, N. Troitin and J. Y. Lallemand, *J. Mag. Res.*, **56**, 163-166, (1984); D. Neuhaus, G. Wider, G. Wagner and K. Wüthrich, *J. Mag. Res.*, **57**, 164-168, (1984).
14.   L. D. Field and B. A. Messerle, *J. Mag. Res.*, **62**, 453-460, (1985).

10

Spin Echoes and J-Spectroscopy

10.1 INTRODUCTION

By now we have seen many experiments which involve spin echoes, for instance INEPT, INADEQUATE, and RCT. In all these examples the echo is used as a means of *eliminating* rather than adding something to the experiment; it takes away the chemical shift dependence of the manipulations of the spins. There remains a class of experiment in which the echo makes a more direct and positive contribution to the result, gathered here under the term 'J-Spectroscopy', which emphasises the fact that the dependence of the eventual spectrum on multiplet structure is the main feature of interest. J-spectroscopy received a great deal of attention during the early development of 2D NMR, and its theoretical and practical aspects have been quite comprehensively analysed. However, for various reasons most of the experiments are not of such general usefulness as the shift correlations we have seen in the previous two chapters. Because of this, and because there is so much literature information available, I will not explore this topic in much detail, but simply present a short survey of spin-echo based experiments and some possible applications for them.

10.2 HETERONUCLEAR *J*-MODULATED SPIN ECHOES

10.2.1 Introduction

This experiment is a good place to start because it is easy to understand using the rotating frame vector model. In section 4.4.4 of Chapter 4 I remarked that if you perform a spin echo sequence on a heteronuclear system, you have the option of whether or not the heteronuclear coupling is refocused, because you may or may not apply a π pulse to the second nucleus. If you choose to pulse the second nucleus, using the sequence:

$$\text{X:} \qquad \left(\frac{\pi}{2}\right)_x - \tau - \pi_y - \tau - Acquire \;\cdots$$

$$\text{Y:} \qquad\qquad\qquad \pi \qquad\qquad (Decouple \;\cdots)$$

where X is the observed nucleus (e.g. ^{13}C) and Y is the other nucleus (e.g. ^1H), then the coupling is not refocused by the echo. Broadband decoupling of Y may or may not be turned on during acquisition as desired, and suitable choice of τ can give interesting effects according to the coupling between X and Y. Using the rotating frame vector model we can understand this quite easily.

Figure 10.1 illustrates the results for XY and XY_2 groups assuming X and Y have spin $\frac{1}{2}$ (typically these would be methine and methylene groups), with τ set to $1/2J$. For an XY group, the X signal is a doublet, so if we set the reference frequency at the centre of this we see two components moving at $\pm J/2$ Hz in the rotating frame. After $1/2J$ s they will have rotated through $\pm 1/2J.J/2$, i.e. 1/4 cycle, so they lie along the $\pm x$ axes.

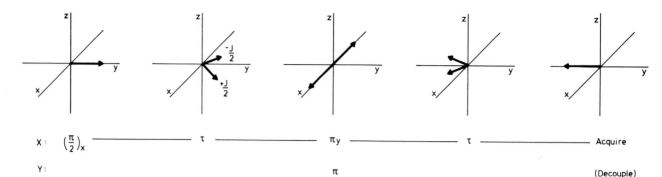

X : $\left(\frac{\pi}{2}\right)_x$ ——————————— τ ——————————— π_y ——————————— τ ——————————— Acquire

Y : ————————————————————————— π ——————————————————————— (Decouple)

Figure 10.1 Spin echo sequence for a heteronuclear AX system.

During the second half of the sequence they keep going in the same direction, as the π pulse on Y interchanges all the α and β states while π_y on X swaps the two components, so at the beginning of the acquisition period they both align along the −y axis. Since they are aligned, we can turn on broadband decoupling of Y with impunity, and record a spectrum of X which contains a singlet. With conventional phase correction such that signals starting along +y are positive, it would be an upside-down singlet.

For an XY_2 group, the X signal is a *triplet* (Figure 10.2). If we put the reference on the centre of this, then we have one line remaining static along the +y axis, while the other two rotate in opposite directions at ±J. As they are going twice as fast as were the doublet components, they find themselves along −y at the time of the π pulse, and back along +y at the start of acquisition. Thus the decoupled signal acquired has *opposite* phase to that from the doublet. The experiment is therefore a means of distinguishing numbers of attached protons, and competes with off-resonance decoupling and spectrum editing methods. If you work through the procedure for X (e.g. quaternary carbon) and XY_3 (e.g. methyl) groups, you should find that X ends up with the same phase as XY_2, and XY_3 with the same phase as XY. In general, all groups with an even number of couplings have opposite phase to all groups with an odd number.

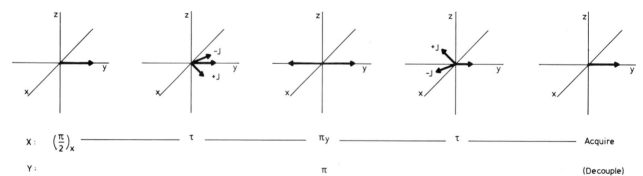

X : $\left(\frac{\pi}{2}\right)_x$ ——————————— τ ——————————— π_y ——————————— τ ——————————— Acquire

Y : ————————————————————————— π ——————————————————————— (Decouple)

Figure 10.2 As Figure 10.1, but for an AX_2 system.

Interestingly, the same experiment can be done without the need for pulsing Y, simply by turning the decoupler on and off (Figure 10.3). Since the decoupling essentially freezes the multiplet components wherever they happen to be in the rotating frame at the time it begins, setting τ = 1/J and switching on decoupling simultaneously with the X π pulse leads to the desired result. An equivalent effect is also achieved by having the decoupler on during the first part of the echo, off during the second part and back on during acquisition (these two experiments are the same only when decoupled signals are acquired; if decoupling is not used during acquisition, then the former leads to exotic results[1]).

All of these sequences give spectra in which odd and even numbers of substituents are distinguished by the phase of the signals (Figure 10.4). In comparison with off-resonance decoupling, the J-modulated spin echo is superior in terms of resolution and sensitivity, because acquisition occurs with broadband decoupling. It contains less information, though, since X and XY_2 groups are indistinguishable, as are XY and XY_3. Properly

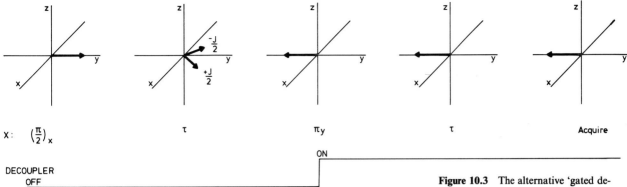

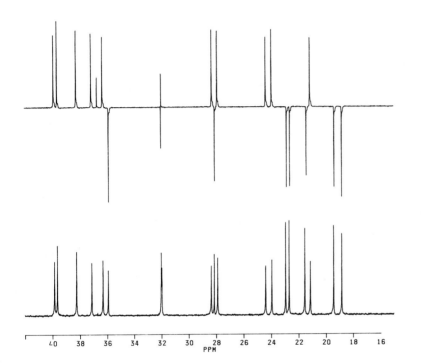

Figure 10.3 The alternative 'gated decoupler method' for bringing about heteronuclear *J*-modulation of spin echoes.

Figure 10.4 A normal ^{13}C spectrum (bottom, partial spectrum of cholesteryl acetate), and the *J*-modulated echo spectrum (top, gated decoupler method with $\tau = 1/J$). The spectrum is phased so that CH$_2$'s and quaternary carbons are positive, with CH's and CH$_3$'s negative.

executed spectrum editing should give better results than either, with distinction between all groups, but requires more attention to experimental detail. The *J*-modulation experiment performed with decoupler switching (to avoid the need for calibrating decoupler pulse widths) can sometimes be a useful quick assignment aid, when an off-resonance decoupled spectrum is too crowded or noisy and setting up DEPT seems like too much bother.

The big problem with this experiment is the requirement that the delay τ be related to *J*. In real situations we encounter a spread of *J* values which may be quite wide, for instance 125–210 Hz for 1-bond proton carbon couplings. Any deviation from the condition $\tau = 1/2J$ (or $1/J$, as appropriate) leads to a reduction in the amplitudes of the detected signals; in the extreme, if $\tau = 1/2J$ for the decoupler switching experiment (or $1/4J$ for the pulse experiment) then all signals but those from unsubstituted nuclei are nulled (Figures 10.5, 10.6). This variation in the *amplitude* of the signals as τ varies should give us a clue to how we can generalise the experiment; I will return to this shortly.

10.2.2 Spin Echo Difference

The phase behaviour of *J*-modulated echoes can be used as the basis for a simple type of editing technique. Consider the specific example of a

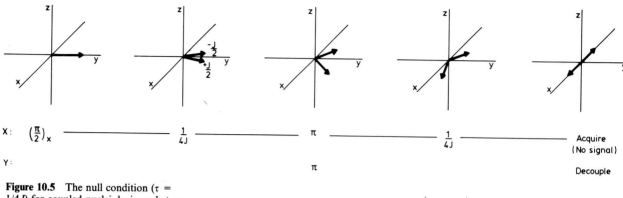

Figure 10.5 The null condition ($\tau = 1/4J$) for coupled nuclei during a heteronuclear *J*-modulated echo.

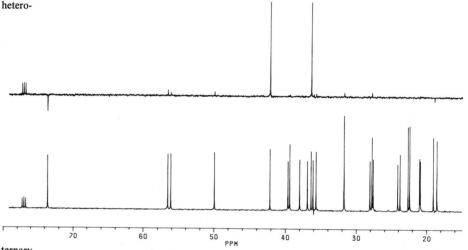

Figure 10.6 Identification of quaternary carbons by use of the null condition (same sample as Figure 10.4). The nulling of protonated carbons depends on exact setting of the τ delay, so in the presence of a range of coupling constants some signals are not completely eliminated.

substance containing a single ^{13}C label. We will do the *J*-modulated echo experiment backwards, using carbon decoupling and proton observation (broad-band carbon decoupling is feasible using composite-pulse methods - Chapter 7; if your spectrometer is not up to this then the pulsed method for the echo can be used, and the proton spectrum acquired with the ^{13}C coupling still present). In this case we only have to consider 'X' groups (i.e. protons attached to ^{12}C) and 'XY' groups (i.e. protons attached to the label). *If* we assume that the proton-proton couplings are so much smaller than the proton-carbon couplings that they have negligible effect during a τ interval set to $1/J_{HC}$, then the analysis is exactly as before. Suppose we do two experiments:

^{1}H: $\left(\dfrac{\pi}{2}\right)_x — \tau — \pi_y — \tau — Acquire \;\cdots$

^{13}C: *Decouple* $\cdots$

and

^{1}H: $\left(\dfrac{\pi}{2}\right)_x — \tau — \pi_y — \tau — Acquire \;\cdots$

^{13}C: *Decouple* $\cdots$

The first of these leads to a normal spectrum, since the proton-carbon coupling is refocused along with everything else. In the second, however, decoupling during the second half of the sequence stops the coupling from being refocused. Thus the signals from the protons attached to the ^{13}C label (and *only* these protons) are inverted in the second experiment. Subtracting the two spectra reinforces the labelled protons while cancelling everything else (Figure 10.7). As with all difference experiments, the level of cancellation will depend on many aspects of spectrometer performance;

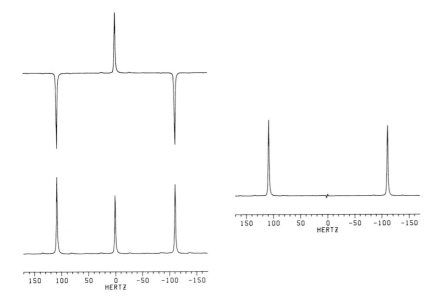

Figure 10.7 The principle of spin echo difference (proton spectra of formic acid with 60% ^{13}C enrichment; the pulse method was used so as not to obscure the ^{12}C line). Two spectra are acquired, one with a ^{13}C π pulse (top left) and one without (bottom left). In the difference spectrum (right) the signals with coupling are reinforced, and those without are cancelled.

a realistic expectation would be for a few hundredfold reduction of the unwanted peaks.

The merit of this experiment is that it combines the selective nature of ^{13}C labelling with the sensitivity of proton detection. Unlike reverse DEPT (Chapter 6), we should have *full* proton sensitivity, subject only to loss of signal through transverse relaxation during the echo (reverse DEPT transfers carbon populations to the protons, thus reducing sensitivity by a factor of $0.75 - 0.25$; also the repetition rate is determined by carbon T_1's). On the other hand, the reverse polarisation transfer experiment gives much better suppression of the protons attached to ^{12}C, because they can be saturated by broadband irradiation between scans. Since the technical requirements of these two competing experiments are rather different, both are worth considering when planning how to tackle a labelling problem. For examples of the application of spin-echo difference, see reference 2.

Simple spin echo difference will probably give insufficient suppression of unlabelled resonances for many applications. Fortunately it is possible to improve the experiment, by replacing the $\pi/2$ pulse with a sequence known as TANGO[3]. This is an excitation sequence which discriminates between nuclei according to whether they are coupled to a heteronucleus or not, acting as a $\pi/2$ pulse for the coupled nuclei and a π pulse otherwise. Once again, this is achieved by means of a spin echo:

^{1}H:
$$\left(\frac{\pi}{4}\right)_x - \tau - \pi_y - \tau - \left(\frac{\pi}{4}\right)_x$$

^{13}C:
$$\pi$$

With τ set to $1/2J_{CH}$, the action of this can be visualised as in Figure 10.8; compare this sequence with the 'bilinear rotation operator' introduced in Chapter 9, section 9.3.1. TANGO alone does not usually give adequate suppression of unlabelled resonances for use in realistic systems, but in combination with spin echo difference the results are comparable with those achieved by reverse DEPT, but with better sensitivity. Figure 10.9, showing a TANGO-spin echo difference spectrum should be compared with Figure 6.19 of Chapter 6, which is the equivalent reverse DEPT experiment on the same compound.

DISTANT PROTONS:

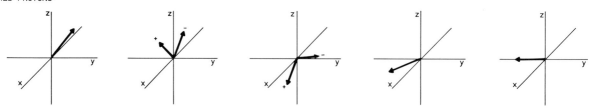

ATTACHED PROTONS:

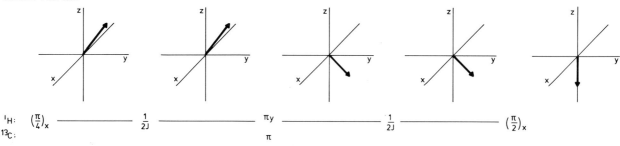

Figure 10.8 The TANGO sequence for selective excitation of nuclei with heteronuclear coupling.

Figure 10.9 Selective detection of protons attached to a ^{13}C label by spin echo difference; the initial $\pi/2$ pulse was replaced with TANGO. Broadband ^{13}C decoupling was applied during acquisition. Suppression of unlabelled signals is similar to that obtained using reverse DEPT, but sensitivity is significantly higher. This is the same sample as was used for Figure 6.19, but the experiment was run at a later date; some signals from ^{13}C containing decomposition products are visible to low field of the main signal.

10.3   THE HETERONUCLEAR *J*-SPECTRUM

10.3.1   Introduction

Consider the *J*-modulated echo sequence using the pulse method, but this time instead of a carefully selected interval τ we will use an arbitrary one $t_1/2$; I will refer to ^{13}C observation and ^1H coupling throughout, but of course it could be any pair of nuclei:

$$^{13}C: \qquad \left(\frac{\pi}{2}\right)_x - \frac{t_1}{2} - \pi_y - \frac{t_1}{2} - Acquire \cdots$$

$$^1H: \qquad\qquad\qquad\qquad \pi \qquad\qquad Decouple \cdots$$

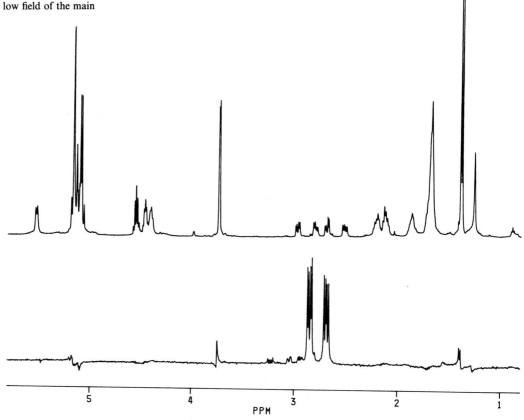

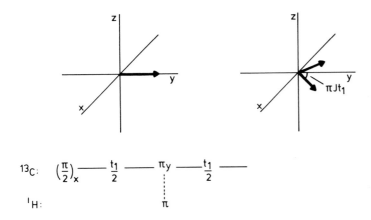

^{13}C: $\left(\dfrac{\pi}{2}\right)_x$ ——— $\dfrac{t_1}{2}$ ——— π_y —— $\dfrac{t_1}{2}$ ———

^{1}H: π

Figure 10.10 Amplitude modulation of the carbon signal from a CH group - the basis for heteronuclear *J*-spectroscopy.

Examining Figure 10.10, we can follow the fate of a doublet. Each component travels through an angle $\pi J t_1$ during the sequence. As we are going to turn on decoupling during acquisition it is the resultant of these two components which is important; projecting them onto the x and y axes, we find $M \cos \pi J t_1$ along the y axis, while the x axis components are in opposite directions and cancel. Thus the vector being detected always begins along the $+y$ axis (i.e. it has constant phase as a function of t_1), but it is amplitude modulated by the coupling. Note that this pure amplitude modulation arises as a result of the symmetrical disposition of the two components of the doublet, which in turn arises because the system shows first-order coupling. In a more complex case involving coupling amongst protons as well as between protons and carbon, systems which are not first-order may occur, in which case the assumption that multiplets must be symmetrical about their centres breaks down; this can lead to a considerable increase in complexity of the resulting spectrum. I am going to assume throughout completely first order coupling; for a more detailed analysis see reference 4.

If you follow through the above sequence for CH_2 and CH_3 groups, you should be able to convince yourself that in general lines experience amplitude modulation during t_1 as a function of their separation from the centre of the multiplet. This means that if we perform a 2D experiment with variable t_1, the v_1 dimension will contain *only* multiplet structure, while v_2 will contain *only* chemical shifts (because acquisition occurs with broadband decoupling). This is the *J-spectrum;* it amounts to a proton coupled carbon spectrum in which the multiplet structure has been unscrambled by rotation into the second dimension. The same experiment can be done using decoupler switching instead of a π pulse on the protons, as described above for the 1D *J*-modulated echo (the 'gated decoupler' experiment). In this case, since the couplings are only evolving during half of t_1, the line separations in that dimension are halved. This may be considered helpful or detrimental according to circumstances.

10.3.2   Examples of Heteronuclear *J*-Spectra

There are two rather extreme situations in which it might be profitable to run a heteronuclear *J*-spectrum: aiming either for very high or very low resolution in v_1. The former case exploits the experiment as a means of separating patterns of lines which would overlap in a conventional spectrum, and also takes advantage of the inhomogeneity refocusing effect of the spin echo. In the latter case, it is speed and sensitivity of determination of multiplicities that is of interest. For the majority of structural problems it is the fast, low resolution experiment that will be applicable, but I will discuss both.

With high v_1 resolution

As a *high resolution* experiment, the *J*-spectrum offers the interesting possibility of observing multiplets with *natural* linewidths. This is because the signal envelope during t_1 is determined by the true value of T_2, not by T_2^*, as the spin echo refocuses magnet inhomogeneity (subject to limitations imposed by diffusion within the sample, as usual for spin echoes). However, other practical problems may prevent this feature being fully exploited. Fine digitisation will be required in v_1 if the true linewidths are to be observed, as T_2 values for carbon may be long, easily as much as 10 s.

Acquisition times several times greater than T_2 would be ideal, requiring many increments in t_1, even though the spectral width in that dimension is small. The widest coupled carbon multiplets are quartets arising from methyl groups, and assuming coupling constants less than 200 Hz we require to digitise a spectral range of 600 Hz (for the experiment with proton pulsing). This implies a t_1 increment of about 0·8 ms. To get even a 10 s acquisition time in t_1 we must therefore make *more than a thousand* increments. This is seldom possible, as the repetition rate of the experiment, being determined by ^{13}C T_1's, will not be particularly fast. Sensitivity will also be low, since many experiments need to be accumulated with long t_1 values, so the v_2 acquisition time should not be less than T_2^* even if shorter values would still give acceptable discrimination of lines in that dimension. Combined with the need for a large number of t_1 values, this requirement implies a very large dataset.

The gated decoupler version of the experiment appears to reduce the problem of digitising v_1 (in fact, to halve it) by halving the spectral range. Unfortunately, this also detracts from the very resolution we desire by halving the line separations, so any advantage is likely to be negated in the high resolution application. For the low resolution experiments described below the reduced spectral range is helpful, however, and there are also other significant experimental advantages to using decoupler switching rather than proton π pulses, discussed in section 10.3.3.

Despite the above comments, with patience, and when the problem requires it, such a high resolution *J*-spectrum can be run. Although the dataset will be large, provided it can at least be *stored*, the data processing requirement is not severe as the complete 2D transform need not be performed. Instead it is sufficient to transform in v_2 and then pick out single columns from v_1 corresponding with the chemical shifts of interest; these may be zero-filled and transformed as 1D spectra with a large number of data points. Figure 10.11 shows slices extracted in this manner from the *J*-spectrum of an organometallic compound. Note that, although a complete spectrum would be displayed in magnitude mode for reasons discussed below (section 10.3.3), individual columns can be adjusted into pure absorption phase.

Figure 10.11 Slices through v_1 of a heteronuclear *J*-spectrum of an organometallic compound , with high v_1 resolution. In principle natural linewidths can be obtained in this way.

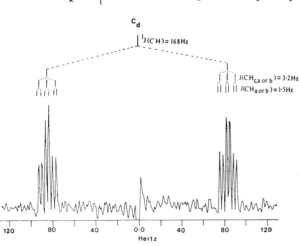
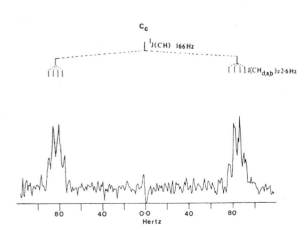

The main obstacle to obtaining high resolution of heteronuclear couplings by use of the *J*-spectrum is the relatively large spectral width necessitated by the presence of large one-bond couplings. If the only feature of interest is the fine structre due to long range coupling (e.g. two- and three-bond couplings in a ^{13}C spectrum), then an attractive alternative approach exists[15]. Replacing the π pulses at the centre of t_1 with the 'bilinear rotation operator' already described in Chapter 9 (section 9.3.1) eliminates the effect of large one-bond couplings, by inverting, and hence refocusing, those multiplet components. Small couplings remain unaffected, and hence modulate the signal as a function of t_1. This reduces the required spectral range to the multiplet width due to long range couplings, making fine digitisation of v_1 feasible.

With low v_1 resolution

A perennial problem in ^{13}C NMR is determining the number of protons borne by a particular carbon. We have by now seen two alternatives to the traditional method of off-resonance decoupling: spectrum editing through polarisation transfer and the use of *J*-modulated echoes. It is possible to bridge the gap between the incomplete information provided by the 1D spin echo experiment and the complete, but experimentally taxing, editing methods, using the *J*-spectrum. Obviously the information we want is there, as the ^{13}C multiplet structures are present in v_1, but if the experiment is to compare favourably with its competitors it must be possible to perform it quickly and with high sensitivity. This requires determination of the absolute minimum number of increments acceptable in t_1 without loss of multiplet resolution.

For this low-resolution experiment we naturally choose the decoupler switching method, for minimum v_1 spectral width and maximum experimental convenience. Recalling the discussion in Chapter 8 (section 8.3.5) about digital resolution, we know we need an acquisition time around $1/\delta v$ to get an effective linewidth δv. Now, the minimum proton-carbon coupling constant will be about 130 Hz, reduced in the decoupler switching experiment to a line separation of 65 Hz. The maximum acceptable linewidth which would not lead to unresolved couplings must be less than this, say 40 Hz to be on the safe side. This implies an acquisition time of 25 ms in v_1. The spectral range is 300 Hz, so we need about 15 increments. A carefully chosen window function will then have to be applied to the severely truncated signal in this dimension to avoid distortions of the peaks. Figure 10.12 shows a spectrum obtained under precisely these conditions; the *total* acquisition time was 30 minutes, which compares favourably with the time needed to accumulate an off-resonance deoupled experiment with acceptable signal-to-noise ratio.

It is even possible to reduce the number of t_1 increments further by taking a different approach to the data processing. The limiting factor in the above discussion was the assumption that the effective linewidth δv_{eff} would need to be increased by the window function so that A_{t_1} was about $1/\delta v_{eff}$ ($3T_2^*$). If this is not done, then it is possible to show that the effective linewidth is about $0.6/A_{t_1}$, but the signals are distorted by truncation. It is possible to remove the resulting 'sinc wiggles' by other means than apodisation, particularly since the form of the distortion is known as it derives from the value of A_{t_1}. One possible way of processing such highly truncated data is by means of maximum entropy methods[5] (Chapter 2), which have inbuilt assumptions about the form of the spectrum being sought. An alternative proposed recently[6], which puts much less demand on available computing power, derives bizarrely from radio astronomy and entails iterative fitting of sinc lineshapes to the observed spectrum to determine the true peak positions. Figure 10.13 illustrates multiplicity determination using only 5 t_1 increments in the *J*-spectrum by this method.

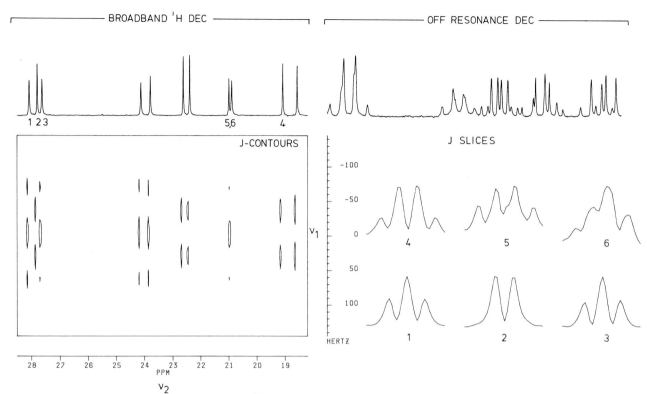

Figure 10.12 With low v_1 resolution the *J*-spectrum provides a quick multiplicity check (partial ^{13}C spectra of cholesteryl acetate, experimental conditions as described in the text). The broadband decoupled spectrum is plotted above the contour representation of the *J*-spectrum, while an off-resonance decoupled spectrum is compared with vertical slices through v_1. Slices 1-4 allow multiplicities to be determined very clearly, while slices 5 and 6 illustrate the effect of insufficient resolution in v_2. Here a triplet and quartet overlap in the 2D experiment, leading to confusing traces; this could be rectified by increasing the v_2 acquisition time slightly, or even just by zero-filling in that dimension.

10.3.3 Experimental Aspects

EXORCYCLE and composite π pulses

Deficiencies in both pulses on the observed nucleus, but particularly in the π pulse, prove to be most detrimental to the quality of *J*-spectra. Detailed analysis of this problem is available[7]; I will simply illustrate a couple of the more obvious problems and present the conventional solutions.

Imagine that the $\pi/2$ pulse at the beginning of the sequence is a little short, either because it was incorrectly calibrated, or inevitably in certain regions of the sample because of B_1 inhomogeneity. This leaves a little z magnetisation at the start of t_1. The π pulse will presumably be short too, so instead of harmlessly inverting this magnetisation, it will send some of it into the x-y plane. Here it precesses during the second half of t_1, with both ^{13}C chemical shifts and proton-carbon couplings in the pulsed version of the experiment. After the 2D transform extra peaks appear; as they have evolved with chemical shifts as well as couplings they will almost certainly be folded in v_1.

Some further quite alarming things happen in the presence of non-ideal π pulses. In the heteronuclear experiment with proton pulsing, deficiencies in the proton pulse cause extra peaks to appear in between the true multiplet components[8]. In a *homonuclear* J-spectrum (see next section) there is even more scope for trouble. Suppose the π pulse were long; we can imagine decomposing the experiment into a perfect part originating from the first 180° of the nominal π pulse, and the remainder. The remainder consists of $\pi/2-t_1/2-\alpha-t_1/2-Acquire$, where α represents the rest of the long 'π' pulse. Aside from the details of the timing, *this is a coherence transfer sequence*, just like COSY. We expect proton magnetisation, evolving during the first part of t_1 with proton chemical shifts and couplings, to be transferred and detected as a modulation of the magnetisation of other coupled protons during t_2. In the narrow v_1 dimension, chemical shifts will be folded many times over, leading to extra peaks in unpredictable places. π pulses are very likely to suffer from imperfect inversion of magnetisation (Chapter 4), so this problem may be severe. In

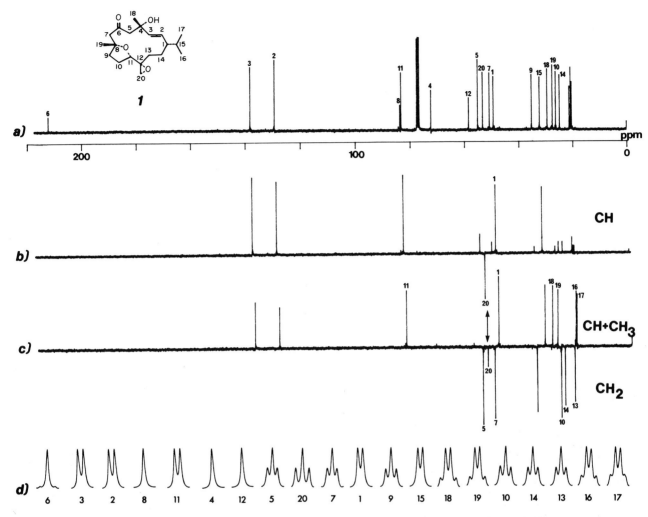

Figure 10.13 *J*-spectroscopy with only 5 t_1 increments, with special data processing (see text). For this compound spectrum editing has given misleading results (upper traces), because of a large $^1J_{CH}$ for C_{20} (arrowed). The *J*-spectrum, however, is quite unambiguous.

the heteronuclear experiment problems with the proton π pulse are simply avoided by choosing the decoupler switching variation; in *homonuclear J-spectroscopy* (section 10.4) this option is not available.

To counteract the tendency for extra peaks to appear in *J*-spectra, it is essential to apply proper phase-cycling. Cycling the ^{13}C π pulse through the sequence $x,y,-x,-y$, with inversion of the receiver phase for each 90° step in the pulse phase, eliminates many problems. This procedure is known as EXORCYCLE[9] (because some of the extra peaks in *J*-spectra are referred to variously as 'phantoms' and 'ghosts'). It can be combined as usual with CYCLOPS by stepping all the ^{13}C phases and the receiver phase together. Problems with the proton π pulse are best tackled in the heteronuclear case by using the gated decoupler method, as mentioned above. For homonuclear *J*-spectroscopy, use of the composite π pulse $(\pi/2)_x\pi_y(\pi/2)_x$ is recommended (Chapter 7).

Sign discrimination and lineshape

Distinguishing the signs of frequencies in v_1 in a *J*-spectrum is often not so problematical as in coherence transfer experiments. For the purely first-order cases we have been discussing, or always in the low resolution experiment, multiplets are symmetrical about $v_1 = 0$, so there is no objection to allowing folding about this line. The resulting spectrum can be phase corrected throughout into pure absorption mode, so there is no need for the use of unusual window functions. However, the absorption mode results from the superimposition by folding of pairs of lines with the objectionable phase-twist shape, and any deviation from perfect symmetry about $v_1 = 0$ will spoil this.

In this case (i.e. for systems which are not purely first order), or simply to restore the familiar multiplet patterns in v_1, combining appropriate quadrants of the transformed data allows discrimination of the signs of the v_1 frequencies[10]. Because an *individual* multiplet component (rather than the resultant of a pair of components) undergoes phase modulation during t_1, the lines then have phase-twist shapes as for echo-selection lines in coherence transfer experiments (Chapter 8). Magnitude calculation and strong resolution enhancement, with their associated penalties, therefore become necessary in order to produce an absorption-like display of the complete spectrum. Even this is not fatal, however, because individual columns from the data can still be adjusted into a true absorption mode. For second-order systems it is also necessary to use the gated decoupler method to acquire the data, because in the proton pulse method the assumption that the π pulse on protons interchanges the labels of all states breaks down, leading to patterns of lines in v_1 which are not the same as the multiplet structures observed directly in a 1D spectrum.

For the greatest convenience and best resolution, procedures are available that allow sign discrimination in v_1 to be combined with a pure absorption mode display for either the pulsed or gated decoupler methods[10,11]. As only the latter correctly reproduces the multiplet structures of second order systems, obviously it should be preferred. The necessary sign discrimination is achieved by adding two experiments using both possibilities for gating the decoupler (i.e. off during the first half of t_1 and off during the second half). The signs of frequencies in v_1 are reversed between the two experiments, so before addition it is necessary to reverse the v_1 dimension of one of them; this causes a reversal in the sense of phase-twist and cancellation of the dispersive components. Although this procedure has been available since 1979, there do not appear to be any reports of its application to a real problem (up to mid 1984); the fact is that measurement of very fine details of proton-carbon couplings is rarely performed anyway, so a new and better way to do it is not too useful.

10.4   THE HOMONUCLEAR *J*-SPECTRUM

10.4.1   Introduction

When the spin echo sequence is applied in the presence of homonuclear coupling, *J*-modulation of the detected signal as a function of t_1 arises automatically since homonuclear couplings are not refocused. A 2D experiment performed by varying t_1 therefore leads to separation of the multiplet structures into v_1, as in the heteronuclear case. There is, however, an important difference: the couplings are still present during t_2; in the heteronuclear experiment we eliminate them by broadband decoupling, but clearly this is not possible for homonuclear couplings. The result is that multiplets do not lie parallel with v_1, but instead are tilted at 45°. Figure 10.14 compares a heteronuclear *J*-spectrum of a methyl group with the homonuclear spectrum of the X part of a proton A_3X system, with equal scales in v_1 and v_2 to illustrate this. In a more realistic case the v_2 plot scale will be much smaller than that in v_1, because it must encompass the complete range of chemical shifts rather than just the width of a multiplet, so the tilt will be less obvious (Figure 10.15).

The tilting of the multiplet structures proves to be something of a problem in certain applications, as we will see later. However, first consider why we might want to perform a homonuclear *J*-spectrum. The advantages that spring to mind are: unscrambling of overlapping multiplets by their rotation into the second dimension, potentially improved linewidth in v_1 because of refocusing of inhomogeneity, and the possibility of distinguishing between homonuclear couplings, which are present in v_1, and any

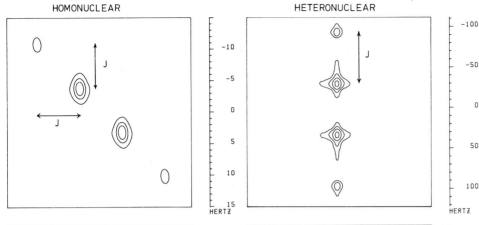

Figure 10.14 A comparison of homonuclear and heteronuclear J-spectra of A_3X systems. In the homonuclear case the coupling appears in *both* dimensions, leading to a 45° tilt of the multiplet.

additional heteronuclear couplings, which from the point of view of this experiment act like chemical shifts and only appear in v_2. Taking the first two points, which are related, the problem is that the experiment only works properly for purely first order systems (in the presence of strong coupling extra lines appear). This means that the additional resolution of multiplets will only be fully effective for the rather limited case of many overlapping signals which are nonetheless first order. Such circumstances do arise, and then the J-spectrum can be informative provided some complications discussed next are taken into account.

10.4.2 Tilting J-spectra

Most of the interest in a J-spectrum lies in the v_1 dimension. We would like to be able to take slices parallel to v_1, as for the heteronuclear experiment, and find that they contain complete multiplets. Unfortunately because of the 45° tilt, we actually find that they only contain single lines (i.e. each signal-bearing v_1 column contains a single component of a multiplet). Another loss due to the tilt is evident if we imagine *projecting* the spectrum onto v_2, parallel with v_1 (i.e. summing the columns of the data). In the heteronuclear experiment columns either contain a multiplet, because they are at an v_2 frequency corresponding with a signal, or they contain nothing (except noise). Projecting onto v_2 therefore reconstitutes

Figure 10.15 Homonuclear J-spectrum of the ubiquitous compound **1** from Chapter 8 (partial spectrum only, normal 1D spectrum presented above the contour plots). The acquisition time in the J dimension was about 2·5 s. The contour representation of such a spectrum is not ideal; with a fairly high contour level, suited to resolve details of the more intense peaks, some resonances are missing (left), while with a low contour level the spectrum becomes confusing (right).

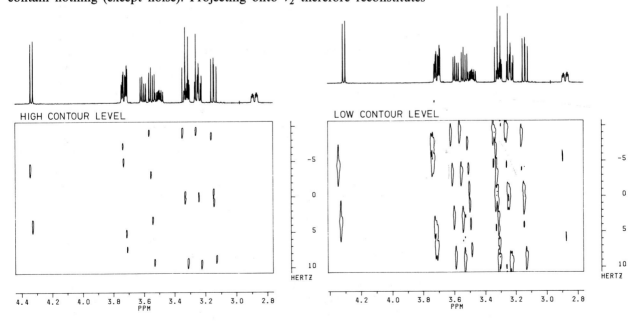

the 1D spectrum of the detected nucleus, without any coupling. In contrast, projecting a homonuclear experiment parallel with v_1 simply gives you a normal spectrum, complete with couplings. It seems at first sight, however, as though projecting along a line at 45° to v_2 would cause all the multiplet components to line up (it does), and so lead to a spectrum containing simply a single line at each chemical shift - the exciting possibility of a *'homonuclear broadband decoupled'* spectrum. Regrettably if the 2D dataset retains its phase information, a property of the Fourier relationship (the *cross-section projection theorem*) indicates that in fact *no* signals will be observed in the 45° projection. This can be understood as being the result of the positive and negative parts of the phase-twist lines aligning in just the right way to cancel in the sum made at 45°. Projecting the spectrum in magnitude mode avoids this problem, but we get all the usual troubles with broad lines and the need for strong resolution enhancement.

The 45° projection can be useful even in magnitude mode, and this and the process of extracting complete multiplets by taking slices through the data are made straightforward by a data processing trick: *tilting* the complete dataset. This involves slewing each successive row across to a greater and greater extent according to its frequency in v_1, and leads to a representation identical to that of the heteronuclear J-spectrum. *Provided the magnitude spectrum is calculated*, then slices parallel with v_1 contain multiplets, and the projection onto v_2 has the appearance of broadband homonuclear decoupling; this is illustrated in Figure 10.16. The disappointing thing about this is that having gone to the trouble of running an experiment which should *increase* resolution, we have then degraded it again by performing a magnitude calculation.

At present there is no way to escape this completely. If the main point of interest is resolution of single multiplets, then individual columns from the tilted dataset can be examined without magnitude calculation. Even so, unlike the heteronuclear case, these do not have a true absorptive ap-

Figure 10.16 Tilted version of Figure 10.15, with the projection onto the v_2 axis (magnitude mode). This has the appearance of broadband homonuclear decoupling, but this magic trick only works for weakly coupled systems. In the presence of strong coupling, extra peaks appear half way between the coupling partners (e.g. at the arrowed positions in this spectrum).

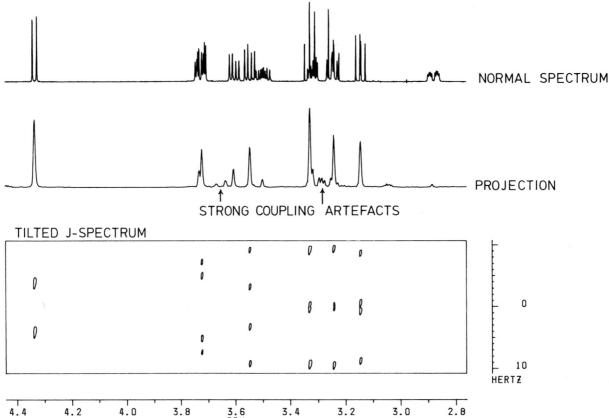

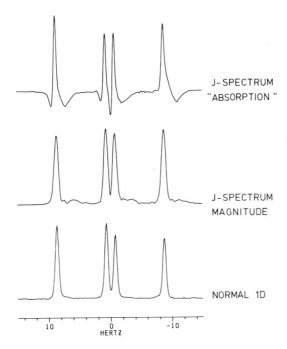

J-SPECTRUM
"ABSORPTION"

J-SPECTRUM
MAGNITUDE

NORMAL 1D

10   0   -10
HERTZ

Figure 10.17 Vertical slices from a tilted homonuclear *J*-spectrum can be adjusted into a pseudo-absorption mode, but the lineshape is severely distorted. This figure compares a slice from the dataset of Figure 10.16 (with and without magnitude calculation) with the same multiplet from a 1D spectrum processed with an equivalent window function.

pearance; the lines show severe distortion around the base as a result of the tilting process mixing in dispersive contributions (Figure 10.17). As the refocusing effect on the linewidth is often not large in practice, the main advantage is simply the separation of overlapping multiplets. If it is the projection that is interesting, then careful attention to choice of window functions in each dimension will be required, but even so very narrow lines will not be obtained. A true pure absorption phase representation of the homonuclear *J*-spectrum would be highly advantageous for both applications, but no entirely satisfactory experimental solution is yet available. A sequence which allows pure absorption phase to be obtained for two-spin systems is known[12], but with more than two spins dispersion contributions reappear. A computational method has also been proposed[13], related to that for processing highly truncated data mentioned in section 10.3.2, but seems quite demanding of processing power and is not to my knowledge presently available in a form suitable to run on spectrometer data systems.

Overall, then, the homonuclear *J*-spectrum is useful only subject to a number of restrictions. We must have pure first order coupling, and either the resolution obtainable in magnitude mode or severe lineshape distortion must be acceptable. We may then be able to use the experiment to disentangle overlapping multiplets, and to obtain the positions of multiplet centres using the projection. If *heteronuclear* coupling is also present, it will appear in the projection, so the experiment can be used to identify it.

10.4.3 The Indirect *J*-spectrum

A final twist to the idea of separating overlapping multiplets is an experiment which has the appearance of a cross between hetero- and homo-nuclear *J*-spectroscopy, with an experimental procedure very similar to heteronuclear shift correlation (Chapter 9). This tackles the problem of truly superimposed multiplets, which will evidently remain superimposed even in a homonuclear *J*-spectrum, by allowing the homonuclear multiplet structure to be carried on the chemical shift of another coupled heteronucleus.

The idea is actually quite simple. The heteronuclear shift correlation experiment in principle contains fine structure in v_1 due to homonuclear coupling between the nuclei from which magnetisation is transferred. In practice this is hard to observe in detail, because it is difficult to digitise v_1

Figure 10.18 The sequence for the indirect *J*-spectrum. Compare this with HSC: a π pulse on ^{1}H is inserted in t_1 to eliminate chemical shifts in v_1 (this also refocuses the heteronuclear coupling, obviating the need for a ^{13}C pulse here); π pulses on both nuclei are required during Δ_1, but are optional during Δ_2.

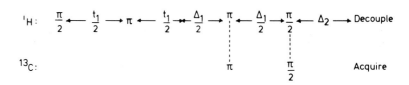

finely enough as it covers a complete chemical shift range, typically the 10 p.p.m. range of protons. If the chemical shifts were eliminated from that dimension, as can be done using a spin-echo, then only the width of the widest multiplet would need to be characterised, making high resolution feasible. Thus the normal shift correlation sequence is altered simply by adding a proton π pulse in the centre of the evolution period (Figure 10.18), eliminating chemical shifts from v_1. For further experimental details, see reference 14. This experiment is subject to the same limitations as homonuclear *J*-spectroscopy, but with lower sensitivity because a heteronucleus is being detected, so it should only be considered for the specific case of exact overlap in the normal *J*-spectrum. Figure 10.19 illustrates the results obtained for a sample of glucose in D_2O. The normal proton spectrum (top right) is ill-resolved, because the linewidth is quite large and the compound exists as a mixture of anomers. The proton *J*-spectrum is a little better, but it is still hard to identify all the multiplets. In the indirect *J*-spectrum, on the other hand, the multiplet structures are clearly visible.

Figure 10.19 Roughly 'equivalent' regions of the homonuclear and indirect *J*-spectra of glucose. Note that the order of proton and carbon shifts is not the same.

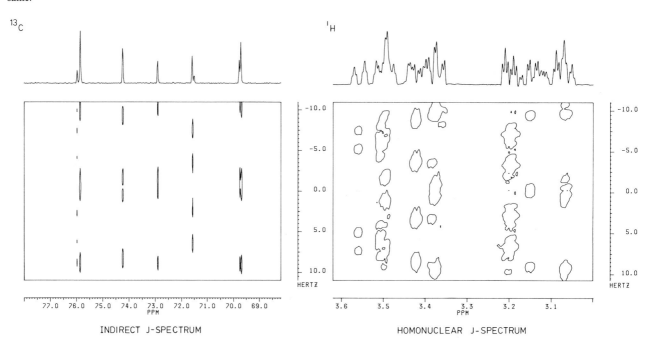

INDIRECT J-SPECTRUM   HOMONUCLEAR J-SPECTRUM

REFERENCES

1. M. H. Levitt, G. Bodenhausen and R. R. Ernst, *J. Mag. Res.*, **53**, 443-461, (1983).
2. R. Freeman, T. H. Mareci and G. A. Morris, *J. Mag. Res.*, **42**, 341-345 (1981); D. M. Doddrell, D. G. Reid and D. H. Williams, *J. Mag. Res.*, **56**, 279-287, (1984); D. M. Doddrell, D. H. Williams, D. G. Reid, K. Fox and M. J. Waring, *J. Chem. Soc. Chem. Commun.*, 218-220, (1983).
3. S. C. Wimperis and R. Freeman, *J. Mag. Res.*, **58**, 348-353, (1984).
4. G. Bodenhausen, R. Freeman, G. A. Morris and D. L. Turner, *J. Mag. Res.*, **28**, 17-28, (1977).
5. P. J. Hore, *J. Mag. Res.*, **62**, 561-567, (1985).
6. J. Keeler, *J. Mag. Res.*, **56**, 463-470, (1984).
7. G. Bodenhausen, R. Freeman, R. Niedermayer and D. L. Turner, *J. Mag. Res.*, **26**, 133-164, (1977).
8. R. Freeman and J. Keeler, *J. Mag. Res.*, **43**, 483-486, (1982).

9.   G. Bodenhausen, R. Freeman and D. L. Turner, *J. Mag. Res.*, **27**, 511, (1977).
10. R. Freeman, S. P. Kempsell and M. H. Levitt, *J. Mag. Res.*, **34**, 663-667, (1979).
11. P. Bachmann, W. P. Aue, L. Müller and R. R. Ernst, *J. Mag. Res.*, **28**, 29-39, (1977).
12. M. P. Williamson, *J. Mag. Res.*, **55**, 471-474, (1983).
13. A. J. Shaka, J. Keeler and R. Freeman, *J. Mag. Res.*, **56**, 294-313, (1984).
14. G. A. Morris, *J. Mag. Res.*, **44**, 277-284, (1981).
15. A. Bax, *J. Mag. Res.*, **52**, 330-334, (1983).

Index

Jeener's COSY

2D. COSY P-190 (COrrelation Spectroscopy)